MANAGING
TECHNOLOGICAL
INNOVATION

MANAGING TECHNOLOGICAL INNOVATION

COMPETITIVE ADVANTAGE FROM CHANGE

SECOND EDITION

FREDERICK BETZ

WILEY

John Wiley & Sons, Inc.

Library of Congress Cataloging-in-Publication Data:

Betz, Frederick, 1937-
    Managing technological innovation : competitive advantage from change / by Frederick Betz.—
2nd ed.
        p. cm.
    Includes bibliographical references and index.
    ISBN 0-471-22563-0 (cloth)
        1. Technological innovations—Management.   2. Research, Industrial—Management.   I. Title.

    HD45.B443 2003
    658.5'14—dc21   2002192426

Printed in the United States of America

10 9 8 7 6 5 4 3 2 1

Dedication

For my wife, Nancy, and my children, Sara, Fred, and David.
From love and perspective,
I have tried to understand the totality of life,
beginning with science.

CONTENTS

INNOVATION CAPABILITY
(*What a Business Needs to Be Innovative*)

TECHNOLOGY STRATEGY
(*How to Plan Innovation*)

PREFACE

The study of the interaction between science and economy is an interdisciplinary topic. The topic spans the different disciplinary areas of: science and engineering and management and economics. To indicate this wide span from science to economy, in the mid-1980s this interdisciplinary topic was named *management of technology* (MOT) (NRC, 1987). Later, in 1994, Cyret and Kumar (1994) commented on the importance of the new field:

> Forty years ago only a relatively few perspicacious observers realized the significance of technology management for our society. . . . Technology management has become increasingly important in the modern age. . . . Organizations have to be in a position to adapt to technological innovation. In order to be in a position to adapt, management must have knowledge of the kind of technological innovation that will come . . . [and] the firm must have a [technology] strategy for searching [for innovation]. . . ." (p. 333)

Despite the importance of the concept of technological innovation, it has not always been well understood or managed. This is because the topic bridges two very different worlds: the technical and business worlds. There is a vast cultural gap between these worlds: a world of matter and a world of money—and the material and financial worlds run on different laws: laws of nature or laws of economy. In the past, concentrating on the laws of economy, business schools mostly ignored the business functions based on the laws of nature: that is, the business functions of research and engineering and information. The academic business schools assumed that the other academic schools of engineering and computer science would take care of this. And they did, but only on the technical side and not on the business side. Engineering schools did not teach their engineering students the laws of economy; neither did science departments teach their science students such laws. Accordingly, after a university education, the practical education of managers and engineers and scientists needed to be completed. What was missing in such incomplete education was a sophistication about the interaction of matter and money. MOT educational programs were developed to provide such sophistication, an understanding of *technological innovation*. Researchers in many different disciplines have studied parts of technological innovation in:

- Philosophy and sociology of science
- History of science
- History of technology
- Theory of economic development
- Management of research and development
- Management of the engineering function in business
- Management of new high-tech business ventures

As these studies had not been integrated, it was integration that became the challenge for MOT. As late as 1987, these topics were divided among different groups of disciplinary researchers who studied different aspects of innovation (Kocaoglu, 1994). One group of economists studied the macro level of innovation process at the national scale, with emphasis on the fact that technological progress made and continues to make important contributions to economic development and industrial productivity. Another group of engineering researchers and of practicing industrial scientists studied the micro level of innovation process in the industrial organization, with emphasis on managing the engineering function and research and development (R&D) function in industrial firms. A third group of business school researchers studied the micro level of new high-tech businesses, with emphasis on entrepreneurship and new business ventures. Two separate groups of historians studied the history either of science or of technology (but not both). A group of political scientists and sociologists and experienced R&D administrators studied the macro level of national support of R&D, with emphasis on how to administer governmental programs of R&D and on how to formulate effective science and technology (S&T) policy.

From an integrative perspective, one could see that these different research groups were describing different slices of the process of technological innovation. Taken together, these could reveal the whole of innovation. This was the challenge, and in the middle of the 1980s, those interested in integration banded together under the banner of MOT. For example, in 1986, Richie Herink (then of IBM) approached Eric Block (then director of the National Science Foundation) to sponsor a workshop on MOT. With NSF support, Herink then chaired a study group of the National Research Council (NRC) of the U.S. National Academy of Science, and the group published of a brief pamphlet entitled *Management of Technology: The New Challenge* (NRC, 1987). They emphasized that the challenge of MOT consisted of integrating knowledge about technological innovation in the interface between the disciplines of engineering and management. Tarek Khalil at the University of Miami began a series of biannual conferences specifically on the management of technology, which later grew into the scholarly society of the International Association for Management of Technology. An earlier program in engineering management had been established in the engineering school at the University of Missouri–Rolla. As such departments grew in engineering schools, the society for engineering management was begun. Dundar Kocaoglu at Portland State University developed an engineering management program at Portland State University and became editor of the *IEEE Transactions on Engineering Management*. Kocaoglu later began biannual conferences as the Portland International Conference on Management of Engineering and Technology (PICMET). George Washington University had a program on engineering administration, and one of its graduates, John Aje, began a graduate program in management of technology at the University of Maryland University College. By 2000, there were over 100 departments in engineering management in the United States and over 80 programs in management of technology throughout the world.

Because of this interdisciplinary and integrative character, MOT as an academic topic has been pursued in two different methodological paths:

- On the one hand, MOT was *descriptive,* with practitioners attempting to describe actual historical patterns in the changes in science, technology, and economy.
- On the other hand, MOT was *prescriptive,* with practitioners attempting to develop useful concepts, techniques, and tools for managing new and future change in science, technology, and economy.

In this book we present both the descriptive and prescriptive approaches of MOT. This second edition traces from a book of integration of technology and management that I published in 1987. At that time, the literature about technological change

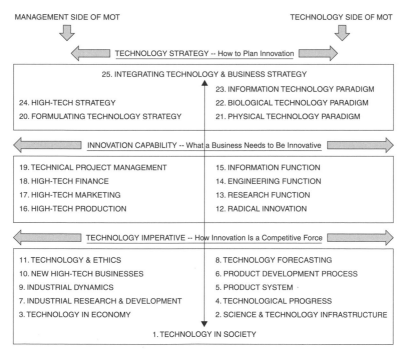

Figure 0.1.

drew principally on management experience with progress in the physical aspect of technologies (i.e., the paradigms of "mechanism" and "logic"). For the first half of the twentieth century, technological progress was driven primarily by these paradigms in materials and energy technologies. But the world keeps changing. In the second half of the twentieth century, dramatic new technological progress came from the science and technologies of information and molecular biology, and it was in the last quarter of the twentieth century that practical experience in managing innovation in information and biological technologies began to accumulate. So the major change in MOT in the last decades of the twentieth century has been the accumulation of practice, cases, and theory for managing progress in the information and biological technologies.

This new edition has been expanded to provide a balance of focus across all technologies: material, power, biological, and informational. We summarize the theory for the whole of technological innovation, illustrated in actual cases that illustrate innovation.

In the early report on MOT (NRC, 1987), the MOT working group emphasized that there were two sides of MOT needing to be integrated—management and engineering. This book includes both sides of MOT—emphasizing either the managerial aspects of MOT or the technical aspects of MOT. As shown in Figure 0.1, the different chapters of the book elaborate the three major issues of MOT (technology imperative, innovation capability, and technology strategy) while emphasizing in an issue, either the management or technical side of MOT.

FREDERICK BETZ

1

TECHNOLOGY IN SOCIETY

INTRODUCTION

The grand theme of managing technological innovation is the story of technological change and its impacts on society. Historically, this story is both dramatic and ruthless. The drama is the total transformation of societies in the world from feudal and tribal to industrial. The ruthlessness in technological change has been its force, which no society could resist and which has been called the *technology imperative*. For the last 500 years, technological change has been irresistible in military conflict, in business competition, and in societal transformations.

Two technological innovations mark the beginning of the modern era: the gun and the printing press. The gun ended the ancient dominance of the feudal warrior, and the printing press secularized knowledge. This combination of the rising dominance of a mercantile (or *capitalist*) class and the secularization of knowledge (or *science*) are hallmarks of a modern society. From the fourteenth through the twentieth centuries, the political histories of the world are stories of the struggles between nations and peoples, wherein the determining factor has been the superiorities made possible by new *scientific technologies*. The direct connection between science and economy is through technology; and the study of the interaction of *science* and *economy* is the topic of *technological innovation*. We begin by providing a comprehensive overview of the role of technological change in society—covering the basic ideas of *strength* of technological change, *function* of technology in societies, *interaction* of technology with nature, and *leadership* in technological change.

CASE STUDY:

Dot.Com Stock Bubble of the 1990s

A clear case of the direct connection between science and economy was the Internet stock bubble of the late 1990s. Technological innovation of the Internet and its impact on commerce made that decade a very interesting and challenging—

even scandalous—time: "The Internet decade has seen the unscrupulous rewarded, the dimwitted suckered, the ill-qualified enriched at a pace greater than at any other time in history. The Internet has been a gift to charlatans, hypemeisters, and merchants of vapor . . . and despite all that, it still changes everything" (Nocera and Carvell, 2000, p. 137).

This case illustrates how great an impact a technological innovation can have on an economy. Figure 1.1 shows the average value of the U.S. Nasdaq stock market in 2002, with the Internet stock bubble. The growth of the Internet had been exciting and rapid. For example, in 1996 in the United States, 14% of the population used the Internet, which jumped to 22% in 1997, 31% in 1998, 38% in 1999, and 44% in 2000. In 2000, the average time spent monthly online per user was 19 hours. U.S. consumer spending online grew from a few million in 1996 to $3 billion in 1997, $7 billion in 1998, $19 billion in 1999, and $36 billion in 2000. Of the $36 billion spent in 2000, $11.0 billion was for travel, $7.7 billion for personal computers, $2.4 billion for clothes, $13.4 billion for books, and $13.4 billion for other merchandise. The Internet stimulated an economic expansion in the U.S. economy. This pattern of technological innovation stimulating an economic expansion has been a general pattern in modern economies.

The Internet was created as a technology to enable computer networks—from scientific research in computer science. The Internet began in the United States as ARPAnet, became NSFnet, and was transferred to the commercial sector as the Internet. This innovation occurred in the United States because (in the 40 years of the Cold War, during the second half of the twentieth century) the U.S. government funded computer research to advance military capabilities, Modern progress in technology arises from research, and advanced research requires large and sustained funding. In addition to research funding, technological innovation requires intensive capital funding for commercialization of the new technology.

A coincident event in the middle of the 1990s was the deregulation of telecommunications business in the United States. In this and the new Internet, many en-

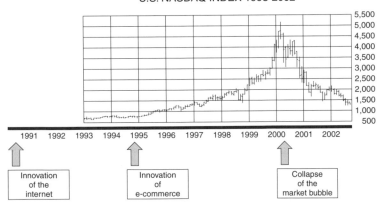

Figure 1.1. E-commerce stock bubble: U.S. Nasdaq Index, 1993–2002.

trepreneurs saw new business opportunities. They launched new telecommunication businesses and many new businesses in electronic commerce (i.e., *dot.com businesses*). Venture capital firms poured huge amounts of capital into hundreds of ideas for new dot.com businesses, and immediately took many of these public as IPOs (initial public offerings). From 1997 to 2000, the excitement over the Internet and the new dot.com businesses drove stock markets to new highs. Business pundits then wrote that the Internet was creating a *new economy* for which one need not need worry about the profitability of a new business, only market growth. In the year 2000, this stock bubble burst. The "new economy" fell back to the principles of the old economy, which always required profitable businesses. The many new unprofitable dot.com businesses collapsed. The Nasdaq average dropped 65% early in 2000. Following the Internet stock bubble bursting in 2000, there was a rash of major corporate failures due to greed and misdeeds stimulated by the bubble. The failed companies included Enron, WorldCom, Global Crossing, Andersen, and others.

Historically, the Internet stock bubble was just one of many examples of excess financial *enthusiasm* about technological innovation. Another example of an earlier financial bubble was the U.S. railroad bubble, which occurred 100 years earlier, late in the nineteenth century. There are important lessons that one can learn about innovation from history:

1. Scientific research is necessary to invent new basic technologies.
2. New basic technologies create new economic opportunities.
3. The path to wealth in economic opportunities is usually rocky.
4. However rocky the entrepreneurial path, economic development from technological innovation becomes a permanent feature of societal structures (despite the stock bubble, the Internet and electronic commerce are in the world to stay).
5. Technological innovation and economic development are neither simple nor inevitable.
6. If progress in technology is to occur and be implemented successfully in business opportunities, technological innovation needs to be managed carefully.

TECHNOLOGY AND SCIENCE

We see in the Internet case that the connection between science and economy is through technology. The Internet was new technology arising from new science (research in computer science to enable computers to communicate). The economic impact of the new technology was its implementation in new products and services (enabling Internet users to communicate through their computers and phone connections). Since technology always provides a direct connection between science and economy, we need to be precise in using the terms *technology* and *science.*

The historical derivation of the term *technical* comes from the Greek word *technikos,* meaning "of art, skillful, practical." The portion *ology* indicates "knowledge of" or "systematic treatment of." Thus, the derivation of the term *technology* is literally "knowledge of the skillful and practical." This meaning of *technology* is a common definition of the term but is too vague to express the exact interactions between science, technology, and the economy. We use a more precise definition:

Technology is *knowledge of the manipulation* of nature for human purposes.

Knowledge of the skillful and practical is knowledge of the manipulation of the natural world. Technology is useful knowledge—knowledge of a *functional capability.* In all technologies, nature is being manipulated.

Also, since nature is "discovered" by science, we need to define science precisely. Derivation of the term *science* comes from the Latin term *scientia,* meaning "knowledge." However, the modern concept of scientific research has come to indicate a specific approach to knowledge, which results in discovery and explanations of nature. For this reason, we use the following definition:

Science is the *discovery and explanation* of nature.

In science, the term *nature* is used to indicate the essential qualities of things that can be observed in all of the universe. Accordingly, we use the term *nature* in the following way:

Nature is the *totality* of the essential qualities of the experienciable phenomena of the universe.

There are difficult concepts in this definition, such as "totality," "essential qualities," "experienciable," and so on, and we examine these as we proceed. But it is sufficient for now to emphasize that the basic connection of science with technology is that they both deal with nature.

Science *understands* nature, and technology *manipulates* nature.

CASE STUDY:

Technology in Archaeology

The fundamental ideas of nature and technology are nicely illustrated in the case of archeology. In the modern discipline of archeology, one can see two types of forces in human history: biological evolution and cultural evolution. In cultural evolution, technological change has been imperative. For example, if one takes an encyclopedia of archcology (such as Scarre, 1998) and lists key archeological events, the role of technology in human culture stands out. Table 1.1 shows such a list as a sample of major human events that have been deduced from archeological evidence.

TABLE 1.1 Technology in Archaeology

Approx. Dates	Material Technologies	Biological Technologies	Information Technologies	Energy Technologies
22 Million years		Earliest hominids (Africa)		
4 Million years		Bipedal hominids (Africa)		
2.5 Million years	Earliest stone tools (Africa)	Homo hobilis (Africa)		
1.5 Million years	Diversification of stone tools (Africa)	Homo erectus		
1 Million years				Fire (Africa)
200,000 B.C.	Pointed stone tools attached to wooden handles (Africa)			
120,000 B.C.		Homo sapiens (Africa)	Neanderthal burials (Europe and West Asia)	
100,000 B.C.	Mines for red ocher (Africa)			
42,000 B.C.			Cave art and Venus figurines (Europe)	
30,000 B.C.			Microliths (Europe)	
14,000 B.C.	Pottery			
10,000 B.C.	Copper			
9,000 B.C.				
8,000 B.C.		Agriculture		
7500 B.C.				Irrigation
5500 B.C.				Sailing ships
4500 B.C.				Wheeled vehicles
3500 B.C.	Bronze			
3500–2000 B.C.			First civilizations	
3500 B.C.			Writing	
1500 B.C.			Alphabetic writing	
1200 B.C.	Iron (steel)			
1000 B.C.			Empires	

From the perspective of biology, the human species evolved, as have all other species. Genetically, human beings are 97% similar to chimpanzees and 96% similar to gorillas. The evolutionary assumption is that modern humans and apes shared a common ancestor millions of years ago called a *hominoid.* Fossils of the earliest representative of the hominoids, the Proconsul, have been found in African Early Miocene deposits made 22 million years ago (Scarre, 1998). Fossils of another hominoid, *Kenyapithecus* (similar in appearance to the modern ape), were found from 15 million years ago: "The common ancestor of chimpanzees, gorillas and humans probably lived in open woodland or wooded savanna habitats in the tropics of Africa. It would have been a medium-sized active primate adapted to a mixture of tree-climbing and ground-living" (Scarre, 1998, p. 52). About 4 million years ago, the species that was to be a direct ancestor of humans appeared, with upright mobility: "Evidence that this fundamental adaptation had evolved in Africa by at least 4 million years ago has been found in Laetolil, northern Tanzania, where footprints preserving the unmistakable outlines of the human foot were discovered in 1978" (Scarre, 1998, p. 54).

The oldest fossils yet found, called *Homo* (man), date from about 2.5 million years ago, found first at the Olduvai Gorge in Tanzania. A skull named *Homo habilis,* was found which was distinguished from other early skulls by a larger brain cavity and a somewhat human face. From this same period, the first evidence of stone tools was found in Hardar in Ethiopia in the form of sharpened flakes, dating from 2.5 million years ago, that look as if they were used to butcher animals and to smash bones to obtain marrow.

With the development of tools and their later changes, we can see a curious shift in archaeology from a tale of purely biological evolution toward a cultural evolution of humanity through improved toolmaking. A new technology in stone tools appeared alongside biological changes in humanity's ancestors: "About 1.5 million years ago significant changes in tool types coincided with the appearance of a new hominid. *Homo erectus* replaced the earlier, smaller-brained *Homo habilis.* The most important change in stone implement technology at this period was the development of the Acheulian industry, characterized by hand axes and cleavers worked on both faces (Scarre, 1998, p. 56).

Hand axes and cleavers continue to appear in the fossil records for the next million years, with *Homo erectus* spreading from Africa into Europe and Asia. Other examples of cultural evolution next appear in the record, such as the use of fire at campsites over 1 million years ago (traces can be found at some early archaeological sites in Africa, such as at Kalambo Falls and Chesowanja). By 120,000 years ago, the entire set of technologies in stone age cultures existed that provided evidence of a common human culture, including hunting and gathering, the uses of fire, and stone and wood technologies.

About that time, the first evidence of a species related to the human, the Neanderthal, appears in the fossil record: "The Neanderthals . . . first known some 120,000 years ago . . . lived in western Eurasia until some 35,000 years ago. The Neanderthals are distinguished from fully modern man by a strong and heavy skeleton and a projecting face, with a large, broad nose and large teeth. . . .

Neanderthals are the first type of human known to have buried their dead, and display a variety of ritual behavior" (Scarre, 1998, p. 64).

Also from 100,000 to 35,000 years ago in Africa appear the fossils of Cro-Magnons, direct ancestors of modern humans, *Homo sapiens sapiens*. About 42,000 years ago, Cro-Magnons mined ocher, probably used for body decoration, on a large scale at the Lion Cave in Swaziland. Painted rock slabs appear from about this time. In Europe, cave paintings appear in France around 30,000 B.C. The famous "Venus figurine" dates from 20,000 B.C., and carved mammoth ivory was found at Lespugue, France. The aesthetic aspects of the Cro-Magnon appear in these figures, along with cave wall paintings and ornaments, such as amber swan pendants from Russia and cut beads of ivory and antler.

Also at this time, a technology called *microliths* appeared in the fossil records. Microliths are small stone tools shaped into points or barbs, used on arrows or harpoons, and these appear in sites from the last ice age, which occurred at the end of the Upper Paleolithic Era (35,000 to 12,000 years ago). Traces of sites indicate large huts and storage pits. These sites are sometimes located at entrances to valleys that could be used by hunters to intercept migrating herds.

The shift in the story of human evolution during the last 100,000 years is one from biological evolution to cultural evolution. This is a particularly curious fact about the human species—first evolving partly biologically and partly culturally and then wholly culturally! Biological evolution was driven by nature and the necessity to adapt to nature. Cultural evolution consists of acts of human innovations that alter nature.

Cultural evolution consists of deliberate human acts of technological innovation.

TECHNOLOGICAL IMPERATIVE

We pause in this story of early humanity to note an interesting coincidence. The Neanderthals disappear from the archeological record about 35,000 years ago, and in about the same era, microlithic tools (such as arrowheads) begin to appear (and they increase in numbers notably about 16,000 years ago). There has been speculation about reasons for the disappearance of the Neanderthals. Were they killed off by our ancestors, or did they interbreed with our ancestors? This is not yet resolved. In the 1990s, DNA from Neanderthal bones was recovered and examined. That first study of Neanderthal DNA concluded that it was too different from ours for interbreeding to have occurred. This leaves the speculation that it might have been that the new and lethal *microlithic technology* of the bow and arrow which helped our Cro-Magnon ancestors kill off their Neanderthal competitors. We will probably never know for certain. But in later history, there have been many instances of superior weapons (such as iron weapons or the gun) enabling one human tribe to dominate or exterminate other human tribes.

It is a particularly striking fact about human history that technological change has always forced societal change. Whenever a superior technology in human competition has entered the historical record, its use in enabling competitive dominance has been harsh.

> The *imperative* in technological innovation is that superior technology of a competitor cannot be ignored by other competitors, except at their peril.

Technological change has been irresistible in military conflict, in business competition, and in social forms. The technology imperative is not a kind of "historical law"; rather, it is simply a form of human nature, the social biology of the human species to use technology in their internal competitions over territory, government, economy, and culture.

CASE STUDY CONTINUED:

Technology in Archaeology

Returning now to the case of archeology, the late record becomes one solely of cultural evolution without biological evolution. About 10,000 years ago, as the Earth warmed from the last ice age, agriculture appeared as a new human technology. Whereas hunting and gathering technologies of animals and humans take from the environment, agriculture is a technology that alters the environment. Agriculture began in the Near East around 8000 B.C. and in China and Mesoamerica by 6000 B.C. In the new agricultural cultures, other new technologies were innovated, such as pottery. Pottery was invented independently in different parts of the world and at different times. The earliest examples of pottery are found in Japan from about 10,000 B.C. The pottery was made by coiling ropes of moist clay into a shape and then baking it. Next appeared the technologies of irrigation (first evidence about 7500 B.C.), sailing (first evidence about 5500 B.C.), and wheeled vehicles (first evidence about 4500 B.C.).

Then another major change in cultural evolution occurred, the material technology of metals. Metals can be taken from ores by cracking the ore rock and heating to melt out the metal. The metal can be reheated and pounded into useful shapes. Copper was used for making tools and weapons as early as 7000 B.C. But copper is too soft to hold a hard edge for strong tools and weapons. In the third millennium B.C., bronze was invented. Bronze is a heated combination of copper and tin, very important because of its hardness. Bronze could be used for weapons: the bronze sword or bronze-tipped arrow and spear. Bronze was probably invented by a potterymaker who had melted copper and tin for a glaze and found hard nuggets of bronze from that combination. Bronze weapons were so important that the period in prehistory from 3500 to 1500 B.C. in the Mediterranean area is called the *bronze age:* "[Bronze] was discovered in western Asia between 4000 and 3000 B.C. . . . Its spread was helped by the development of long-distance trade good routes for metals, especially as deposits of tin . . . were rare" (Scarre, 1998, p. 120).

TAXONOMY OF TECHNOLOGIES

With the precise definitions of technology, science, and nature, we can now classify all the possible kinds of technologies in the world. We do this by classifying the kinds of "nature" of the world in which we live. As a species, we humans live in two kinds of natural worlds, worlds of things and worlds of ideas: physical and conceptual worlds. The *physical world* in which we exist includes breath, food, warmth, body, matter, forces, energy, and so on. The *conceptual world* within which we exist includes mind, spirit, communication, money, and so on. Earlier, these two kinds of natural worlds were called worlds of *body* and *spirit,* or of *matter* and *mind.* This *matter–mind* dichotomy is useful because we can distinguish between classes of technology that aid the manipulation of nature as either the physical or conceptual worlds of nature—*technologies of matter or mind.*

Also in the physical world, we distinguish between matter that is inanimate or animate. Examples of *inanimate matter* include atoms, molecules, liquids, gases, solids, planets, and stars; examples of *animate matter* include bacteria, viruses, multicellular organisms, and humans. Accordingly, we can also classify technologies that aid the manipulation of either inanimate or animate matter: *technologies of inanimate or animate matter.*

These two dichotomies (matter or mind and inanimate or animate) provide the basis of a comprehensive taxonomy of technologies, as shown in Table 1.2. Technologies that manipulate the nature of inanimate matter are those that facilitate materials, such as plastics, steel, and asphalt. Technologies that manipulate the nature of animate matter are those that facilitate biological processes, such as agriculture, recombinant DNA, and surgical procedures. Technologies that manipulate the nature of inanimate and pure form without substance (mind) are those that facilitate processes of power, such as electrical power, flight, and transportation. Technologies that manipulate the nature of animate mind are those that facilitate functional capabilities of cognition, such as perception, information, thinking, decision, communication, computation, and control. The usefulness of this topology is its inclusiveness. All things in nature are either inanimate or animate matter, and all things in nature are either of matter or of mind. Accordingly, all technologies for

TABLE 1.2 Typology of Technologies

	Inanimate	Animate
Matter	Material technologies	Biological technologies
Mind	Power technologies	Information technologies

manipulating nature are either material, biological, power, or information technologies (or a combination such as neuroengineering, using biological, energy, and information technologies).

Technology in Archaeology

Civilization becomes the next episode in the archeology of the human species. It was the combination of irrigation, bronze weapons, wheels, and sailing that provided the technologies for the next advance in human society—civilization. Civilizations were urban-centered agricultural territories and rose separately in different parts of the world: southern Mesopotamia (3500 B.C.), the Egyptian Nile valley (3200 B.C.), the Indus valley (ca. 2500 B.C.), China (1800 B.C.), and Mesoamerica (1500B.C.). Most early civilizations were developed in fertile alluvial plains along rivers that provided the rich and annually renewable agricultural basis for a feudal structure of a civilization.

The form of the new agricultural civilization was feudal. A feudal structure consisted of a dominant tribe subduing less dominant tribes for agricultural cultivation. A tribal form of human society consists of a group of related families occupying and defending a territory for hunting or agriculture. A feudal form of society consists of a militarily dominant tribe controlling a territory over subservient tribes engaged in agriculture. The militarily dominant tribe in a feudal structure is called an *aristocracy* and the subservient agricultural population is called a *peasantry*. What made feudal social structures widespread (as human history evolved) were agriculture and the invention of bronze weapons. Agriculture gave economic value to a fixed residence of a tribe; while bronze weapons gave dominance to a military aristocracy over tribes reduced to peasantry. Bronze weapons were superior to stone weapons in warfare and were difficult and expensive to obtain. A conquering aristocratic tribe could maintain control over a much larger population of peasants through monopoly of military power based on bronze weapons.

Next, the technology of "writing" was developed independently as a religious and administrative tool in each of these civilizations: "Writing was a crucial development in the administration of early civilizations, and with it emerged a new class of specialist the scribe. . . . Writing was invented to keep business accounts in the Near East in the late 4th millennium B.C." (Scarre, 1998, pp. 122–134). The first writings were on clay tablets to record transactions, and the first texts were only numbers. Then representations of domestic animals were listed with numbers to tell what the numbers were about. Gradually, a pictographic script developed to communicate such things as stored or traded objects. More abstract ideas were then recorded with symbols from concrete objects where the spoken words for the ideas and symbols contained similar syllabic sounds. Thus, a syllabic writing style evolved from the early pictographic scripts. The Near Eastern scripts were developed as the idea spread from one area to another. The Chinese and Mesoamerican scripts were invented independently.

In the Near East another major technical progress occurred in writing as script was changed from transcribing the syllabic sounds in speech to phoneme sounds within a syllable—alphabetic writing. The first alphabet script appears in Semitic writings around 1500 B.C., transcribing only consonant phonemes. Then a full alphabet with both consonant and vowel sounds appeared in Greek writings. This Greek written language with a complete alphabet enabled a literate aristocracy, and the ancient Greek classics on philosophy, history, and mathematics appeared.

TECHNOLOGY, HUMANITY, AND THE UNIVERSE

We noted how the grand story of human history can be seen as technological progress in control over nature, and we show this as sketched in Figure 1.2. The taxonomy of technologies (material, power, biological, and information) provides the means for humanity to manipulate nature in the universe in which it lives. From this manipulation of nature, the universe provides a kind of feedback to humanity. It is a *fundamental characteristic of modern culture* that both humanity and the universe are described as being within *nature*. Prior historical cultures used the ideas of "myth" and "god" to describe the totality of the universe; but science uses "nature" for this totality.

An example of how technologies used by humanity interact with the universe is the Aswan dam, built in Egypt in the 1950s. The dam was a flood control technology

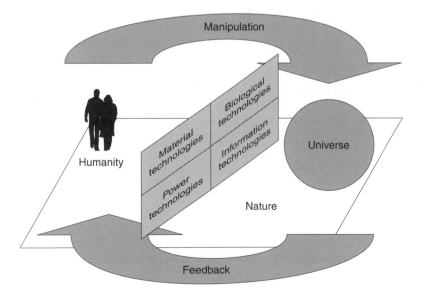

Figure 1.2. Humanity interacts with nature in the universe through technology.

(material and power technology) that controlled the level of annual flooding of the Nile. Since ancient times, the annual Nile flood provided both renewal of the soil for Egyptian agriculture and in some years, disastrous floods. The Aswan dam mitigates against flood disasters and provides a major source of electricity for the country. But as a feedback, the dammed Nile no longer replenishes the soil, and chemical fertilizers must be used to keep the soil fertile. The fertilizing silt now accumulates behind the dam in a broad and increasingly shallow lake. Moreover, increased levels of disease flourish in the Nile delta, undisturbed by Nile flooding. While humanity interacts with its universe through technology, the universe interacts back with humanity (natural feedback).

Humanity is that part of nature which manipulates nature through technology.

CASE STUDY CONCLUDED:

Technology in Archaeology

After the invention of technology of writing in the human story, we next see a transition from archaeology to history. In the early history of the Near East, their civilizations are stories of military struggles for control of the Levant (the eastern shore of the Syrian and Arabian desert). The Levant region consisted of small but prosperous city-states which controlled the sea trade in the Mediterranean. To the west of the Levant was the Egyptian kingdom based on control of the Nile river valley. To the east was the Mitanni kingdom, centered on control of the river valleys of the Tigris and Euphrates. To the north was the Hittite kingdom, based on the control of agriculture along rivers begun on the Anatolian plateau. Phoenicians dominated the bronze age trade until the twelfth century B.C. Invasions of new peoples came from the Eurasian steppes, disrupting trade and ending the bronze age: ". . . onslaughts from maritime raiders of uncertain origin, the so-called 'Sea Peoples.' In the face of these attacks the Levantine empires crumbled and the Hittite state was destroyed, while Egypt survived only in a seriously weakened condition. . . . This [was] a Dark Age" (Scarre, 1998, p. 142).

That Dark Age disrupted the Mediterranean trade needed for bronze production (e.g., tin was imported all the way from what is now Austria and what is now Cornwall in Britain). A substitute material for bronze was needed. Although iron production had been known, iron weapons had been brittle and broke against a bronze weapon. Then a new way to process iron into steel was invented, probably in the Anatolian region. Iron was heated repeatedly in smoky furnaces and hammered into steel (the reheating in smoky fires added carbon to the hot iron, which was then beaten to form a steel surface around the iron). Thus, in the Near East the age of bronze ended and the age of iron began. Steel was discovered by hammering iron in carbon rich (smoky) atmospheres (steel is iron containing a small percentage of carbon atoms).

After this event, we are down to the time when *history* begins. Then the story is one of the rise and fall of empires. Beginning in 1000 B.C. there was in succession the Assyrian empire, the Archemenid empire of Persia, the sort-lived empire of Alexander the Great, and then the long-lived Roman empire. In other parts of the world, such as China, there was a succession of kingdoms.

> **When technologies are stable, history turns only upon political and economic events.**

NATURE AND ECONOMY

In the case of key events in archaeology, we saw the essential role of technological change in societal change. The early evolution of hominids branching into a human species was, of course, not technological but biological. But even in this biology, the hominid inventions of stone tools and fire were significant technological events. Then further cultural evolution of *Homo sapiens* depended on key technological innovations such as stone arrow tips, pottery, copper, agriculture, irrigation, sailing, wheels, bronze, writing, steel, and so on. Civilization and the feudal kingdoms and empires of the human race that were to follow were all dependent on these technological inventions.

> **The story of the human species turns on two themes of evolution— *biological evolution* and *cultural evolution*—with cultural evolution dependent on technological progress.**

Huge periods of history have been named for dominant technologies, such as the stone age, bronze age, iron age, and industrial age. Technology has divided earlier from later cultures. Even when the twentieth century ended, contemporaries talked of a new division, an age of information (meaning, of course, the pervasive impacts upon society occurring from information technology).

Since the technology in a society connects nature to its economy, we will use a meaning of the term *economy* that indicates this. The common use of the term is to indicate the administration or management or operations of the resources and productivity of a household, business, nation, or society. But we use the term *economy* in the following way:

> **Economy is the *social process* of the human use of nature.**

In modern economies, the direct connection between technology and economy is in the *products and processes and services* that embody the technologies that a business uses in an economy. The personnel in a business who design these are *engineers.* Thus, we also need to define the term *engineering* generically:

> **Engineering is the *design of economic artifacts* embodying technology.**

Engineers are the modern technologists. Engineers are employed by business (and government) to design products, production, and operations. Engineering designs enable businesses to use nature in adding economic value through its activities. Technological innovation requires the three essential roles of scientist, engineer, and entrepreneurial businessperson.

Energy in a Modern Economy

Next we look at an example of nature and economy as the sources and uses of energy in a modern economy. This illustration shows how general is the idea of nature to all economies and all societies. Modern industrial societies are all based on the economic utility of energy extraction from nature, and Figure 1.3 sketches how the energy we use on Earth ultimately derives from nature in the hydrogen fusion reactions of the sun. Radiant energy from the sun powers the weather cycles on Earth that transfer water in a hydrological cycle from the oceans to land and back to the oceans again. Radiant energy from the sun plus rainfall from the hydrological cycle powers the growth of biomass. Biomass, ancient and modern, provides energy sources to society in the form of coal, petroleum, gas, and wood. In addition, the hydrological cycle provides energy in the form of hydroelectric power as rivers return water to the ocean. Wind and wave motion can also provide energy sources.

Figure 1.4 shows how a modern economy can use these cycles of nature to provide energy to a society. This involves a sequence (chain) of industrial sectors to

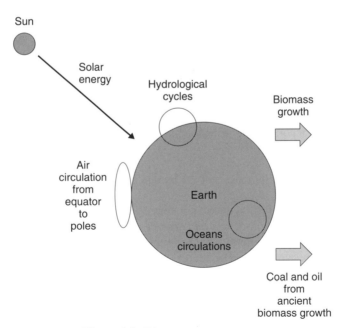

Figure 1.3. Biomass energy sources.

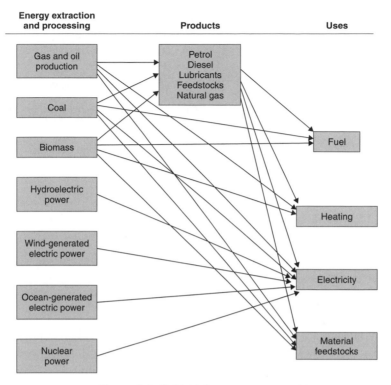

Figure 1.4. Industrial energy sectors.

acquire energy, process energy, and distribute energy to consumers in an economy. Industrial sectors acquire energy for economies through mining of coal and uranium, production of gas and oil, logging of timber, or wind or wave generation of electricity; and thus the first industrial sectors in the energy chain are the energy-extraction industrial sectors of coal mining, petroleum exploration and production, timber, and wind and wave farming. Other industrial sectors process energy for an economy in the form of electrical utilities that produce electricity from burning, coal, uranium, or petroleum, and oil refineries that process petroleum into gasoline, diesel fuel, and petroleum lubricants. Then another set of industrial sectors distributes energy in economies through electrical power transmission networks, gasoline and diesel petroleum distribution stations, fuel oil distribution services, and natural gas distribution networks. Through this complicated scheme of natural cycles and industrial sectors, economies acquire energy from nature. Figure 1.5 shows the known world reserves by region in the year 2000, which totaled 1004 billion barrels. The largest known source of reserves was in the Middle East, which then held about 65% of the world's oil. In this case of a modern society obtaining energy from nature, one can see directly how science and technology affect both the *totality of nature* and *economic value*.

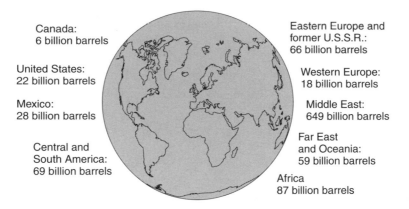

Canada:
6 billion barrels

United States:
22 billion barrels

Mexico:
28 billion barrels

Central and
South America:
69 billion barrels

Eastern Europe and
former U.S.S.R.:
66 billion barrels

Western Europe:
18 billion barrels

Middle East:
649 billion barrels

Far East
and Oceania:
59 billion barrels

Africa
87 billion barrels

Figure 1.5. Known world oil reserves in 2000. (Data from the Energy Information Administration.)

FUNCTIONAL CAPABILITIES OF SOCIETY

Since all humans are biological animals, requiring the satisfaction of physical needs to survive (e.g., air, water, food, clothing, shelter, safety), technology is essential to humanity as the knowledge of how to obtain the things from nature needed to satisfy the physical needs of human societies. Of course, there are many needs of society other than physical needs (e.g., social needs for communication, justice, education). Technology also affects the ability of society to fulfill its social needs (e.g., military technology or communications technology). Accordingly, our conception of nature includes human society itself. From a biological perspective, human society can be conceived of as one example of the social nature of animals (compared to bee society or wolf society). In this broad sense of nature (as both the physical and social aspects of the human animal), technology is the knowledge of manipulating both physical and social nature to satisfy human purposes.

> Technology provides the knowledge bases for all the *functional capabilities* of a society.

Figure 1.6 sketches a model of any modern society in terms of four modes of association within a society: territory, culture, government, and economy. The human species is a territorial animal, organizing within cultural and governmental groups to control a territory. Thus, the basic territorial units of modern societies are nations, with control over specific territories. Within a territory, human occupants can be distinguished according to demographics and wealth. These provide important characteristics in analyzing mass markets. Also, the organization of human use of the environment and ecology of a territory provide important ways of describing territory as a

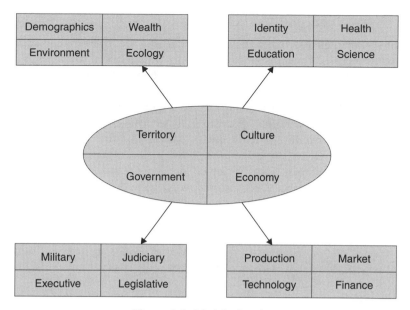

Figure 1.6. Model of society.

societal construct. Culture is another important construct for describing a society, providing the ways that people form identity groups (e.g., familial, tribal, occupational, religious). Education, public health systems, and science also provide important descriptors of culture in modern societies. Government provides the means for ruling populations within a territory, and modern governments are organized into four branches: legislative, executive, judiciary, and military. Economy is the fourth major construct of societal description, and within an economy, the production of goods and services for markets provides the primary societal activities of an economy. Technology and finance provide the means for production and trade in the market.

Accordingly, the capabilities of a modern society include economic technologies, governmental technologies, cultural technologies, and territorial technologies. Change in the technical knowledge in a society is necessary to change the capabilities of that society, and this change can alter that society's relative power within its fellow societies. This is why historically, tribes and nations that developed superior military technology ahead of their neighbors were able to use such improved military capability to conquer their neighbors. This is also historically why firms that innovated new products with superior technical capability have put their competitors out of business.

Accordingly, change in technologies can alter societal structures when they affect a society's (1) power structure (government), (2) means of production (economy), (3) organization and communication (culture), and (4) distribution of wealth and ecology (territory).

Power Structure

For example, the gun affected the power structure of societies from feudal forms by changing the demand for skills in using weapons. In the feudal form of a dominant warrior caste of aristocracy, the barons and kings had almost equal ability to field feudal armies equipped with swords, bows and arrows, lances, and siege weapons. Swords, bows and arrows, and lances require a great deal of skill to use successfully, along with superior fitness. Training in these skills takes years. A skilled feudal warrior could defeat several relatively less skilled warriors. In contrast, the skills to use guns can be attained in a few weeks; and a skilled gunmen cannot take out several semiskilled gunmen. Raw firepower counts over marksmanship. Thus, peasants could be trained into an effective army when given guns instead of swords and bows and arrows. A monarch's ability to raise money by taxes and then to pay and equip a gun-armed peasant army became the base of new power when the feudal age ended in Europe. In the sixteenth and seventeenth centuries, monarchies in England, France, and Russia began systematically to reduce their barons into a dependent service aristocracy without independent military capability.

Means of Production

The invention of steam-powered textile machinery provides an example of how technology can alter both production means and organization in societies. Prior to powered textile machinery, thread was spun and cloth woven by hand on spinning wheels and looms in the cottages of spinners and weavers. Periodically, a jobber would bring them more materials to be worked and would pick up the worked pieces, paying a piece rate. Jobbers played the role of contractors, and cottagers played the role of individual producers. Both the roles and the organization of production were altered with the introduction of the factory system for the new means of production by powered textile machinery. This was the beginning of the industrial revolution in England in the late eighteenth century.

Organization and Communication

It was the high cost of powered textile machinery that necessitated putting the machinery in a central factory and hiring people to come to the factory and attend the machinery. They were then paid for their time. Factory-based industries changed societal organization from the guild form of feudal organization of production into modern capitalistic forms of production. People left farmland and moved into cities to find employment in the new factories. Capitalist owners and managers became "bosses" and peasants became "labor." Societal classes changed from aristocracy and peasantry to capital and labor.

The railroads affected the communications of people and goods. Prior to railroads, the time and cost of transporting people and goods across land inhibited the development of national economic infrastructures and the growth of national-scale markets. For example, in the nineteenth century, the European colonization of the

North American continent expanded rapidly beyond the east coast of North America into its west when the railroad promoted economic interchanges between the manufacturing capability in the east and the agricultural capabilities of the midwest of the United States.

Distribution of Wealth and Ecology

The uneven industrialization of the modern world from the eighteenth century to the twenty-first century created both wealthy and impoverished nations and regions. Technological progress has greatly affected the distribution of wealth in societies and the ecologies of the Earth.

CASE STUDY:

Invention of Xerography

We conclude this introduction to technological innovation by looking at a fundamental question: Just how is a new technology innovated? Technological innovation is a complex process, as nicely illustrated by the famous case of the invention of xerography by Chester F. Carlson. Carlson was born in Seattle, Washington on February 8, 1906. His father had tuberculosis and arthritis and moved the family from Seattle to California. During high school in San Bernadino, California, Carlson worked in part-time jobs in a newspaper office and in a small printing business. He became interested in "the difficult problem of getting words onto paper or into print" (Hall and Hall, 2000, p. 15). After graduating from high school, Carlson first graduated from the two-year Riverside Junior College and then the California Institute of Technology in Pasadena, California, graduating in 1930 with a B.S. in physics. While at Caltech, Carlson (Figure 1.7) began to think of himself as an inventor: "I had read of Edison and other successful inventors and the idea of making an invention appealed to me as one of the few means to accomplish a change in one's economic status, while at the same time bringing to focus my interest in technical things and making it possible to make a contribution to society as well" (Hall and Hall, 2000, p. 17). Carlson got a job with the Bell Telephone Company in New York City, working in their patent department to assist Bell's patent attorneys. It was the time of the Great Depression in the United States and Carson was laid off from Bell in 1933, but found a similar job in a New York law firm and then a year later a job at P.R. Mallory and Company. While at Mallory, Carlson graduated from New York University Law School and eventually became head of Mallory's patent department.

Accordingly, Carlson had a technical background in the physics and chemistry of carbon (the powder he was to use in his invention), printing (his invention was a printing invention), and intellectual property rights (he understood the commercial value of a good basic patent). As a patent lawyer, he was frustrated by the errors in copying a patent for public dissemination and with the trouble in

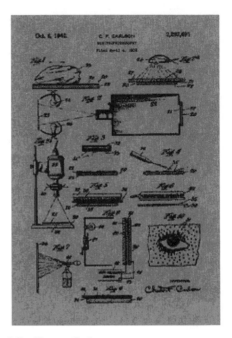

Figure 1.7. Chester Carlson.

making large numbers of copies whose quality decreased continually with number. Then one had to type multiple copies with multiple carbon papers—any typing error requiring an equal multiple of corrections, and the quality of the printing decreasing according to the number of copies. As a potential customer for his invention, he clearly understood the market's need for the invention.

In 1935, Carlson began experimenting evenings and weekends with ways to create a new copying process. His idea was (1) to project the image of a typed paper onto a blank sheet of paper coated with dry carbon, (2) to hold the carbon temporarily on the spaces occupied by letters using static electrical charges induced by the light, and (3) by baking, to melt the ink onto the paper in the patterns of the projected letters. This would produce a quick, dry reproduction of a typed page.

In the fall of 1938, Carlson moved his apparatus from his kitchen in his apartment to a one-room laboratory in Astoria on Long Island, New York. He hired Otto Kornie, a recent Austrian immigrant and a physicist, to help him in the invention. On October 22, 1938, they used static electricity and a photoconductive, sulfur-coated zinc plate to transfer a written phrase, "10-22-38 ASTORIA," from a glass plate to paper. It was the first demonstration of what would later become called *xerography*. It was a crude image, but it reduced his idea to practice, and he filed for a patent.

Yet like all new inventions, it was still not commercially efficient, cost-effective, or easily usable. It required research and development (R&D). The development of a new technology usually costs a great deal of money, takes time, and requires skilled resources. All inventors face similar problems—first conceiving the invention, reducing it to practice, obtaining a patent, and then obtaining support for its development and commercialization.

From 1939 to 1944, Carlson went from company to company seeking support. He was turned down, again and again, by 20 major companies. Each of the companies that turned him down missed one of the great commercial opportunities of that decade. (That story you may have heard about the world beating a path to the door of the inventor of a better mousetrap—not true! A newly invented mousetrap that uses new technology is seldom capable of catching a real mouse until after much costly R&D.)

Finally, in 1944, Carlson's patent work for Mallory brought him into contact with Russell Dayton, who then worked at Battelle Memorial Institute in Ohio. Some researchers at Battelle found Carlson's idea interesting and signed a development agreement with Carlson on October 6, 1944, in return for a share in royalties from the invention. Battelle Memorial, a nonprofit R&D organization, proceeded to make several improvements in the technical process of the invention.

The same year, 1944, an article on Carlson's patent (by Nichols Langer) was read by John Dessauer, the director of research at Haloid Company. The president of Haloid, Joseph Wilson, was informed of the invention by Dessauer. Wilson was looking actively for new technology for his company, for new products. At the time, Wilson's main customer was the giant Kodak, which could at any time eliminate his small business if they chose. Wilson watched Battelle's research progress and in 1947 signed a license agreement with Battelle and Carlson. Finally, all the innovative pieces for Carlson had fallen in place: inventions, patents, R&D, commercialization.

Wilson funded Battelle for the rest of the development and then commercialized the first copiers, which Wilson called *Xerox,* and Wilson later changed the name of the company to *Xerox.* The rest became commercial history in the second half of the twentieth century. Xerox created the new industry of office copying and was one of the fastest-growing companies in the world during the 1950s and 1960s. However, in the year 2000, Xerox came close to bankruptcy. No innovation lasts forever; and no high-tech company can stand still. Carlson derived very substantial income from royalties, as did Battelle; Carlson died in 1968.

Some interesting questions arise from this case: Why did many companies first seeing the new technology not have the vision to grab and exploit it? Why can some R&D outfits, such as Battelle, have better technological vision than many commercial companies but not be able to commercialize products themselves? What kind of leadership qualities do innovative, risk-taking managers such as Joseph Wilson possess? What was the power of the innovation that it enabled the new firm, Xerox, to be dominant for 50 years? Why did Xerox begin to fail after 50 years? These are some of the questions that we shall pursue.

TECHNICALLY SAVVY MANAGERS

Next, we give a precise definition of *innovation:*

> **Technological innovation is both the *invention* of a new technology and its *introduction* into the marketplace as a *new high-tech* product, process, or service.**

The idea of technological innovation has two complex ideas within it: *invention* and *innovation.*

> **Invention is the *creation* of a functional way to do something, an idea for a new technology; innovation is the *commercialization* of new high-tech products, processes, or services.**

A technological entrepreneur such as Joseph Wilson is a manager who seizes a technical opportunity and commercializes it. Wilson added commercial innovation to the technological invention of Carlson. These kinds of entrepreneurial managers play a critical role in technological innovation. They are technically savvy managers.

What made Wilson an outstanding business leader was that he had two kinds of skills: technical savvy and financial savvy. But how much technology does a modern executive need to know to be technically savvy? Actually not much, for it is an understanding of the innovation process that counts.

> **Technological innovation is a business process; and technically savvy managers use principles of innovation.**

Principles of innovation capture the general patterns of technological progress. But they require context to provide a competitive advantage. All *principles* may appear simple but attain complexity in contexts. One must *specifically* understand the context of the application of any *general* principle. To apply the general principles of innovation, one must understand the specific contexts of technological progress. These contexts include technology strategy, high-tech businesses, science and technology infrastructures, engineering function, research function, information function, new product development, high-tech marketing and finance, and so on. It is in the dependency of principles in contexts that one establishes *theory.*

> **Any general theory is both *proved* in specific contexts and *acquires significant meaning* only in specific contexts.**

Good theory should be grounded in principles wherein the context is explicit. Management theory of innovation must be grounded in the principles of technological change in the context of science and economy. Actual cases of innovation provide the grounded context for the principles of innovative management. Accordingly, we will study the grounded truths for managing technological change through actual cases of innovation—innovation theory as summarized in this book.

Innovation theory addresses the *why, what, and how* of technological innovation:

1. Why does technology affect competition?
2. What does a business need to be competitive?
3. How can one effectively plan innovation?

Innovation theory answers each question in turn by elaborating:

1. The context of technological competition
   —*Technology imperative*
2. The context of business innovation
   —*Innovation capability*
3. The context of technology planning
   —*Technology strategy*

SUMMARY AND PRACTICAL IDEAS

In case studies on

- dot.com stock bubble of the 1990s
- Technology in archaeology
- Energy in a modern economy
- Invention of xerography

we have examined the following key ideas:

- Types of technologies
- Technological imperative
- Technological innovation
- Technology and nature
- Functional capabilities of societies
- Technically savvy leadership

These theoretical ideas can be used practically in understanding how modern business change depends on technological innovation. The power of technological innovation in business strategy is a kind of imperative on business competition: to match the technology or perish. The ability to manage technological innovation as a competitive strategy requires a basic understanding of the connections between science and engineering and technology and nature and economy:

- Nature   totality
- Science   discovery
- Technology   manipulation
- Engineering   design
- Economy   utility

FOR REFLECTION

Identify an industry begun after 1850. What were the key technologies important to its establishment? What are the science bases of these technologies? What early firms grew to dominance in the industry? Do they still exist?

2

SCIENCE AND TECHNOLOGY INFRASTRUCTURE

INTRODUCTION

We have seen how changes in technology have affected the cultural evolution of the human species as stages of civilization. Modern civilization now depends on progress in science and technology. We live not in the age of bronze or in the age of iron but in the age of science. Let us look at how modern civilization creates continuing scientific and technological progress through a *science and technology* (S&T) *infrastructure*. We will present this S&T infrastructure in terms of both the *methodological* and *institutional* forms of modern *scientific and technological progress*.

The boundary line between the ancient world and the modern world was at the beginning of science in the seventeenth and eighteenth centuries. After that time, technological innovations become frequent and continuing. It was in the nineteenth and twentieth centuries that there occurred an incredible cultural evolution of social institutions to foster both scientific and technological progress. In modern times, new technologies are created by scientific technology: technological progress enabled by scientific progress. It is an S&T infrastructure that makes scientific technology possible. Now let us look deeply into science, at the specific instance of the origin of molecular biology. This will illustrate the complexity and challenges and drama that is modern science.

CASE STUDY:

Origin of Biotechnology

The case of the origin of biotechnology from progress in a new scientific discipline of molecular biology nicely illustrates the modern interrelation between scientific and technological progress. It also emphasizes how the university part of a modern S&T infrastructure establishes the science base for radical innovations that create new industries. The historical setting was the century of biological research from the 1870s to the 1970s that established the science base for the new biotechnology industry, which began after 1970.

As we saw in the case of the Internet and the stock bubble of the 1990s, new industries begin on the innovation of a new and radical basic technology. The rate of occurrence of these has depended on the rate of scientific progress, which often has taken a long time. For example, the final scientific event in creating the science base for the new biotechnology was a critical biology experiment performed by Stanley Cohen and Herbert Boyer in 1972 that invented the technique for manipulating DNA—recombinant DNA. But the scientific ideas that preceded this experiment began 100 years earlier. These ideas were directed toward answering a central question in science: How is life reproduced? This answer required many stages of research to be performed, including:

1. Investigating the structure of the cell
2. Isolation and chemical analysis of the cell's nucleus, DNA
3. Establishing the principles of heredity
4. Discovering the function of DNA in reproduction
5. Discovering the molecular structure of DNA
6. Deciphering the genetic code of DNA
7. Inventing recombinant DNA techniques

Structure of the Cell

By the early part of the nineteenth century in the then new scientific discipline of biology, scientists were using an eighteenth-century invention, the microscope, to look at bacteria and cells. Cells are the constituent modules of living beings. The scientists saw that cells have a structure consisting of a cell wall, a nucleus, and protoplasma contained within the wall and surrounding the nucleus. In 1838, Christina Ehrenberg was the first to observe the division of the nucleus when a cell reproduced. In 1842, Karl Nageli observed the rodlike chromosomes within the nucleus of plant cells. Thus by the middle of the nineteenth century, biologists had seen that the key to biological reproduction of life involved chromosomes that divided into pairs when the nucleus of the cell split in cell reproduction (Portugal and Cohen, 1977).

Discovery and Chemical Analysis of DNA

Scientific attention next turned to investigating the chemical nature of chromosomes. Science is organized into disciplines; and the techniques and knowledge in one discipline may be used in another discipline. Chemistry and physics, as scientific disciplines, provided tools and knowledge to the scientific discipline of biology. In 1869, a chemist, Friedrich Miescher, reported the discovery of DNA, by precipitating material from the nuclear fraction of cells. He called the material *nuclein*. Subsequent studies showed that it was composed of two components, nucleic acid and protein.

While these studies were occurring, there were continuing improvements in microscopic techniques. The more detail one wishes to observe, the more the means of observation of science need to be improved. For the microscope, specific chemicals were found that could be used to stain the cell selectively. In the 1860s, Paul Ehrlich discovered that staining cells with the new chemically derived coal-tar colors correlated with the chemical composition of the cell components. (This is an example of how technology contributes to science, for the new colors were products of the then new chemical industry.)

In 1873, A. Schneider described the relationships between the chromosomes and various stages of cell division. He noted two states in the process of mitosis (which is the phenomenon of chromosome division, resulting in the separation of the cell nucleus into two daughter nuclei). In 1879, Walter Flemming introduced the term *chromatin* for the colored material found within the nucleus after staining. He suggested that chromatin was identical with Miescher's nuclein.

At this time, studies of nuclear division and the behavior of chromosomes were emphasizing the importance of the nucleus. But it was not yet understood how these processes were related to fertilization. In 1875, Oscar Hertwig demonstrated that fertilization was not the fusion of two cells but the fusion of two nuclei. Meanwhile, the study of nucleic acid components was progressing. In 1879, Albrecht Kossel began publishing on nuclein in the literature. Over the next decades, he (along with Miescher) was foremost in the field of nuclein research; and they and Pheobus Levine (1869–1940) finally laid a clear basis for the determination of the chemistry of nucleic acids.

As early as 1914, Emil Fisher had attempted the chemical synthesis of a nucleotide (a component of nucleic acid, DNA), but real progress was not made in synthesis until 1938. Chemical synthesis of DNA was an important scientific technique necessary to understand the chemical composition of DNA. One of the problems was that DNA and RNA were not distinguished as different molecules until 1938. This is an example of the kinds of problems that scientists often encounter—that nature is more complicated than originally thought. By the end of the 1930s, the true molecular size of DNA had been determined. In 1949, C. E. Carter and W. Cohn found a chemical basis for the differences between RNA and DNA. By 1950, DNA was known to be a high-molecular-weight polymer with phosphate groups, linking deoxyribonucleosides between positions 3 and 5 of sugar groups. The sequence of bases in DNA was then still unknown. Thus, by 1950 the detailed chemical composition of DNA was finally determined but not yet its molecular geometry. Almost 100 years had passed between the discovery of DNA and the determination of its chemical composition.

Principles of Heredity

From 1900 to 1930, while the chemistry of DNA was being sought, the foundation of modern genetics was being established. Understanding the nature of heredity began in the nineteenth century with Darwin's epic work on evolution and with Mendel's pioneering work on genetics. Modern advances in genetic research began in 1910 with Thomas Morgan's group researching the heredity in the fruit fly, *Drosophila melanogaster.* Morgan demonstrated the validity of Mendel's analysis and showed that mutations could be induced by x-rays, providing one means for Darwin's evolutionary mechanisms. By 1922, Morgan's group had analyzed 2000 genes on the four *Drosophila* fly's chromosomes and attempted to calculate the size of the gene. Muller showed that ultraviolet light could also induce mutations. (In the 1980s, an international human genome project would begin, with the goal of mapping the entire human gene set.)

Function of DNA in Reproduction

While the geneticists were showing the principles of heredity, the mechanism of heredity had still not been demonstrated. Was DNA the transmitter of heredity, and if so, how?

Meanwhile, other scientists were studying the mechanism of the gene, with early work on bacterial reproduction coming from scientists using bacterial cultures. R. Kraus (1897), J. A. Arkwright (1920), and O. Avery with A. R. Dochez (1917) had demonstrated the secretion of toxins by bacterial cultures. This had raised the question of what chemical components in the bacterium were required to produce immunological specificity. The search for the answer to this question then revealed a relationship between bacterial infection and the biological activity of DNA.

There had also occurred (as early as 1892) the identification of viruses and their role in disease. In 1911, Peyton Rous discovered that rat tumor extracts contained virus particles capable of transmitting cancer between chickens. In 1933, Max Schlesinger isolated a bacteriophage (virus) that infected a specific kind of bacteria. Next, scientists learned that viruses consisted mainly of protein and DNA. In 1935, W. M. Stanley crystallized the tobacco mosaic virus, which encouraged scientists to study further the physical and chemical properties of viruses.

Meanwhile, in 1928, Frederick Griffith had shown that a mixture of killed infectious bacterial strains with live noninfectious bacteria could create a live infectious strain. In 1935, Lional Avey showed that this transformation was due to the exchange of DNA between dead and living bacteria. This was the first clear demonstration that DNA did, in fact, carry the genetic information. By 1940, work by George Beadle and Edward Tatum further investigated the mechanisms of gene action by demonstrating that genes control the cellular mechanisms of gene action by demonstrating that genes control the cellular production of substances by controlling the production of enzymes needed for their synthesis. The scientific stage was now set to understand the structure and of DNA and how DNA's structure could transmit heredity.

Structure of DNA

We have reviewed the many long lines of research necessary to discover the elements of heredity (genes and DNA) and its function (transmission of heredity). Yet before technology could use this kind of information, one more scientific step was necessary—understanding the mechanism. This step was achieved by a group of scientists that were later to be called the *phage group* and would give rise directly to the modern scientific specialty of molecular biology (Judson, 1979).

In 1940, M. Delbruck, S. Luria, and A. Hershey, founders of the phage group, began collaborating on the study of viruses. One of their students was James Watson, who studied under Luria at the University of Illinois. Watson graduated in 1951 with a desire to discover the structure of DNA, as this was then a great goal in biology. He heard that the Rutherford Lab at Cambridge University was strong in the x-ray study of organic molecules, and he knew that an x-ray picture of DNA would be necessary. With a postdoctoral fellowship from the U.S. government, he asked Luria to arrange for him to do his work at the Rutherford Lab.

Meanwhile, the Rutherford Lab had a researcher using x-rays to try to determine the structure of DNA, Rosalind Franklin. She was a young, bright scientist from Portugal and had joined the lab to study x-ray diffraction of crystallized DNA, under the supervision of a senior scientist, Maurice Wilkins. Prior to the work at the Rutherford Lab, x-ray crystallography had been developed. It was a technique for sending x-rays (high-energy photons) through crystals and inferring the structure of the crystals from the diffraction patterns that the x-rays produced from the structure. (An analogy would be to have a line of pilings near the shore and watch a big wave come in and produce smaller waves from the pilings; and from watching the smaller waves, calculate backwards to measure the spacing between the pilings.) The diffraction of x-rays by crystals was first suggested theoretically by Max von Laue in 1912. Lawrence Bragg and William Henry Bragg established experimental x-ray crystallography as a scientific technique. William Astbury applied the technique to the study of organic fibers in 1928. Dorothy Crowfoot Hodgkin published the first x-ray study of a crystalline protein in 1934. Scientific instrumentation and instrumental techniques are critical to the progress in science. Just as the microscope was essential to observing the cell and its structure, x-ray crystallography was essential to observe the structure of DNA.

Once at the Rutherford Lab, Watson found a collaborator in Francis Crick, then a graduate student working on a physics degree. Crick was a bit older than Watson because his graduate studies had been interrupted by service in World War II. While Watson brought a knowledge of biology and organic chemistry to their collaboration, Crick brought a knowledge of physics. Both were necessary for the job of constructing a molecular model of DNA.

But a critical piece of information they needed was a good x-ray diffraction picture of a crystalline DNA. Franklin was working on this, and it was not easy. There were two crystalline forms of DNA, and only one of these would yield a good picture. Moreover, it had to be oriented just right to get a picture that would be interpretable as to structure. Franklin would finally get a good picture of the right form. She was an excellent scientist.

Meanwhile, Watson had learned of Linus Pauling's and Robert Corey's work on the structure of crystalline amino acids and small peptide proteins. From these Pauling published a structural description of a first example of a helical form of a protein. Pauling was one of the most famous organic chemists in the world.

Watson feared that Pauling would soon model DNA structure, robbing the young and ambitious scientist of fame and scientific immortality. Watson saw himself in a scientific race with Pauling to be the first to discover the structure of DNA. Pauling then did not know that Watson existed. (But this is the general model of human competition, the young racing to exceed their elders; and scientists are only human.)

Watson then conjectured: What kind of x-ray diffraction picture would a helical molecule make? If DNA were helical, Watson wanted to be prepared to interpret it from an x-ray and asked another young expert in diffraction modeling for a tutorial. He was told how (if the picture were taken "head-on" down the axis of the helix) to measure the angle of the helix. Watson was thus equipped to interpret an x-ray picture of DNA if he could only get his hands on one.

Meanwhile, Watson and Crick had scoured the chemical literature about DNA and been trying to construct ball-and-wire-cutout models of DNA. At first they had tried a triple-helix model, and it didn't work. Finally, Watson heard that Franklin had got a good picture, but he feared that Franklin would not show it to him. Franklin was as fierce a competitor as Watson and was not willing to show her picture before she had time to calculate its meaning; so Watson sneaked a peak at the picture without Franklin's permission. There it was! Watson saw it. Clearly a helix, and a double helix!

Quickly, Watson measured the pattern and rushed to Crick with the information on the angle of the helix. Watson and Crick put their model together in the form of a double helix, two strands of amino acid chains, twisting about each other like intertwined spiral staircases. They used the angle for the helix as measured from Franklin's picture. All the physical calculations and organic chemistry fit together in the model beautifully. Without a doubt! This was the holy grail of biology—the double-helix structure of DNA.

Moreover, the structure itself was informative. It clearly indicated the molecular action of DNA in the mitosis of cell reproduction. DNA was structured as a pair of twisted templates, complementary to one another. In reproduction, the two templates untwisted and separated from one another, providing two identical patterns for constructing proteins, that is, reproducing life. In this untwisting and chemical reproduction of proteins, life was inherited biologically.

In 1995, Watson and Crick and Wilkins were awarded the Nobel Prize in Physiology. Unfortunately, Rosalind Franklin was not so honored, the reason given being her untimely death before the prize was awarded. (This is not a nice story because she should have received appropriate recognition for her essential contribution.)

Genetic Coding

By the early 1960s, it was therefore clear that the double-helical structure of DNA was molecularly responsible for the phenomenon of heredity. Proteins serve as structural elements of a cell and as catalysts (enzymes) for metabolic processes in a cell. DNA provides the structural template for protein manufacture, replicating proteins through the intermediary templates of RNA. DNA makes RNA, RNA makes proteins: a two-step biological process. DNA structures the synthesis of RNA, and RNA structures the synthesis of proteins. What was not yet clear was how the information for protein manufacture was encoded in the DNA. In 1965, Marshall Nirenberg and Philip Neder deciphered the basic triplet coding of the DNA molecule. The amino acids that composed the DNA structure acted in groups of three acids to code for a segment of protein construction. Thus, in 100 years, science had discovered the chemical basis for heredity and understood its molecular structure and mechanistic function in transmitting heredity information.

Recombinant DNA Technique

After understanding the mechanism of the gene, the next step was trying to manipulate the gene. Several scientists began trying to cut and splice genes. In 1965, Paul Berg at Stanford planned to transfer DNA into *Escherichia coli* bacteria, using an animal virus (Svrp lambda phage). *E. coli* bacteria can live in human intestines, and the SV40 virus is a virus of monkeys that can produce tumor cells in cultures of human cells. Because of the dangerous nature of the SV40 virus, Berg decided not to proceed with the experiment, publishing a design for hybridizing bacteria in 1972. Berg organized a historic meeting on safety, the Conference on Biohazards in Cancer Research in California on January 22–24, 1973 (Olby, 1974). This stimulated later U.S. government action to set safety standards for biotechnology research.

Also at this time, Peter Lobban, a Stanford graduate student, had been working on a similar idea for gene splicing. Lobban was studying under Dale Kaiser of the Stanford Medical School. (Kaiser had been one of the phage school group, which also had spawned Watson.) When ideas are ripe, there have often been scientists competing hotly for the same scientific goal. Scientists compete for the fame of being the first to discover or to understand—second place gets no recognition.

A colleague at the University of California responded to Berg's request for some EcoRI enzyme, which cleaves DNA (and leaves the "sticky ends" of the cut DNA). Berg gave the enzyme to one of his students, Janet Mertz, to study the enzyme's behavior in cutting DNA. Mertz noticed that when the EcoRI enzyme cleaved an SV40 DNA circlet, the free ends of the resulting cut and linear DNA eventually re-formed into a circle. Mertz asked a colleague at Stanford to look at the action of the enzyme under an electron microscope. They learned that any two DNA molecules exposed to EcoRI could be recombined to form hybrid DNA molecules. Nature had arranged DNA so that once cut, it respliced itself automatically.

Another professor in Stanford University's Medical Department, Stanley Cohen, learned of Janet Mertz's results. Cohen had also thought of constructing a hybrid DNA molecule from plasmids using the EcoRI enzyme. Plasmids are the circles of DNA that float outside the nucleus in a cell and manufacture enzymes the cell needs for its metabolism (the DNA in the nucleus of the cell are used principally for reproduction). In November 1972, Cohen attended a biology conference in Hawaii. He was a colleague of Herbert Boyer, who had given the EcoRI enzyme to Berg (and Berg's student Mertz). At a dinner one evening, Cohen proposed to Boyer that they create a hybrid DNA molecule without the help of viruses. Another colleague at that dinner, Stanley Falfkow of the University of Washington at Seattle, offered them a plasmid, RSF1010, to use that confers resistance to antibiotics in bacteria so that they could see whether the recombined DNA worked in the new host.

After returning from the Hawaii conference, Boyer and Cohen began joint experiments. By the spring of 1973, Cohen and Boyer had completed three splicings of plasmid DNAs. Boyer presented the results of these experiments in June 1973 at the Gordon Research Conference on Nucleic Acids in the United States (with publication following in the *Proceedings of the National Academy of Sciences,* November 1973). Cohen and Boyer applied for a patent on the new technique.

After 100 years of scientific research into the nature of heredity, humanity could now begin to manipulate genetic material deliberately at the molecular level—and a new industry was born, biotechnology. Boyer and Cohen would win Nobel prizes. Boyer would be involved in the first new biotechnology company (Genentech) to go public and would become a millionaire. The days for biologists to become industrial scientists had begun.

SCIENCE BASES FOR TECHNOLOGY

Science has provided the knowledge bases for scientific technology in a pattern:

1. Scientists pursue research that asks very basic and universal questions about what things exist and how things work. (In the case of genetic engineering, the science base was guided by the questions: What is life? How does life reproduce itself?)

2. To answer such questions, scientists require new instrumentation to discover and study things. (In the case of genetic research, the microscope, chemical analysis techniques, cell culture techniques, x-ray diffraction techniques, and electron microscope were some of the important instruments required to discover and observe the gene and its functions.)

3. These studies are carried out by different disciplinary groups specializing in different instrumental and theoretical techniques: biologists, chemists, and physicists. (Even among the biologists, specialists in gene heredity research differ from specialists in viral or bacterial research.) Accordingly, science is pursued in disciplinary specialties, each seeing only one aspect of the existing thing (much like the tale of the blind philosophers who never saw an entire elephant but went around the elephant feeling each part and puzzling how it all went together). Nature is always broader than any one discipline or disciplinary specialty.

4. Major advances in science occur when sufficient parts of the puzzling object have been discovered and observed and someone imagines how to put it all

together properly (as Watson and Crick modeled the DNA molecule). A scientific model is conceptually powerful because it often shows both the structure and the dynamics of a process implied by the structure.

5. Scientific progress takes much time, patience, continuity, and expense. Instruments need to be invented and developed. Phenomena need to be discovered and studied. Phenomenal processes are complex, subtle, multileveled, and microscopic in mechanistic detail. (In the case of gene research, the instruments of the microscope and electron diffraction were critical, along with other instruments and techniques. Phenomena such as the cell structure and processes required discovery. The replication process was complex and subtle, requiring determination of a helix structure and deciphering of nature's coding.)

6. From an economic perspective, science can be viewed as a form of societal investment in possibilities of future technologies. Since time for scientific discovery is lengthy and science is complicated, science must be sponsored and performed as a kind of overhead function in society. Without the overhead of basic knowledge creation, technological innovation eventually stagnates for lack of new phenomenal knowledge for its inventive ideas.

7. Once science has created a new phenomenal knowledge base, inventions for a new technology may be made either by scientists or technologists (e.g., scientists invented the recombinant DNA techniques). These radical technological inventions start a new technology. This is the time to begin investment in a technological revolution and to begin new industries based on it.

8. When the new technology is pervasive across several industries (as genetic engineering is across medicine, agriculture, forestry, marine biology, materials, etc.), the technological revolution may fuel a new economic expansion. The long waves of economy history are grounded in scientific advances that create basic new industrial technologies.

9. There are general implications for management. Corporations should be supportive of university research that focuses on fundamental questions underlying core technologies of the corporation. Corporations need to perform some active basic research in the science bases of their core technologies to maintain a window-on-science for technological forecasting.

SCIENTIFIC METHOD

Science is organized into many scientific disciplines. Major disciplines of science include mathematics, physics, chemistry, biology, computer science, social sciences, environmental sciences, and astronomy. Each of these is divided into subdisciplines and specialties. The kinds of observational or experimental techniques used in different disciplines vary, as does how theory is socially constructed and methodologically validated. But all the scientific approaches fall within the general approach of *scientific method,* and the goals of all disciplines are to discover and understand nature. Nature is discovered by observation and experimentation that results in the development of theory. Scientific knowledge accumulates through observation and experimentation, abstracted and generalized into scientific theory. Theory is then validated by further observation and experimentation.

> **The way of conducting observation and experimentation to develop and verify theory is called the scientific method.**

Scientific method is an interaction of the activities of (1) experimental observation of nature, (2) construction of theoretical representations of nature, (3) prediction of the sequence of natural states of nature, and (4) formulation of a disciplinary perspective paradigm: experiment and theory, prediction, and paradigm. One can envision a model of science as a set of activities, as shown in Figure 2.1. Any model of science requires at least two scientists (S_1 and S_2) and two things (T_1 and T_2) of nature, about which the scientists are observing and constructing theory and predicting nature. At least two scientists are needed in a model of science, since science is a community activity; and at least two things are needed as science describes relationships between things in nature.

Experiment and Observation of Nature

For experimentation, scientists need to observe and measure nature. For observing nature, scientists invent means of observation, scientific instruments, to see the phenomena of nature. (For example, the invention of the recombinant DNA technique was a scientific technique to experiment with genes.) Instruments enable discovery of new nature. Instruments enable scientists to make careful comparisons of both the quality and quantity of things observed. (For example, the instruments of the microscope and electron diffraction were essential to gene research.) The idea of quantity as well as the quality of things in nature is a fundamental aspect of scientific method. The quantitative aspects of nature must be in a particular formal language of mathematics. Modern mathematics operates as a specialized kind of language to express the quantitative forms of natural observations.

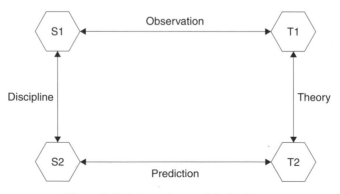

Figure 2.1. Information model of science.

Generalization and Theory of Nature

Scientists construct a general theory of nature to express the qualitative and quantitative observations of natural things (T_1 and T_2). Such theory construction adds new specialized terms to common languages (such as *nucleon, DNA*) to express the objects and properties of the observed things along with new terms (such as *mitosis*) to express relationships between phenomenal things. These new terms of observable objects, properties, and relationships form a scientific subset of common language, a *specialized semantic language,* called a *scientific theory.* A scientific theory provides a precise language for describing the abstract and general description of nature.

Scientific theories of nature are created from observations as semantically specialized languages (with terms about the precise objects, properties, and relations in the field of nature observed). If these semantically specialized languages can be expressed additionally in a syntactical language of mathematics, scientific theories provide the capability to express not only ideas about the qualitative aspects of nature but also facts as to the quantitative aspects of nature.

Models and Prediction of Nature

The validity of the theory that scientists construct to explain nature is tested and validated by predictions from models expressed in the theory of the sequence of natural states. In a quantitative model, a future state of nature can be predicted theoretically from a present state of nature, and then further observations confirm or discredit the theory by testing the accuracy of predictions. The theoretical relationships between the objects T_1 and T_2 of nature are said to be *causal relationships;* and the prediction from theory is possible if theory realistically captures nature's causal relationships among the things in nature. To do this, a quantitative model of a natural thing is constructed in the semantically specialized language of the theory. For example, a model of an atom is expressed in the theory of quantum mechanics. Quantitative models of natural things allow validation of the theory about nature through precise predictions of the model and the testing of these predictions by experiments.

Disciplines and Paradigms of Nature

The complexity of nature is viewed in different perspectives by separate communities of scientists through a unit called a *scientific discipline.* For example, the disciplinary community of physicists views everything in nature as physical phenomena in the paradigm of mechanism. Biologists view all living things in nature in the paradigm of mechanism and function. Sociologists view all human beings in nature in the paradigm of social communities. Physiologists view all individual human beings in nature in the paradigm of conscious/unconsciousness. Economists view all human transactions in the paradigm of rational trades of value. All scientists divide into separate disciplinary groups that view nature in a distinct perspective of distinct concepts—*disciplinary paradigms.*

> **Scientific method is an integrated approach of experiment, theory, prediction, and paradigm.**

CASE STUDY:

Origin of the Biotechnology Industry

S&T infrastructures have a major impact on the economy when technological revolutions are begun on scientific discoveries. The opportunity to start new firms and the ability to position corporations in a new technology occur in the early years of the technology. The biotechnology industry was created directly from scientific discoveries in genetics. Two of the first new biotech firms were Genentech and Cetus.

As we noted earlier, in 1973 Cohen and Boyer applied for a basic patent on recombinant DNA techniques for genetic engineering. Subsequently, Boyer founded Genentech, and Cohen joined Cetus. In the years from 1976 to 1982 in the United States, over 100 other research firms were formed to commercialize the new biotechnology. In 1980, Genentech and Cetus both went public, and Boyer and Cohen became millionaires.

Genentech was founded by a scientist, Herbert Boyer, and an entrepreneur, Robert A. Swanson. Swanson heard of the new DNA technique and saw the potential for raising venture capital to start a genetic engineering firm. The story is that Swanson walked into Boyer's office, introduced himself, and proposed that they start a new firm. They each put up $500 in 1976 and started Genentech. Early financing in Genentech was secured from venture capital funds and industrial sources. Lubrizol purchased 24% of Genentech in 1979; Monsanto bought about 5%.

Founded earlier, in 1971, Cetus had been established to provide a commercial service for fast screening of microorganisms. In 1976, Cetus changed its business to designing gene-engineered biological products. For this, Cetus retained Stanley Cohen as one of its 33 scientific consultants; and then Cetus hired Cohen as head of Cetus Palo Alto (*Business Week,* 1984b). Further investment in Cetus came from companies interested in the new technology. A consortium of Japanese companies owned 1.59% of Cetus; Standard Oil of Indiana purchased 28% of their stock; Standard Oil of California bought 22.4%; National Distillers & Chemical purchased 14.2%. Corporate investors wanted to learn the new technology.

Both Genentech and Cetus offered stock options to their key scientists. Genentech and Cetus were the first biotechnology firms to go public. Genentech realized net proceeds of $36 million. At the end of fiscal year 1981, it had $30 million in cash but required about $1 million yearly for its R&D activities.

In its public offering, Cetus raised $115 million at $23 a share. Of this, $27 million was intended for production and distribution of Cetus-developed product processes, $25 million for self-funded R&D, $24 million for research administrative facilities, $19 million for additional machinery and equipment, and $12 million for financing of new-venture subsidiaries.

For new firms it is important that early products create income. In 1982, Genentech's product interests were in health care, industrial catalysis, and agriculture. In 1982, early products included genetically engineered human insulin, human growth hormone, leukocyte and fibroblast interferon, immune interferon, bovine and porcine growth hormones, and foot-and-mouth vaccine. Genentech's human insulin project was a joint venture with Eli Lilly aimed at a world market of $300 million in animal insulin. Genentech's human growth hormone project was a venture with KabiGen (a Swedish pharmaceutical manufacturer), a world market of $100 million yearly. The leukocyte and fibroblast interferon was a joint venture with Hoffmann–La Roche, and the immune interferon with Daiichi Seiyaku and Toray Industries. The bovine and porcine growth hormones were a joint venture with Monsanto, and the foot-and-mouth vaccine, with International Minerals and Chemicals.

In comparison in 1982, Cetus was interested primarily in products in health care, chemicals, food, agribusiness, and energy. Their commercial projects included high-purity fructose, veterinary products, and human interferon. The high-purity fructose project was a joint venture with Standard Oil of California. In 1983, Cetus introduced its first genetically engineered product, a vaccine to prevent scours, a toxic diarrhea in newborn pigs. But product income was not raising fast enough for Cetus. In 1983, Cetus replaced its president with a new president, Robert A. Fildes. He developed a five-year plan to provide Cetus with a positive cash flow by 1987. His plan narrowed the focus to the two major health care markets, diagnostics and cancer therapeutics. Both monoclonal antibody and recombinant DNA technologies were employed. He also reduced employment to 480 people (with 280 scientists). The new president then hired people experienced in developing drugs and taking them to market. Hopes then were high for quick profits.

SCIENCE AND TECHNOLOGY INFORMATION TRACKS

In the model of science, one sees that disciplinary communication between two scientists is recorded as a disciplinary paradigm in archival scientific journals, scientific conferences, and scientific textbooks. This communication, which can be called a science information track, is open and without proprietary rights. However, communication between scientists and technologists (engineers) and between technologists is along a different information track, a technology information track. Communications between scientists and between engineers occur in two distinctly different tracks of information, publications and conferences and journals (Figure 2.2). Research progress in science and in technology operates on two separate communication tracks: a science information track and a technology information track. Figure 2.2 also summarizes how these two research tracks interact in creating knowledge for innovation. In the science track, the current state of *scientific knowledge* is archived in scholarly journals of disciplinary-focused scientific and professional societies. This knowledge is summarized and codified in textbooks and communicated to the next generation of scientists, engineers, and other professionals through education courses at the undergraduate and graduate level and as continuing education.

From an understanding of parts of the current state of *scientific knowledge,* researchers in scientific, engineering, and professional *disciplines* pose *fundamental*

SCIENCE INFORMATION TRACK

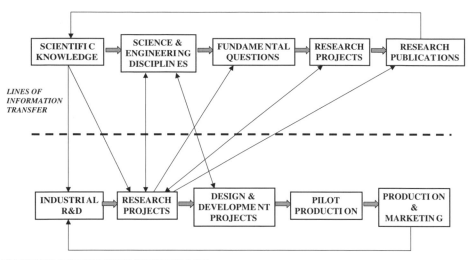

TECHNOLOGY INFORMATION TRACK

Figure 2.2. Science and technology information tracks.

questions to advance the state of knowledge in their disciplinary specialties. These fundamental questions are framed as a *research project* with methods to obtain answers, and from successful research projects come *research publications*. These research publications in scholarly journals then add to the previous scientific knowledge. The research publications are put into the open literature of international science so that all basic knowledge is available to everybody: to all competitors, to all nations, and to future generations. This is the reason that all technological knowledge eventually diffuses throughout the world—because the science bases of technology are in the public domain. No nation or firm can hope to retain for long an intellectual monopoly on any knowledge. Only temporary legal monopolies are possible for technology because of patent laws, and patients are always for a limited period: usually, 17 years. It makes the knowledge base of technology essentially open over the long term, which is why industrialization is open to all nations.

New technology arises in the technology track from invention occurring in research projects in this track. In the technology track, *industrial R&D strategy* provides the basis for focusing and funding most *technological research projects*. These projects use information from the current state of *scientific knowledge* and from *science and engineering and professional disciplines* and from *scientific research publications*. The arrows connecting the science and technology research tracks indicate these knowledge contributions of science to technology. Textbooks and handbooks from the general state of scientific knowledge and from particular disciplines provide the underlying knowledge base of facts, theories, instrumentation, and methods that are used in technological research projects. The latest results in scientific progress are found in the science research publications.

Successful *technological research projects* do not usually result in research publications but in inventions that are developed further in *design and development projects*. Technological projects do not usually result in publications because technology is a proprietary value in improved products, production, or services. Novel and useful new ideas in technology are sometimes published in the form of a patent, because a patent provides temporary proprietary rights. Technology development projects are aimed at new or improved products, services, or production processes. As appropriate, new products or services need to be produced at a *pilot plant level* and *tested* before *initial production* and *marketing* begin.

One can see that the goal of the science research track is to increase public knowledge about nature, whereas the goal of the technology research track is to increase private economic benefits. There are many intellectual interactions between these two tracks. The science and engineering disciplines take many problems for research from industrial research projects, and industrial research projects take much information from science and engineering disciplines and from scientific research publications.

Particularly on the scientific information track, the scientific infrastructure extends beyond national borders to an international infrastructure of scientific activities. From the beginning, science has been international, and scientific ideas have always spread rapidly throughout the world, and this was facilitated by a major change in the university structure beginning in 1800. Before 1800, universities were organized as training in the four colleges of theology, law, medicine, and natural philosophy. Then in Germany, Wilhelm Humboldt advocated reforming the medieval universities in Prussia into the form of the modern disciplinary science departmental structure and the professional schools of the modern university. The Prussian reforms spread through the German universities and provided a model for graduate research-based education that was emulated in the remainder of Europe and in the United States. Public universities designed as research-based institutions (such as the universities of Michigan at Ann Arbor, Wisconsin at Madison, and California at Berkeley) were established in the U.S. land-grant college system after the Civil War. In the same period, American seminary institutions (such as Harvard, Yale, and Princeton) were reorganized to provide research-based graduate programs.

In the last part of the nineteenth century, equally important changes were occurring in the new technology-based industries of electricity and chemicals with the establishment of engineering departments in firms. After 1920, many of these firms established science-based corporate research laboratories. Famous examples in the United States include General Electric's corporate research laboratory in New York, AT&T's Bell Laboratories in New Jersey, DuPont's corporate research laboratory in Delaware, Dow's corporate research laboratory in Michigan, and General Motors' technical laboratory in Michigan. After World War II, national governments began supporting basic research in universities and applied research in government laboratories and in some industries.

In summary, science and technology information tracks evolved as a result of cultural evolution in societal sectors:

- In higher education in the reform of the university into a research-based educational institution

- In the establishment of corporate central research laboratories
- In government in the support of research

By the end of the twentieth century, the result of these institutional changes was to bring together and more directly couple scientific advance and technological development.

CASE STUDY CONTINUED:

Origin of the Biotechnology Industry

However, the path to riches for the new biotechnology industry did not go smoothly from the early 1980s to the mid-1990s. The reason was that the complexity of nature turned out to be more than anticipated by the new biotechnology industry, as summarized by Thayer in 1996: "Fighting waves of hype and pessimism—while trying to create products and access markets—tests [biotechnology] firms' ability to endure" (p. 13). By 1996, the biotechnology industry had created 35 therapeutic products which had a total annual sale of over $7 billion. These biopharmaceutical products were used to treat cancer, multiple sclerosis, anemia, growth deficiencies, diabetes, AIDS, hepatitis, heart attack, hemophilia, cystic fibrosis, and some rare genetic deceases. But the industry was not initially as successful as early investors had hoped. Genentech had hoped that producing a protein product called TPA would catapult them into the large-firm status, but the costs of developing and proving products and the relatively small market for TPA put Genentech into a financial crises in 1990. To survive, Genentech sold 60% of its equity to Hoffman–La Roche: "Despite TPA's success today, it took the 20-year-old company many years and many millions of dollars to prove that it had an important product" (Thayer, 1996, p. 13).

In the 1990s, most marketing of new biotechnology therapeutic products was through the older, established pharmaceutical firms rather than the new biotechnology firms that pioneered pharmaceutical recombinant DNA technology. Genentech had partnered with Hoffman–La Roche and Chiron with Ciba–Geigy. Other startups from the early 1980s were firms such as Biogen, Amgen, Chiron, Genetics Institute, Genzyme, and Immunex. Biogen did pioneering work on proteins such as α-interferon, insulin, and hepatitis. But to support themselves, Biogen licensed their discoveries so that other pharmaceutical firms would market their products. Genetics Institute and Immunex were majority-owned by American Home Products. The one exception to this pattern of the fate of early biotechnology firms was Amgen, which in 1995 had become an industry leader in biotechnology and an independent, fully integrated biopharmaceutical producer, with sales of $1.82 billion.

Why had the early hoped-for big profits in biotechnology not occurred, although the biotechnology has survived and continues to develop? The answer to this was rooted in biological science: "Early expectations, in hindsight considered naive, were that drugs based on natural proteins would be easier and faster to develop. . . . However, . . . biology was more complex than anticipated" (Thayer, 1996, p. 17). For example, one of the first natural proteins, α-interferon, took 10 years to be useful in antiviral therapy. When interferon was first produced, there had not been enough available to really permit an understanding of its biological functions. The production of α-interferon in quantity through biotechnology techniques allowed real studies and experiments to learn how to begin to use it therapeutically. This combination of developing technologies to produce therapeutic proteins in quantity and to use them therapeutically took a long time and many developmental dollars.

Cetus spent millions of dollars and bet everything on interleukin-2 as an anticancer drug but failed to obtain the U.S. Food and Drug Administration's (FDA) approval to market for this purpose. Subsequently, in 1992, Chiron acquired Cetus. Even in 1995, interleukin-2, which was eventually approved by the FDA, provided only 4% of Chiron's revenues of $1.1 billion. About this, George B. Rathmann commented: "The pain of trying to get interleukin-2 through the clinic just about bankrupted Cetus and never has generated significant sales" (Thayer, 1996, p. 17).

The innovation process for the biotechnology industry in the United Sates included (1) developing a product, (2) developing a production process, (3) testing the product for therapeutic purposes, (4) proving to the FDA that the product is useful and safe, and (5) marketing the product. In fact, recombinant DNA techniques were only a small part of the technology needed by biotechnology and the smallest part of its innovation expenditures. The testing part of the innovation process to gain FDA approval took the longest time (typically, seven years) and the greatest cost.

Because of this long and expensive FDA process in the United States, extensive partnering occurred between U.S. biotech firms and the larger, established pharmaceutical firms. For example, in 1995, pharmaceutical companies spent $3.5 billion to acquire biotechnology companies and $1.6 billion on R&D licensing agreements (Abelson, 1996). Also, pharmaceutical firms spent more than $700 million to obtain access to data banks on the human genome that were being developed by nine biotechnology firms. The U.S. government role in supporting science was essential to the U.S. biotechnology industry: "The government has a very big role to play [in helping] to decrease the costs. Support of basic research through NIH [National Institutes of Health] is very important to continue the flow of technology platforms on which new breakthrough developments can be based" (Henri Termeer, chairman and CEO of Genzyme and chairman of the U.S. Biotechnology Industry Organization) (Thayer, 1996, p. 19).

In this case study of the early decades of the biotechnology industry, we see that the scientific importance of understanding the molecular nature of biology (the discipline now called *molecular biology*) proved to be the future of the pharmaceutical industry, as an essential methodology in the development of new drugs. Yet making money from the technology of recombinant DNA was harder and took longer than expected. The reason was that biological nature turned out to be more complicated than anticipated. The biotechnology industry's technology depended on and continues to depend on new science. In turn, the technology needs of the biotechnology industry have helped drive discoveries in biological science.

The progress of a new technology depends on the progress of understanding the complexity of nature underlying the technology.

All the expense going into the biotechnology industry has turned out to be worth it, for biology continues to be the future of the pharmaceutical industry. Rathmann summarized the situation nicely in 1996: "It doesn't have to follow that science automatically translates into great practical results, but so far the hallmark of biotechnology is very good science and that now is stronger and broader than ever. . . . The power of the science is ample justification that there should be good things ahead for biotechnology" (Thayer, 1996, p. 18).

NATURE AND SCIENTIFIC TECHNOLOGY

It is the complexity of nature that requires scientific technology to discover, understand, and manipulate nature. Technology uses science to know and understand nature. New technology is sometimes invented by scientists (as in the invention of the recombinant DNA technique) and sometimes by engineers. But wherever invented, it is engineers who use new technology for the design of new high-tech products, processes, and services. For example, early in the new biotechnology, industrial bioengineers designed new production processes for producing proteins using biological cells (*E. coli*). The relationship between science and technology and between scientists and engineers is to use new science for new technology for new high-tech products, processes, or services. The training and practice of engineering is organized by fields of technologies such as mechanical engineers, civil engineers, electrical and electronic engineers, and chemical engineers.

Scientists construct mathematical models of nature by theory and experiment. These models can be used for the prediction of technical performance when nature is manipulated. By predicting technical performance, an engineer can design (prescribe) the degree of performance required for application of the technology. Thus both scientific theory and observation/experimentation are useful to technology and engineers.

We can envision this relationship between scientists and engineers by adding an engineer to the model of science as in Figure 2.3. There, an engineer E_1 organizes a field of technologies, engineering discipline, upon inventions and a scientific knowledge base of nature. Engineer E_1 uses the design principles of a technology to design products P_1 that manipulate nature according to a scientific theory that can predict the states of nature obtained by the manipulation. In this way scientific method eventually provides the ground for scientific technology.

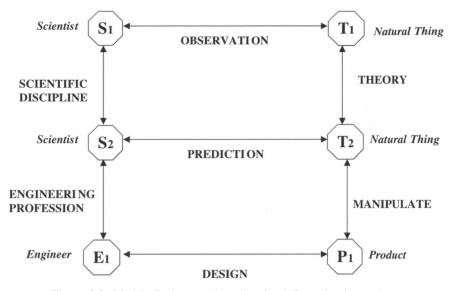

Figure 2.3. Model of science and engineering information interactions.

Scientific methodology has turned out to be critical to the invention of new technologies and to the systematic improvement of technologies. The power of the scientific method enables engineers as technologists to understand and predict the phenomena underlying a technology. Predicting the phenomena underlying a technology enables an engineer or technologist to predict (and thus prescribe) the technical performance of the technology. In modern times, all the major new technologies have been invented based on scientific progress.

> **Scientific technology is a** *manipulation* **of nature for human purpose,** *based on scientific phenomena.*

Technology has existed since humans became toolmakers; but scientific technology needed to wait for the advent of science. The dramatic change from the ancient world to the modern world was due to the rise of scientific technology, requiring (1) the origin of science, (2) the application to economic production of new technological inventions that science made possible, and (3) continuing dynamic interaction between the new science and new technology.

NATIONAL S&T INFRASTRUCTURES

The modern means of scientific technology evolved into three institutional sectors of a national S&T infrastructure: university, industry, and government. The institutional sponsors of R&D projects are principally industry and the federal government (with some support by state governments and private philanthropic foundations). The institutional performers of R&D are industrial research laboratories, governmental research laboratories, and universities. Industry has become the principal producer of technological progress, and universities, the principal producers of scientific progress. Government laboratories participate in varying degrees by country in performing some technological and scientific progress. Government became the major sponsor of scientific research and a major sponsor of technological development in selected areas. In the second half of the twentieth century the S&T infrastructure of nations changed dramatically, due to increased direct participation of governments in the support of research.

Accordingly, in a modern S&T infrastructure, there are two levels of national innovation: a macro and a micro institutional level. At the macro level, innovation occurs in international, national, and industrial research activities. The international level of innovation involves the sharing knowledge and cooperation in creating knowledge across nations and throughout the world. The national level involves sharing knowledge across the institutional sectors of a nation. The industrial level involves sharing knowledge across the firms in an industry. At the micro level, individual firms innovate by performing research and embodying new technologies in their products, services, production, and operations. It is in the interaction between the micro and macro levels of innovation that international competition in high-tech industries is won or lost.

The first issue to address about technological innovation in an S&T infrastructure is the question of when and how industry needs scientific progress. Industry uses technology directly and science indirectly. Technology is embedded in the design of products, services, and production. But as all technology matures, industry needs new science to make progress in technology and to create a scientific basis for new technology. So industry does need science when (1) technological progress in an existing technology cannot be made without a deeper understanding of the science underlying the technology, or (2) new basic technologies need to be created from new science.

The second problem about an S&T infrastructure arises from the fact that the university, not industry, is the primary performer of science progress. Industry must look to the university for progress in science, but universities traditionally have advanced science not in forms directly usable by industry nor in a timely manner.

Accordingly, the practical S&T policy issues of a nation's S&T infrastructure to run such an infrastructure effectively are:

1. How can firms use universities to stay technically competitive in a world of rapidly changing science and technology?

2. How can universities obtain funds from both industry and government to support the advancement of science and respond to industrial needs for science in appropriate forms and in a timely manner?

3. How can a government best direct its R&D support to facilitate partnerships between universities and industries?

SUMMARY AND PRACTICAL IDEAS

In case studies on

- Origin of biotechnology
- Origin of the biotechnology industry

we have examined the following key ideas:

- Science bases for technology
- Scientific technology
- Science and technology information tracks
- National R&D infrastructures
- Macro and micro levels of innovation

These theoretical ideas can be used practically in understanding how science can create new technology through understanding nature and/or creating means of manipulating nature. Moreover, scientific and technological progress are long-term societal endeavors, requiring a complex and expensive infrastructure of universities, government support of science, and industrial research. In the modern world of scientific technology, progress has come neither easily nor cheaply. Still scientific technology changes everything!

FOR REFLECTION

Obtain from the U.S. National Science Foundation a recent copy of their description of the U.S. R&D system (*http://www. nsf.gov*). What are the current trends in the support and performance of basic research (e.g., science)?

3
TECHNOLOGY IN ECONOMY

INTRODUCTION

We have emphasized that technological innovation has been the major force for change in the modern world; and we have seen how new science can create the basis for new technology. As in the case of the biotech industry, a new basic technology creates a new industry. This pattern has driven economic development: new science, new technology, and new industry. This economic pattern occurs in distinct spurts—periods of economic expansion, each due to a new basic technological innovation. This has been called a *long wave of economic dynamics*. We will provide a modern overview of how technological change fosters economic change, covering the basic ideas of *economic cycles, pervasive innovations, international technology transfer,* and *globalization.*

CASE STUDY:

Long Waves of Economic Cycles

We look at the case of the first empirical economic study of long-term change in modern economies by a Russian economist, Nicoli Kondratieff. Kondratieff studied the pattern of long-term economic development in the English economy. He did this in the 1930s in the Soviet Union under the dictator Stalin. Then a key tenet in the communist doctrine of Marxism was that modern history was supposed to be a struggle between capitalists and labor, and that when capitalists dominated, they would starve labor, which would lead to the inevitable decline and collapse of capitalism. Marxists assumed that capitalists would never allow labor to have sufficient wages to create a variable internal market that could sustain capitalism over the long term. Kondratieff measured the long-term economic activity of the capitalist country of England to test Marx's prediction. But he found (instead of the inevitable decline of capitalism predicted by Marx) that there had been actual recurrent cycles of economic expansion and contraction in England's economy.

Moreover, the net result was increasing economic activity in England, not decreasing activity. Kondratieff asked the question: How was capitalism in England renewed periodically and expanding overall? Kondratieff's answer was: technology. He plotted a correlation between times of basic innovation to times of economic expansion. *Overall, the capitalistic economies were expanding rather than contracting because of periodic innovations of new technologies.*

When he published his findings in the USSR, Stalin's police arrested Kondratieff and shipped him off to Siberia to work as a slave laborer in the mines of that socialist economy. As did 20 million other people that Stalin murdered in the Siberian gulags, Kondratieff perished from hunger or disease or brutality—no one really knows. Kondratieff was killed because he had violated a cardinal rule of dictatorships—instead of reiterating party dogma, he told the truth. In the 1940s, Kondratieff's work was rediscovered by an American economist, Schumpeter: "The work of Schumpeter . . . put emphasis on innovation and on the subsequent burst of entrepreneurial investment activity as the engine in the upswing of the long cycle, a la Kondratieff. He was hesitantly supported by Kuznets, who added . . . [that] the investment burst must stem from innovations with a far reaching impact across the whole economic system . . ." (Ray, 1980, pp. 79–80).

The "far-reaching impact" of technology on entire economic systems can occur when *new basic technologies* create *new functionality* through new industries or across existing industries. In 1975, Gerhard Mensch in Germany again revived Kondratieff's ideas (Mensch, 1979), as next did Jay Forrester, Alan Graham, and Peter Senge in the United States (Graham and Senge, 1980). They all argued that basic inventions and innovations underlay the beginnings of long economic cycles. In 1990, Ayers brought up to date Kondratieff's earlier empirical correlation between European industrial expansion and contraction and the occurrence of new technology–based industries (Ayers, 1990). Table 3.1 summarizes the empirical data.

The invention of the steam-powered engine required the science base of the physics of gases and liquids. This new scientific discipline of physics provided the knowledge base for Newcomen and Watt's inventions of the steam engine. Coal-fired steel required knowledge of chemical elements from the new science

TABLE 3.1   New Technology Industries and Economic Expansion

1770–1800	The beginning of the industrial revolution in Europe was based on the new technologies of steam power, coal-fired steel, and textile machinery.
1830–1850	The acceleration of the European industrial revolution was based on the technologies of railroads, steamships, telegraph, and coal-produced gas lighting.
1870–1895	Contributions to continuing economic growth were made by the new technologies of electrical light and power, telephone, and chemical dyes and petroleum.
1895–1930	The new technologies of automobiles, radio, airplanes, and chemical plastics contributed to further economic growth.

base of chemistry, which begun to be developed in the middle of the eighteenth century. The new disciplines of physics and chemistry were necessary for the technological bases of the industries of the first wave of economic expansion from the beginning of the industrial revolution in England.

The second economic expansion, to which the telegraph contributed, was based on the new discoveries in electricity and magnetism in the late eighteenth and early nineteenth centuries.

In the third long economic expansion, it was again the new physics of electricity and magnetism that provided the science bases for the inventions of electrical light and power and the telephone. Also, advances in the new discipline of chemistry provided the science base for the invention of chemical dyes. Artificial dyes were in great economic demand because of the expansion of the new textile industry. With these dyes, the gun powder industry began to expand into the modern chemical industry.

The fourth long economic expansion was fueled by the invention of the automobile, which depended on the earlier invention of the internal combustion engine. This invention required knowledge bases from chemistry and physics. Radio was another new invention based on the advancing science in physics of electricity and magnetism. Chemical plastics were invented through further experimentation in the advancing scientific discipline of chemistry.

MODEL OF THE ECONOMIC LONG WAVE

The major economic expansions in the world were derived from major new science-based technologies. But Kondratieff also observed the successive economic contractions in England, as summarized in Table 3.2. In these historic long periods of economic expansion due to technological innovation, why did economic contractions occur between these economic expansions based on new technology industries? Why didn't the English or European economies continue to grow smoothly

TABLE 3.2   Economic Cycles in England from 1792 through 1913

- The English economy expanded from 1792 to 1825 but contracted from 1825 to 1847. Kondratieff assigned iron, steam power, and textile machinery as the basis of the expansion but cited temporary excess capacity as triggering economic contraction.
- There was a second period of economic expansion in England from 1847 to 1873, followed by a contraction from 1873 to 1893. The expansion was due to the beginnings of new industries in railroads, steamships, telegraph, and coal gas, but again, temporary excess capacity triggered a following economic contraction.
- There was a third period of expansion from 1893 to 1913, due to the new technologies of chemical dyes, electrical lighting, telephone, and automobile, followed by continuing expansion after World War I, followed by a worldwide depression in the 1930s. World War I began and England's economic output turned to war materials. A decade after World War I, England and all Western economies fell into a deep depression from 1930 to the beginnings of World War II. The source of this depression was complicated, partly temporary excess production capacities and partly excess war debt on the German economy.

as the new technologies created new industries? The answer that Kondratieff proposed was the following. Even in a new technology–based industry, as the industry is expanding, production capacity can temporarily outstrip market demand, inducing excess-capacity contractions.

This type of long period expansion and contraction (economic long wave) is now called a *Kondratieff economic long wave*. The stages in this process consist of:

1. Science discovers phenomena that can provide nature for a new manipulation by technological invention.
2. New basic technology provides business opportunities for new industries.
3. A new high-tech industry provides rapid market expansion and economic growth.
4. As the new industry continues to improve the technology, products are improved, prices decline, and the market grows toward large volume.
5. Competitors enter the growing large market, investing in more production capacity.
6. As the technology begins to mature, production capacity begins to exceed market demand, triggering price cuts.
7. Excess capacity and lower prices cut margins and induce business failures and raise unemployment.
8. Turmoil in financial markets may turn recession into depression.
9. New science and new basic technologies may provide the basis for a new economic expansion.

History is not deterministic. There is no historical inevitability in Kondratieff's long-wave pattern. It can begin both (1) only after scientists discover new nature and (2) after technologists invent new basic technologies. Then basic technological innovation can provide business opportunities for economic expansion. *But there is no guarantee that new science and new technologies will always be invented.*

What is likely economically are excess competition and lowered prices as any new technology-based industry matures. So the important point about the long-wave pattern is that one should expect eventual overproduction, even in a new high-tech industry as technology matures and many competitors enter the new market. This will always cut profit margins, even in a relatively new industry. High-tech industry will never enjoy a high profit margin for long—only until technology begins maturing and competition intensifies.

ECONOMICALLY PERVASIVE INNOVATIONS

Long-term economic waves are created by new basic technologies that have a pervasive effect on an economy. For example, the Internet was an example of a basic technological innovation that changed business—*pervaded* the operations of all kinds of businesses. The idea that a technological innovation can pervade an economy or a business was proposed first by Abernathy and Clark (1985). They introduced the term *transilience* of innovation to emphasize the pervasive impact that an innovation may have on the operations of a business. By *transilience,* they meant the capability of passing through a system. Basic technological innovations may affect (pass through) the functioning of a business by making changes in the kinds of products and the way that a firm produces products, and the kinds of customers and markets that a business serves.

Accordingly, Abernathy and Clark also classified innovations as to how they conserved core business competencies or caused them to become obsolete:

1. A *regular innovation:* a technological innovation that conserved both existing production and market competencies
2. A *niche-creation innovation:* a technological innovation that conserved existing production competency but altered market competency
3. A *revolutionary innovation:* a technological innovation that made existing production competency obsolete but preserved existing market competency
4. An *architectural innovation:* a technological innovation that made both existing production and market competencies obsolete

Historically, many firms have usually exploited regular or niche-creation innovations successfully, for they sustain current operations. But many large firms have perished during periods of revolutionary or architectural innovation in their industry. For example, Christensen (2000) examined reasons why large firms in the United States have often failed to profit from revolutionary or architectural innovations:

1. Resources are often deployed conservatively in large firms, influenced by investors and current customers.
2. Early in a period of emerging markets for radical innovations, market size is perceived as too small for a big firm's growth needs.
3. The ultimate use of radical innovations is often not known early in the cycle.
4. The performance and features of radical new innovations are often not attractive to current markets.

DYNAMICS OF ECONOMIES

The Kondratieff economic long wave is only one pattern in the dynamics of modern economies. There is also the more frequent and ordinary short-term economic cycles. Modern economies are dynamic systems with cycles of economic expansion and contraction, with some of the cycles of long and some of short duration. Many factors contribute to economic dynamics, such as money supply and quality, government regulations, international trade, resource availability and cost, labor supply, cost and skill, market infrastructure, communications and transportation infrastructures, and educational infrastructures. The short-term business cycles are influenced principally by two factors: the cost of capital and the level of business inventories. When capital is in short supply or too costly, this can influence business investment and foster a contraction in an economy. The first recession of the twenty-first century in the United States began with a contraction of capital due to an overly invested stock market, the dot.com market bubble.

A second major factor in the short-term cycles has been the managing of product inventory. Another traditional form of the short-term business cycle has been related to business inventory levels. There is a rapid buildup of production when sales are growing and then an excess of product when the trend toward increasing sales turns to a trend toward decreasing sales. Production is then reduced and workers laid off until excess inventory is sold off and sales begin increasing again.

Even in this new age of information, the inventory cycle still exists. For example, when the economic slowdown began in 2000, many manufacturers were surprised by high inventory levels: "Despite some claims that the new economy could effectively eliminate the boom-and-bust business cycle, many high-technology companies have been caught by surprise in the slowdown. As a result, they have built up large stocks of unsold goods recently, adding to the overall rise in inventories versus monthly sales" (Schwartz, 2001a, p. C1). It had been hoped that the Internet would improve control of this kind of short-term business expansion–recession cycle. But improved control over economic decisions that information technologies provided still did not eliminate normal short-term business cycles: "The Internet . . . has allowed business managers to peek into every link of the supply chain that feeds their manufacturing processes, and to change direction with nimbleness . . . [but] it is no surprise that the technologies that speed up business decision-making and operations can be a double-edged sword" (Schwartz, 2001a, p. C1).

Moveover, long- and short-duration economic cycles can interact, as was seen in the Internet stock market bubble of the late 1990s, which we described earlier. We recall how from 1995 to 2000 the then-new technology of the Internet fostered a major U.S. stock market boom, based on investments in new e-commerce businesses and expansion of communication services. Fueling this economic expansion was money in the form of venture capital investments. These were funds that provided startup capital to new firms, which from 1995 to 1999 were quickly taken public or sold to larger firms, with rich returns to the venture funds.

An example was a new firm, Flycast Communications, founded in 1996 by Richard Thompson and Larry Braitman. Flycast was sold in 1999 to CMGI for $689 million: "It was a classic tale of success in the Internet age, and the two men decided . . . to leverage their success by creating a venture capital firm" (Cortese, 2000, p. 1). In fact, starting new venture capitals firms became the rage from 1995 to 1999, as the Internet age was like the California gold rush a century and a half earlier. Many rushed in to make their fortunes and the first that got there prospered, whereas the laggards lost. In 1995 there were about 500 venture capital firms in the United States, and in 1999 there were 1000 venture capital firms. In 1995, the amount of venture capital investments was $5 billion, growing to $100 billion in 2000. Venture capital investing in 2001, after the stock market decline of 2000, dropped back to $10 billion. The average return to venture capital funds in 1995 was 40% and jumped to 160% in 1999. But in 2001, the average return dropped to -20%. An observer then wrote: "Most vulnerable are the funds that were raised and invested at the height of the bubble, in 1999 and 2000, when 70 percent of all high technology venture capital for the last two decades was invested. The competition for deals was so fierce that venture capitalists virtually threw money at startups, may of which had no clear plan for making money" (Cortese, 2000, p. 11).

Another example was the Softbank Corporation, a firm that was an early investor in Yahoo. It created a $606 million fund in 1999 and invested it quickly, only to have nine of the Internet businesses in the fund go bust in the next year. At the end of 2000, the fund had lost 29.8% of its value (Cortese, 2000, p. 11). And what happened to the venture fund that Thompson and Braitman established after their fortune made on Flycast? "Just as the 20th century gave way to the 21st, [Thompson and Braitman] opened Signia Ventures in San Mateo, Calif. Mr. Thompson, 47, and Mr Braitman, 43, put up $50 million of the money they received for Flycast and started to pour it into new companies. Now, less than two years after opening, Signia Ventures office closed, its Web site blank and its phones unanswered. Mr. Thompson said he and Mr. Braitman had decided not to raise money . . . or invest in new deals until the market comes back" (Cortese, 2000, p. 1).

In this case one can see the kind of investment enthusiasm that can and has occurred in businesses founded on a new basic technological innovation. We saw in the Kondratieff list of economic expansions that each basic innovation in the twentieth century did foster economic expansions: automobiles, airplanes, plastics, radio, computers, television, biotechnology, Internet. It was proper investment in the new firms in industries based on these technologies that provided the sources of the greatest wealth growth in that century. The Internet-stimulated economic expansion was just the last Kondratieff wave of the twentieth century.

THREE HUNDRED YEARS OF INDUSTRIALIZATION

The pervasiveness of basic innovations in Kondratieff waves altered the world's economies over a long period of time, all toward an industrialized form of societies. However, this transition of the world from feudal forms to industrialized forms went neither smoothly nor evenly. When we think about the industrialization of the world, we need to think over a long period of turbulent times, a period of at least 300 years, from roughly 1765 to 2165. From about 1765 to about 1865, the principal industrialization occurred in the European nations of England, France, and Germany. From 1865 to about 1965 (the second hundred years), other European nations began industrializing; but the principal industrialization shifted to North America. By the middle of the twentieth century in the 1940s, the U.S. industrial capacity was alone so large and innovative as to be a determining factor in the conclusion of the second world war of that century.

For the second half of the twentieth century, U.S. industrial prowess continued, and European nations rebuilt the industrial capabilities that had been destroyed by that war. From 1950 to the end of the twentieth century, several Asian countries began emerging as globally competitive industrial nations: Japan, Taiwan, South Korea, and Singapore. Other Asian countries were also moving toward globally competitive capability: the Philippines, India, China, and Indonesia. We should note that historically, Asian industrialization actually begun in Japan in 1865 but was diverted principally to a military-dominated society. After World War II, a reindustrialization of Japan occurred.

In summary, one can project a pattern of 300 years of world industrialization in which different regions of the world began to develop globally competitive industrial industries: first hundred years (1765–1865), Europe; second hundred years (1865–1965), North America; and third hundred years (1965–2065), Asia.

The way the various regions industrialized was very different. In the case of the industrialization of Europe, the politics of colonialism played the essential role. English trade with nonindustrialized regions was enforced militarily as a colonial empire of the British nation. England accumulated wealth by trading manufactured goods to colonized, nonindustrialized countries for raw materials or agricultural products. Industrial technologies gave England the economic advantage of a higher technologically based productivity in this exchange of goods. Advanced military technologies gave England the power to enforce this trade.

Colonialism was, in fact, an age-old way for one nation to exploit another nation by subduing it militarily. All ancient empires were built on military conquest: Egyptian, Hittite, Persian, Macedonian, Roman, Chinese, Mongol, and so on. From 1800 to 1960, European nations were the dominant military colonizers in the world, simply because they had the dominant military power in the world. Accordingly, the first 100 years of industrialization from 1776 to 1876 in Europe used colonialism as the export arm of the industrializing European economies.

Whereas Europe continued to industrialize from 1876 to 1976 (and fight major wars among themselves), North American began an industrialization period from about 1850 to 1950. U.S. and Canadian industrialization did not use a colonialist model. They had a vast internal landscape, resources, and markets with which to grow. Moreover, European settlers had, indirectly and directly, already massacred many of the indigenous American peoples. North American industrialization did not use colonialism but *internal markets* for economic expansion. Industrialization could proceed on internal markets and did not need the militaristic exploitation of colonies.

After 1950, industrialization efforts in Asia scaled up in scope and quality: first with Japan rebuilding its war-destroyed economy, then with Chinese exiles industrializing Hong Kong and Taiwan, and then with South Korea and Singapore industrializing. Although Japan began modern industrialization in the twentieth century, the military culture of its society led its first industrialization cycle through a military economy dedicated to colonial expansion. After 1950, Japan rapidly reindustrialized into a commercially strong economy and strong international economic competitor. In the 1990s, industrialization of India, the Philippines, Indonesia, and Thailand was proceeding vigorously. In the 1980–1990s, mainland China began to industrialize, with capital inputs from Hong Kong and Taiwanese Chinese. Asian industrialization from 1950 to 1990 proceeded first on a model of importing technology from the United States and Europe and exporting manufactured goods to the United States and Europe. This was encouraged by the United States because of its focus on fighting the Cold War against communist governments from 1948 to 1988. However, in the twenty-first century, economic growth in Asia will be through the development of internal markets. Also, when the twenty-first century began, a wave of anger and frustration erupted in Middle Eastern nations (carved out from a prior Arabic–Turkic civilization) regarding a lack of participation in the wealth of industrialization.

INTERNATIONAL TECHNOLOGY TRANSFER

With respect to this industrialization of the world in the second half of the twentieth century, one of the historically prominent science and technology policies of world organizations was the policy of *technology transfer* from developed to less developed nations—but historically, this was a failed policy. After World War II, most of the previous colonies of European nations gained independence, and there was a long struggle between the United States and the USSR for the allegiance of the new nations to either democracy/capitalism or communism/planned economy. The nonindustrialized nations were then called *less developed countries* (LDCs). Economists then defined the problem of developing LDCs as a problem of technology transfer from developed nations to LDCs. Much money was spent in foreign aid by international organizations to facilitate technology transfer. But little economic development occurred in the LDCs from technology transfer monies, and much of those monies was stolen.

The only countries that did really begin a globally competitive level of industrialization from the 1950s by the 1980s were the four Asian "tigers": Japan, Hong Kong, Singapore, and South Korea. But in each case, the beginnings of industrialization did not occur from a program of technology transfer but from the transfer of manufacturing operations from developed nations to the four tigers by international firms, due to the lower cost of labor in these four regions. What made the transfer of manufacturing operations possible was that each of these regions had established strong educational infrastructures, creating a literate population to provide low-cost manufacturing labor.

This factor in the cost of manufacturing labor continued to spread industrialization to LDC regions. In the last decade of the twentieth century, the liberalizing communist government in China was encouraging manufacturing investments in the once socialist nation, and China was then on the way to becoming one of the largest manufacturing nations in the world. Moreover, in the decade of 2000, not only was direct manufacturing labor in China relatively inexpensive, but so was engineering. In the second half of the twentieth century, industrialization in the world continued to spread, not by government programs of technology transfer but by transfer of manufacturing operations to regions of relatively low-cost labor and high-quality education.

CASE STUDY:

Whirlpool Becomes a Global Company

The historical setting of this case was the late 1980s and early 1990s, when finally there was a growing corporate appreciation of the new dynamics of international competition in a global marketplace, as the world continued to industrialize. This shift of strategy from a national or international company toward a global company had implications for the organization and management of a firm. An illustration of this kind of strategy is the case of a U.S.-based firm, Whirlpool, which decided in 1989, to become a global company.

The strategic change in Whirlpool's enterprise vision began when David R. Whitwam become CEO of Whirlpool in 1987. Then Whirlpool was a North American company, one of four major competitors in the North American consumer major home appliances market: refrigerators, washing machines and dryers, dishwashers, and ranges. In 1987, the industry was a mature technology industry with a saturated market in North America, sales being for replacements and new family startups. It was a substantial market but with low profit margins and intense competition. As Whitwam recalls: " 'Even though we [Whirlpool] had dramatically lowered costs and improved product quality, our profit margins in North America had been declining because everyone in the industry was pursuing the same course and the local market was mature. The four main players, Whirlpool, General Electric, Maytag, Electrolux, were beating one another up every day' " (Marcua, 1994, p. 136). The first thing Whitwam did as a new CEO was convene his managers to plan Whirlpool's future: " 'When we sat down to plan our future in 1987, it was the first time Whirlpool had ever asked itself what kind of company it wanted to become in the next decade or the next century. This lack of self-scrutiny isn't as surprising as it might sound. Whirlpool was successful, profitable, and reasonably secure in a domestic market that was already eliminating the marginal competitors. The world hadn't broken down our doors the way Japanese auto makers had stormed Detroit, for example.' " (Marcua, 1994, p. 137).

In a large, successful company, planning usually means continuing to do more of the same thing that made it successful. That is reasonable except when new competitors storm the castle or when management cannot project continuing as successfully as they had been in the past. Whitwam saw three choices for Whirlpool's future: " 'We could ignore the inevitable—a decision that would have condemned Whirlpool to a slow death. We could wait for globalization to begin and then try to react, which would have put us in a catch-up mode, technologically and organizationally. Or we could control our own destiny and try to shape the very nature of globalization in our industry' " (Marcua, 1994, p. 138).

The first step for Whirlpool to become global was taken in 1989, when Whirlpool purchased N.V. Philips' floundering European appliance business for $1 billion. This acquisition jumped Whirlpool into the leading position worldwide in the appliance business. But the acquisition of Philips' appliance business was merely the first challenge to becoming a global company: " 'When we acquired Philips . . . Wall Street analysts expected us to ship 500 people over to Europe, plug them into the plants and distribution systems, and give them six months or a year to turn the business around. They expected us to impose the 'superior American way' of operating on the European organization. If you try to gain control of an organization by simply subjugating it to your preconceptions, you can expect to pay for your short-term profits with long-term resistance and resentment.' " (Marcua, 1994, p. 139).

Whitwam set about to build a global company. Six months after acquiring the European operation, Whirlpool sent 150 senior mangers to Montreux, Switzerland to spend a week developing Whirlpool's global vision. Then they had the assignment to educate all of the then 38,000 employees of Whirlpool around the world: " 'We have many, many employees in our manufacturing plants and offices who have been with us 25 or 30 years. They didn't sign up to be part of a global experience. . . . And a lot of our Italian colleagues didn't join Philips to work in the United States. . . . Suddenly we give them new things to think about and new people to work with. We tell people at all levels that the old way of doing business is too cumbersome. Changing a company's approach to doing business is a difficult thing to accomplish . . .' " (Marcua, 1994, p. 139).

Whirlpool's decision to globalize followed three principles:

1. Identify common and world-class technologies for the global company.
2. Customize products based on these technologies for local markets.
3. Organize on an interactive, global basis.

A product may use common technology, but products need to be customized to the regional market: " 'Washing technology is washing technology. But our German products are feature-rich and thus considered higher-end. The products that come out of our Italian plants run at lower RPMs and are less costly' " (Marcua, 1995, p. 136).

To organize on a global basis, Whirlpool developed new procedures for the company based on international participation. For example, one of the assignments to the managers at Montreux was to commission 15 projects on management and technology that each company in Whirlpool was to develop for the whole of Whirlpool, which they called "one-company challenges." Common procedures for the new global Whirlpool was a first step in globalizing Whirlpool. The vision of Whirlpool becoming a global company was based on a desire to access all the markets of the world and to create and maintain a dominant position in each local market.

GLOBALLY COMPETITIVE INDUSTRIALIZATION

When the twentieth century ended, there was a growing appreciation of how the continuing industrialization of the world was affecting international business competition, and this began to be called *globalization,* indicating that world markets were globally available and production was distributed globally. In about 1980, global trade accounted for about 17% of economic activity, and by 2000, it had increased to 26%. But this increasing global trade tied together more closely the economic cycles of the nations of the world. For example, in 2001, when the first economic recession of the twenty-first century began, it was seen that the downturn in national economies was tied together as the United States, Europe, and Asia economies all declined simultaneously. As the entire world was industrializing, it was important to make clear the difference between globally effective and ineffective industrialization. For example, Porter (1995) identified several factors in effective national competitive structures, which include: political forms, national and industrial infrastructures, domestic markets, and firm strategies.

During the twentieth century it became clear that a democratic form of government is necessary for long-term economic development. The totalitarian forms of government, communism and fascism, were able to provide only short-term gains in industrialization, and at a high human price, precluding the development of competitive industrialization. Dictatorships foster intellectual dishonesty, economic corruption, and physical brutality. *Intellectual integrity and truth are essential for scientific and technological and economic progress.* Economic corruption impedes the development of an effective marketplace—substituting graft and shoddy products and inefficiency for competitive prices, product quality, and efficient productivity. Physical brutality degrades human relationships and creates a culture of terror and cowardliness, which is contrary to the safe environment and courage required for a competitive economy and free society.

An effective national infrastructure is also a necessary condition for effective industrialization. Elements of necessary national infrastructure include educational systems, police and judicial systems, public health and medical systems, energy systems, and transportation and communication systems. Strategic interactions between universities and high-tech companies is an important feature for industrial competitiveness. For example, Gwynne (1993) described some of the science and technology parks developed in Singapore, South Korea, and Taiwan to build their science and technology infrastructure for high-tech industries. The model for such science and technology parks was the history of the Silicon Valley region in northern California in the building of the chip and personal computer industries. The universities of Stanford and of California at Berkeley played an important role, along with venture capital firms, in growing high-tech chip, computer, and multimedia industries.

Related to national infrastructure (but distinctly different) is the economic requirement to also have an effective industrial infrastructure. An industrial infrastructure requires such elements as a financial industry, power industry, transportation industry, communications industry, health industry, education industry, and so on. Each of these industries must have an industrial structure that interacts and is competitive with international industrial structures. Also, successful industrialization has occurred only when industrialization provided sufficient employment opportunities to create an effective home market for industrial goods *in addition to* creating successful export products. This has been one of the more puzzling aspects of industrialization—how to create domestic demand and exports at the same time.

Finally, the strategies of individual firms and the nature of their competition also affect industrialization. For example, the socialization policies of India from the 1950s through the 1980s fostered inefficient and internationally noncompetitive firms by encouraging local monopolies and the purchase of market share by political influence and bribery. As another example, Dunning (1994) pointed out the important role for multinational enterprise in building the technological and industrial capacities of developing nations and argued that the motivation of a multinational enterprise was an important factor in whether or not the multinational contributed positively to a nation's development.

In summary, economic industrialization and development is a complex problem. Technological progress makes it possible for some nations to capitalize and implement economic development. Which nations succeed in utilizing technology for economic progress has depended on a set of factors, including political system, national and industrial infrastructures, domestic demand, and local firm competitiveness.

For the period 1950 through 1970, some economists advocated a kind of low-tech progress for less developed countries, then called *appropriate technology*. The idea was that the level of technology (low tech versus high tech) used in an application in an nonindustrialized nation should be appropriate to the application need. Although all technology should be appropriate to the application, using only low-tech technology (e.g., older technology) did not help any country competitively industrialized after World War II of the twentieth century. The cases of successful industrialization of Asian countries after that war—Japan, Taiwan, South Korea, Malaysia—all used high-tech technology, the newest and most innovative technologies. The lesson of this was that competition on a global scale always requires the latest and best technology—high tech.

Accordingly, by the time the twenty-first century began, any large company trying to survive competitively managed with both global and local considerations. The famous slogan for this came from SONY management in the 1990s, who had urged: "Think globally and act locally." This meant that to be competitive in the global market required world-class technology, world-class finance, world-class production, and world-class communications and transportation. At the same time, the conditions of local considerations should be given to products that are at the same time both world class and locally focused, local markets, and local distribution.

In addition, production can be localized to employ local people and local management. This is important, because local employment provides the basis for consumer incomes to purchase the products of a company. Integrating local management into an integrated global firm is a principal challenge. For example, MacCormack and co-workers (1994) examined international trends in strategy for locating manufacturing sites in a global market and suggested that the best sites will become decentralized manufacturing planes in large, sophisticated regional markets.

For example, Sugiura (1990) described Honda Motor Company's four principles for localization in a global economy: products, profits, production, management. Localization of products means to develop products suited to the actual and potential needs of customers in a specific region. Localization of profits means reinvesting some profits in the local market to regard each regional unit as a kind of local company. Localization of production means to provide employment in the regional markcts to which a global company sells. Localization of management means to develop local employees and management that share corporate philosophy but work locally.

To gain a competitive advantage from technology, one must not only acquire the best technology in the world but also improve on it. This means that technology transfer from one country or firm to another is insufficient to produce global competitiveness. When a technology is transferred, it is usually what is competitive today, not what will be competitive tomorrow. To provide a competitive difference, technological information must both be timely and be utilized. For example, Kokubo (1992) emphasized the need not only to look around for technology but also to integrate that information search with innovativeness: "Japan's large corporations have a well-deserved reputation for aggressiveness in gathering competitive technology information worldwide. . . . In many industries, the Japanese are steadily changing from imitation to innovation. . . . In keeping with this switch, the intense global R&D competition that is emerging will require companies to adopt a more integrated style of information gathering. The search for technology information must be more closely joined to competitive intelligence activities for R&D" (p. 33).

Competitive intelligence for technology strategy requires that technical information be gathered on a worldwide basis, distributed internally in a company, and used for technology strategy. Kokubo (1992) also emphasized the integration of intelligence into technology strategy: "Competitive intelligence activities must enable a company to evaluate its own technological competitiveness. . . . This analytical drive spreads a dramatically new R&D philosophy throughout a company" (p. 34).

One important result of this globalization is that there is now a trend for global companies to disperse their R&D efforts throughout the world. For example, Serapio studied the growth of R&D investments by U.S. and Japanese firms outside their home countries, citing reasons for this globalization if R&D were to improve product focus on regional markets, monitor technology developments, and acquire new technology (Serapio, 1995). In another example, Chiesa (1996) examined the international dispersion of R&D laboratories of 12 companies in the early 1990s and found that R&D global dispersion was increased for labs taking long-time horizons.

In the 1990s, the trend toward internationalization of a firm's R&D was still relatively new and small. For example, Patel (1996) examined the patenting records of 539 firms in several industries and found that the overwhelming majority of technology invention still occurred in laboratories in the company's home country. But this was the trend—dispersion of the capability of innovating new technology around the world—not only a global market but also a global R&D system. What one could begin to see was a new phase in technological progress, the globalization of technology, in contrast to previous progress in technology in purely national contexts. When the twenty-first century began, technological progress had not only changed the forms of societies and economies, it was even beginning to change the forms of governments.

Figure 3.1 summarizes the major historical milestones of changes in science, technology, and economy. Science began in European civilization in the seventeenth century when Newton combined new ideas of physics (due to Copernicus, Brahe, Kepler, and Galileo) with new ideas in mathematics (due to Descartes and others)

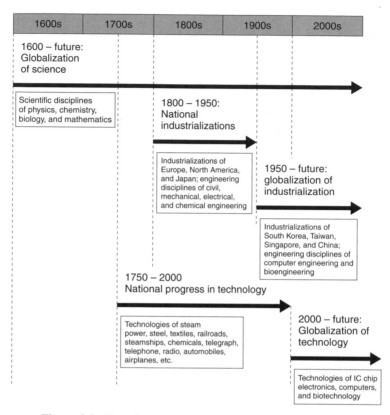

Figure 3.1. Time lines of science, technology, and economies.

and invented the mathematical theory of space, time, and forces (the Newtonian paradigm of physics). In the eighteenth century, these new ideas were developed into the new scientific disciplines of physics, chemistry, and mathematics. The nineteenth and twentieth centuries found dramatic advances in these disciplines, along with the founding of the scientific discipline of biology. By the end of the twentieth century, the physics of the small parts of matter and the largest spaces of matter was established, the chemistry of inanimate and animate matter was established, the molecular biology of the inheritance of life was established, and the computational science of mind and communication was being extended. All this began and occurred in an international context from the very beginnings of science, so that one can see the 400 years of the origin and development of science as a period of the *globalization of science.*

In contrast to this international context of the development of science, the economic and technological developments occurred in national contexts. Each nation industrialized on a national basis and in competition with other nations. England began the industrial revolution in the late eighteenth century, followed by industrializations in the nineteenth century in France, Germany, and other European countries and in North American. This pattern of industrialization in a national pattern continued through the twentieth century in Japan and other Asian countries. But as the twentieth century ended, companies began shifting from principally a national basis to a global basis, organizing for business competition in a worldwide context, *globalization of industry.*

Along with industry, the patterns of developing technological progress also occurred upon a national basis, with technology viewed as a national asset. The technological inventions that began the industrial revolution were invented in England—the steam engine, coke-fueled steel production, and textile machinery and factories, among others. These provided the basis for the industrial revolution begun in England. Thereafter, many new technological innovations—chemical, materials, communications, transportation, electronics, computers, and so on—were innovated in national contexts. In the twentieth century, innovation in new technology became the basis of national economic competition. However, the quickness with which modern technology became international, transferred, and developed began to change in the second half of the twentieth century, so that when the twenty-first century began, a new pattern of change in the modern world emerged as the beginning of the *globalization of technology.*

SUMMARY AND PRACTICAL IDEAS

In case studies on

- Long waves of economic cycles
- Internet and economic cycles
- Three hundred years of industrialization
- Whirlpool becomes a global company

We have examined the following key ideas:

- Economic long waves
- Economically pervasive innovations
- Centuries of industrialization
- Technology transfer
- Globalization

The practical use of these ideas is in seeing long-term economic development as an expansion and contraction cycle, depending on new technological innovation. The expansion of the U.S. economy in the late 1990s with the subsequent stock bubble and recession was a typical Kondratieff wave built on commercial applications of the Internet innovation. Moreover, one can see that modern economic history was hundreds of years of worldwide industrialization and economic integration—

globalization. Basic new technologies affect economies through the products and services needed for customers to apply a new technology in an application system. Business opportunities abound in the times of the early development of a new economic functional system.

FOR REFLECTION

Choose a modern nation and describe its transformation into an industrialized economy from 1800 to 1990. What have been the political and economic and social problems in this transformation? Identify the types of impacts that technologies made on these problems.

4
TECHNOLOGICAL PROGRESS

INTRODUCTION

We have seen that new technologies arise from new science in the modern world, and we have seen that new basic technologies create long-term economic development. But exactly how does progress in a technology occur? Can one plan deliberately to improve a technology? If one cannot plan technological progress, one cannot plan to use innovation as a competitive advantage. We will see that technological progress can be planned because all technologies are types of systems. The idea of a technology as a system of manipulation is the key to all technological progress. We will provide a comprehensive systems overview of technology and its application, including the basic ideas of *technological systems, application systems,* and *economic functional systems.*

CASE STUDY:

Sikorsky's Helicopter

To understand technology as a system, we look at the invention of the helicopter, rotating-wing flight technology. Powered flight was not possible until the invention and innovation of the gasoline-fueled internal-combustion engine. The artist Leonardo da Vinci sketched a rotating-wing helicopter (Figure 4.1), showing that others had envisioned such a flight. But the difference between science fiction and technology is that inventions must be real and must work. The historical setting of this case was the early twentieth century, soon after the Wright brothers demonstrated the first powered fixed-wing flight in 1904 at Kitty Hawk, North Carolina. In the following years, they demonstrated their invention around America and in Europe. Then many other inventors and developers began improving the airplane and inventing other schemes, and one such person was Igor Ivanovich Sikorsky.

Sikorsky was born in Kiev, Russia in 1889 (Wohleber, 1993). As a child, he had showed the technical bent of an inventor, and at the age of 14, Sikorsky

da Vinci's Helicopter   Sikorsky's Helicopter

Sikorsky's   Igor Sikorsky
Pan-Am Clipper

Figure 4.1. Development of flight technology.

entered the Imperial Russian Naval Academy in St. Petersburg. After three years, he decided that he did not wish to become a naval officer and quit to enroll in the Kiev Polytechnic Institute. In the summer of 1908, he visited Germany and learned of Count Zeppelin's dirigibles and of the powered flights of Orville and Wilber Wright. The Wrights were touring Europe with their amazing demonstration of powered flight. Sikorsky determined that his life's work was to be flight. Flight was truly a new functional capability for humanity.

From the beginning, Sikorsky wanted to do something different from the Wright brothers' kind of plane. He resolved to design a flying machine that would be capable of "rising directly from the ground by the action of a lifting propeller." His idea arose a childhood memory of his mother's descriptions of Leonardo's da Vinci's designs for flying machines. In a German hotel room in 1908, he put a 4-foot propeller on a small engine and found that it could achieve a lifting force of 80 pounds for each horsepower the engine could provide.

Back in Kiev, he built two helicopter prototypes, but the French-built, three-cylinder 25-horsepower engine he had purchased did not provide sufficient power to lift the helicopter. He turned then to constructing fixed-wing aircraft. With his fifth prototype, he got one to fly. After this, Sikorsky built a sixth prototype, a three-seater biwing plane that flew at 70 miles an hour, breaking the world record for speed of a plane carrying a pilot and two passengers. He built the world's first four-engine aircraft, and in May 1913 he flew the 4½-ton plane above St. Peters-burg. He then built a larger, refined version of this, and when World War I began, he was put in charge of building military versions for use in long-range recon-naissance and bombing raids. By 1917 he had produced 70 planes of his Il'ya Muiromet.

CONCEPT OF A SYSTEM

We pause in this case study to review briefly the basic concept of a *system*. The de-sign of Sikorsky's bomber aircraft required a design of the complete system of the plane: wings for lift, four engines for power, tail for control, fuselage for crew, and so on. All technologies are systems, and all products (embodying technology sys-tems), such as the Il'ya Muiromet, are also systems.

The concept of a system involves looking at a thing, an object, with a view to see-ing it as a dynamic totality—displaying change and encompassed in an environment.

A description of a thing as a system captures its totality with dynamic changes represented as transformations of system states of the thing, oper-ating within a relevant environment.

One of the most famous examples of the system concept is the solar system. We recall that in the fifteenth century, Copernicus proposed a new model of the uni-verse in which the sun was the center about which the Earth and other planets (Mars, Venus, and Jupiter) orbited in circles. At the time, church officials in the Roman Catholic Church objected to Copernicus's model on the theological grounds that it removed the human being (which was created in God's image) from the center of the universe. To avoid prosecution, Copernicus published his treatise as a mere math-ematical artifact. Later Galileo built a telescope and looked at the moon, seeing Earth-like craters, and publicized the Copernican model as a real system. Catholic Church officials forced Galileo publically to recant the Copernican model. Tycho Brahe improved the measurements of planetary motions and hired a mathematician, Johann Kepler, to calculate the orbits of the planets in a Copernican model of the solar system. Kepler determined that the planets moved not simply as circles but as ellipses. Then Newton developed new mathematics (calculus) and put together a new physical paradigm (using Copernicus's, Galileo's, and Kepler's scientific ad-vances) to represent the solar system as a central gravitational field extending from the sun, which pulled the planets inward so that their inertial momentum kept cir-cling the sun in elliptical forms (Figure 4.2). Thus the solar system was:

- A stellar "thing"
- Composed of components of a star and its planets
- With dynamic states of the solar system as subsequent positions of planets or-biting the sun
- In the environment of space

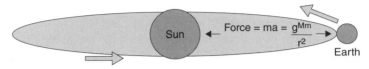

Figure 4.2. Sun–Earth system: a planetary system surrounded by the environment of the sun's galaxy.

System is a very general concept that can describe anything in the natural world as a totality with dynamic change. There are two general forms of systems: closed and open (Betz and Mitroff, 1972). A *closed system* does not have significant outputs from its environment and may not also have significant inputs. A closed system is described principally by transformations between internal states of the system. For example, in the earlier case of energy sources, we saw the Earth as an isolated planet sustaining life and illuminated by the sun. This Earth can be viewed as a closed system. The only significant input is energy from the sun, with the terrestrial system describable as closed cycles of physical and living processes (except when occasionally struck by asteroids).

An *open system* has both inputs from its environment and significant outputs into its environment. Both businesses and technologies are open systems. A business viewed as a system is a legal and financial entity, purchasing resources from its economic environment and selling products/services into its economic environment. A technology viewed as a system is a functional totality, transforming inputs from its purposive environment into outputs of its purposive environment.

All systems have a boundary and an environment and internal states, as illustrated in Figure 4.2. Within the boundary, the system will contain subsystems, parts, and connections between parts that perform a state-to-state transformation of the system. There will always be at least two subsystems within the system, a transformation process subsystem and a control subsystem. A system is defined by its dominant process or processes. The boundaries of an open system are the points at which inputs are received and outputs exported from and to its environment. The functionally dominant process is the transformation that converts inputs to outputs.

CASE STUDY CONTINUED:

Sikorsky's Helicopter

In 1917, the czar of Russia, Nicholas, abdicated his throne. Soon a small group of communists seized power in Russia. Many Russians were forced to flee the communist dictatorship, and Sikorsky left in February 1918 (leaving behind the money he had made). He traveled to France and presented them with a plan to build bombers. The French government accepted them, but World War I ended before Sikorsky could begin production.

In March 1919, Sikorsky traveled to the United States. For the next few years, Sikorsky barely earned a living as a teacher. In 1923 he established a new airplane firm and built an all-metal passenger plane with two 300-horsepower engines. (It was eventually sold to a film maker, and disguised as a German bomber, it went down in flames in a Hollywood film.) In 1928, Sikorsky built a large amphibious plane, the S-38, which sold well. The largest customer was Pan-American Airways, which flew 38 of these on their Latin American routes. Pan-American called these the American Clipper. Sikorsky then built larger versions, and the S-40 was the first to cross the Pacific (Figure 4.1). Sikorsky sold his company to United Aircraft.

In 1938, Sikorsky went back to his first technical love and began working on a design for a helicopter for United Aircraft. He patented a design that used a larger horizontal propeller for lift and a small vertical one on the tail to control torque. The technical problem then was to provide control. (The fixed-wing aircraft, properly designed, provides relatively stable flight dynamics compared to the helicopter.) It wasn't until 1940 that Sikorsky developed a prototype that could hover steadily. Then the U.S. Army got interested in the project.

On May 6, 1941, Sikorsky demonstrated a new prototype, the VS-300, to reporters and military representatives. Sikorsky took the helicopter to a stationary hover and remained there for 1 hour and 32 minutes. But the helicopter still flew too slowly forward. Yet it was enough for a military R&D contract. The Army wanted a higher forward speed and insisted that all control be in the main rotor and that there be an enclosed cockpit with a passenger seat.

By 1942, Sikorsky demonstrated a military version, the XR-4, that had a 165-horsepower engine and could hover and travel forward at a speed of 75 miles per hour. Then Sikorsky began producing helicopters for the Army as model R-4, R-5, and R-6 (Figure 4.1). By the end of World War II, 400 military helicopters had been produced. In March 1944, an R-4 was used for a first military rescue mission in lifting a U.S. pilot and three wounded British soldiers from a crash site in a rice paddy in Burma. Later, during the Korean War, helicopters would be used extensively for rescue, and during the Vietnam conflict, helicopters would also be used to deploy ground troops under combat conditions. Sikorsky retired from Sikorsky Aircraft in 1957 at the age of 68 but remained a consultant for the continuing development of helicopters until he died in 1972.

For the innovation of the helicopter, the following conditions were required: (1) the prior invention of the gasoline engine, (2) the prior invention of fixed-wing powered flight, (3) the development of sufficiently light and powerful engines to power a helicopter, (4) the dedication of an inventor to solve the practical problems of controlled hovering and flight, and (5) the sponsorship of a military customer to fund the R&D required to perfect a practical prototype for application.

TECHNOLOGICAL TRANSFORMATION

Recalling again that we defined technology as knowledge of how to manipulate nature for human purposes, we did so to make evident that if one examines any technology, one can find these ingredients: (1) natural phenomena, (2) manipulation of natural states, and (3) a schematic logic of human purpose.

All technologies are based on natural phenomena. For example, in flight it is the natural phenomenon of the physics of airflow driven by an overhead propeller (helicopter) or over a horizontal wing (fixed wing) that provides lift. The physics of helicopter lift is simply action–reaction, whereas the physics of airplane wing lift is the differences in air pressure as the air flows over the curved upper part of the wing and over the flat lower part of the wing. Different ways to manipulate natural states can produce different versions of a technology (e.g., helicopter and airplane versions of flight technology).

In addition to being based on nature, all technologies are also based on human purposes. The reason that one manipulates nature in a technology is to achieve a purpose, to achieve a human goal. For example, the purpose of flight technology is for transportation through the air.

> **In technology, human purposes are achieved through a functional transformation of nature.**

A functional transformation of nature occurs through establishing initial states of nature and controlling the subsequent sequence of states to desired final states of nature. For example, consider the solar system shown in Figure 4.2. If one wanted to use rocket technology to send a scientific probe around the sun, one would choose an initial state of nature, which would be a day of the solar year to launch the rocket from Earth aiming toward the sun but taking into account the relative motion of the rocket due to being launched from the Earth. The rocket would follow a parabolic path around the sun and then out into the solar system. The parabolic path would be the intermediate states of nature in the technology, with the near approach to the sun for data measurements to be radioed back to Earth as the final state of nature for the technology. After the probe performed its scientific mission, it would fly endlessly around the solar system unless and until possible capture by a solar body.

The representation of these states of nature altogether would provide a picture of the form of the solar system, and in technology we call such a form a *physical morphology*. The term *morphe* means "form" in Greek and *ology* means "a knowledge of," so *morphology* means "a knowledge of form."

> **The physical morphology of a technology is a knowledge of the forms of the physical nature being manipulated by the technology.**

All physical technologies have physical morphologies that describe the structures and processes of physical nature being manipulated in the technology. For example, in the technology of the helicopter, the physical morphology consists of (1) a structure of the body of the helicopter, turbine engine, large horizontal lifting blades, and small vertical control blades, and (2) a lifting process of the turning of the horizontal blades to provide lift through the action and reaction of forces of blades cutting through the air.

A physical morphology of a technology consists of the structures and processes of physical nature.

The choice of the initial and final states of nature and sequential path of states between in a physical morphology is the *schematic logic* of the technology. The schematic logic maps the purpose of the transformation into the states of nature of the physical morphology. For example, in the technology of the helicopter, the schematic logic consists of (1) fueling the helicopter, (2) entering the helicopter, (3) turning on the turbine engine, (4) controlling the pitch of the whirling blades for lift, (5) flying the helicopter to a destination, (6) landing the helicopter, (7) turning off the engine, and (8) exiting the helicopter.

A schematic logic of a technology consists of the logical steps for the physical manipulation of nature to be interpreted as a functional transformation for human purposes.

All technology systems have these two aspects: a physical morphology and a schematic logic. The inventive creation of a new technology is a devising of (1) a schema, (2) a physical morphology, and (3) one-to-one mapping of the sequences between logic scheme and physical morphology.

A technology is transformational knowledge of a logic scheme (for the transformation) mapped to a physical morphology (manipulated in the transformation).

TECHNOLOGY SYSTEM

Now any physical technology must be expressed in a specific configuration of a physical morphology, and such a configuration is called a *technology system.* For example, both fixed-wing and rotating-wing configurations provide two different configurations of the technology of flight: airplane and helicopter technologies.

A specific configuration of a technology transformation is a technology system.

A technology system is a configuration of parts that operate together to provide a *functional transformation.* Any technology might be invented in different configurations. Any particular technology system is a *specific configuration* of a technology focused by an application. Different technology systems provide similar functional capability but with different features and performance. As in the function of flight, the two different systems to attain flight were different configurations of the flight system: fixed wing and rotating wing.

All technology systems are specific configurations of functionally defined open systems, accepting inputs and transforming inputs into outputs in that configuration.

The boundaries of a technology system are the points of the physical structure that receive inputs and which export outputs into the environment.

> **The type of transformation of generic inputs into generic outputs defines the functional capability of a technology system.**

We can see any physical product as embodying a technology system. For example, take the familiar automobile, which can be viewed as two connected technology transformations: power and motion. The technology system of energy conversion in the automobile can be viewed as an open system with four boundary points of the energy system in the automobile:

- *Inputs:* fuel tank inlet and air filtration input
- *Outputs:* exhaust and radiator

Also, the motion transformation of the automobile can be viewed as an open system with five boundary points of the motion system in the automobile:

- *Input:* four tire contact points with the ground
- *Outputs:* airspace immediately in front of the automobile

From a marketing perspective, the automobile can be viewed as a product, and from an engineering perspective, the automobile can be viewed as two connected systems of power and motion. The technology system of the automobile can thus be defined as a physical device that inputs energy sources from the environment that converts this energy to mechanical power to transform the physical location of the device.

In summary any technology can be viewed as an open system, a transformational system, with two levels of system description:

- *Logic schematic:* logical scheme of the functional transformation
- *Physical morphology:* constructed physical structure whose processes map in a one-to-one manner with the logic scheme

The logic schematic is a step-by-step laying out of the order of the discrete transformations required to take a generic input to a generic output of the technology's functional transformation.

> **A logic scheme for a technology system is represented as a topological graph of the parallel and sequential steps of logical unit transformations.**

The physical morphology of a technology system is the assembly and connection of the physical structures and processes whose physical operation can be interpreted in a one-to-one manner with the logical steps of the schematic.

The physical morphology of a technological device may also be analyzed as
an open system defined by a boundary and containing components, connections between the components, and a control subsystem.

This generic way of looking at technology as a system—a system with both a
logical scheme and a physical morphology—is very powerful because it provides a
way to look at both the completeness and the relevance of technology to economic
activities.

TECHNOLOGICAL INVENTION

Earlier we noted that technological innovation requires both invention and commercialization. As invention, progress in a technology occurs first in the original invention of the technology, and later as progress from further inventions for
improvement in the technology system.

The *invention* of a new technology is the *first creation* of the mapping of a
functional scheme to a physical morphology.

Continuing improvements in a technology occur by way of further inventions to
improve the technological system. This requires identifying potential or actual opportunities for technological advances in either the logic schematic of the functional
transformation or in progress in the physical phenomenon and physical morphology. Technology advance may occur by (1) extending the logic schematic, (2) alternative physical morphologies for a given schematic, and (3) improvement of
performance of a given morphology for a given schematic by improving parts of
the system.

Technical progress may occur by adding features and corresponding functional subsystems to the logic. Technical progress may occur in alternative physical processes
or structures or configurations for mapping to the logic. Technical progress may occur in improving parts of the system: its subsystems or components or connections or
materials or energy sources or motive devices or control devices. Thus viewed as a
system, technical progress can occur from changes in any aspect of the system:

1. Critical system elements
2. Components of the system
3. Connections of components within the system
4. Control subsystems of the system
5. Material bases within the system
6. Power bases of the system
7. System boundary

A technology system cannot be innovated until all the critical elements for
the system components, connections, control, materials, and power already exist

technologically. For example, the steam engine was a necessary critical element before the railroad could be innovated. The gasoline engine was a critical system element before the aeroplane could be innovated. The vacuum electronic tube was a necessary element before radio and television could be innovated. The vacuum tube and the ferrite core memory were necessary elements for the general-purpose computer. Critical elements are the building blocks for a system's components or connections or control or material or power bases.

Technical progress in a system may occur from further progress in the components of the system. For example, over the early 30-year history of the computer technology, rapid and continual progress in the central processing units, memory units, and storage units drove the technological progress of the computer. Technical progress in a system may also occur in the connections of the system. For example, in the telephone communications industry, fiber-optic transmission lines began to revolutionize that industry in the 1970s. As noted previously, for computers in the 1980s the development of local area network connections and protocols and the fiber-optic transmission systems also contributed to the continuing revolution of the computer industry.

Technical progress in a system may also occur in the control systems of a technology. For the automobile industry in the 1980s, technical progress was occurring primarily into the development of electronic control systems. In computers, major technical progress occurred in developing assembly languages, higher-level cognitive languages, compiler systems, interactive compiling systems, artificial intelligence languages, and so on.

The material bases in the framework of a system, in the devices of a system, in the connections of a system, or in the control of a system are also sources of technical change and progress. An example of technical progress in the frame was the beginning in the 1980s of the substitution of organic composite materials for metals in the construction of airplanes. And example of advances in materials in a device was the substitution of silicon for germanium in the new transistor in the 1950s, and in the 1990s the attempts to substitute gallium arsenide for silicon in semiconductor chips.

Finally, energy and power sources may provide sources of advance in a technology system. The classic example is the substitution of steam power for wind power for ships in the nineteen century, and in the twentieth century the substitution of nuclear power for diesel power in the submarine.

SCALES OF TECHNOLOGICAL INNOVATION

What is seen by the customer of an innovation is how big an impact the innovation makes on customer applications. The idea of the size of economic impact has been called the *scale of the innovation,* and early in the studies of technological innovation, Marquis (1960) distinguished different scales of innovation impact: radical, incremental systems. However, what Marquis called a radical systems innovation has not proven to be a useful distinction, since all technologies are systems. A more useful distinction is that of a next-generation technology (Betz, 1991). Accordingly, we distinguish three scales of technological innovation:

1. *Radical innovation:* a basic technological innovation that establishes a new functionality (e.g., steam engine or steam boat)
2. *Incremental innovation:* a change in an existing technology system that does not alter functionality but improves performance, features, safety, or quality incrementally or lowers cost (e.g., governor on a steam engine)
3. *Next-generation technology innovation:* a change in an existing technology system that does not alter functionality but changes or improves, performance, features, safety, or quality dramatically or lowers cost and opens up new applications (e.g., substitution of jet propulsion for propellers on airplanes)

Basic inventions occur infrequently in the history of technological innovation, for most innovations are improvements on existing technologies. Although infrequent, basic innovations are extremely important because they do create new industries. Basic inventions can be of two kinds: component inventions and system inventions. A *component invention* is used within a system, and a *system invention* invents new ways of incorporating components into a system. However, even components are systems. In this way there occurs a hierarchy of technology systems, containing technology systems as components.

All basic innovations, component or system, require improvement through incremental innovations. Over time, most innovations are incremental and add up to major improvements in the basic innovation. Occasionally, a technology system is improved not just in part but in whole. When this happens, the new whole replaces the older system, and this is called a *next generation of the technology* (NGT). Next generations of technology systems provide dramatic improvement in performance and features. Incremental innovations provide small but definite improvements in performance or features.

However, the perceived impact of an innovation can depend on the perspective of the user. For example, Afuah and Bahram (1995) pointed out that what may be an incremental innovation in a supplier may make a larger impact on the supplier's customer.

For example, in the nineteenth and twentieth centuries there was a long period of technological progress—from the railroad's basic invention using the steam engine to the eventual technical obsolescence of the steam engine. The key invention to begin the industrial revolution in the eighteenth century was the steam engine. The steam engine provided the first source of mechanical power and was a radical invention. (Also as a component, the steam engine had several initial applications: pumping water from mines, powering textile machinery, powering railroads, powering steamships—each application also became a new radical innovation.)

The steam engine had many technical improvements, one of which was a two-stroke steam-cycle engine that doubled the horsepower of an engine's size. Early steam engines on railroads had steam pushing the piston on only one side; later, improved models had steam pushing the piston alternately on both sides. This was an incremental innovation improvement that advanced the performance of the railroad. Steam-powered locomotives dominated into the early twentieth century as a technology, until replaced by the diesel-engine locomotive. The diesel-engine locomotive is an example of a next generation of a technology system innovation.

Both radical and next-generation technological innovation can have dramatic impacts on industrial structure. For this reason, it has been useful to talk about these innovations as kinds of *discontinuities* in technological progress.

> **Discontinuous technological change creates industrial structures and may alter existing industrial structures.**

This is in contrast to the kind of smaller change provided by incremental technological innovations that also tend not to alter industrial structure. For this reason, incremental technological innovation is also called *continuous technological progress*.

> **Continuous technological change reinforces an existing industrial structure.**

CASE STUDY:

Systems Control in Helicopters

All technology systems require a control subsystem. For example, a helicopter control subsystem consists of lifting, steering, acceleration, flight, and descent. Controlling actions consist of a planned route, ignition, starting, flying, and landing. To accomplish this location transformation, the hellicopter has two separate and parallel control subsystems that need to be coordinated at all times of operation: the energy transformation process and the power-applied and flight processes. As the helicopter is lifted, hovered, flown, and landed, the energy transformation must be controlled in exact coordination as fuel and air are injected into the engine, ignited, and provide torque for the lifting propeller and the lateral-stabalizing propeller.

TECHNOLOGY SYSTEM CONTROL

All technology systems require control of the transforming function of the technology.

> **System control requires at least three parallel subsystems of control:**
> 1. **To input material resources for fuellng the transformation activities**
> 2. **To control the transformation activities**
> 3. **To direct the purposes of the transformation**

Any or all of the parallel control subsystems may, in addition, contain a hierarchy of subsystems. For example, in the automobile, the gasoline and air subsystems provide energy for process, the power subsystem transforms energy to locomotion, and the driver and steering subsystem direct the motion of the automobile. In the computer example, the electrical power subsystem provides energy for computer functioning, and the three hierarchical levels of control (system, application, micro-instruction) control the computational execution. In addition, the programmer who has written the application program (and perhaps also an online user who interacts in real time with the system) provides the directing purpose of the computational activities.

In both these examples, we see a technology system in which partial control lies in the purposive direction of an application programmer or an interactive user (e.g., automobile driver). For this reason there is also a human component in any technology. To include this human operator component, we must also see that all technologies are embedded within a sociotechnical system.

The design of the control system for any technology system must therefore also be encompassed within a larger design of the sociotechnical system. It is here that the market perspective and technology perspective collide head to head. Any technology system that does not perform well in the sociotechnology system will be a commercial failure.

CASE STUDY:

Monster Guns of the Nineteenth Century

Societal conditions influence when and how an invention occurs. This case study illustrates how the change in performance in one technology in an application may stimulate a need for change in another technology of the application. The historical setting was a time of important technological change in the history of military technology—the innovation of the iron-clad warship. Control over the oceans has always been a critical issue among nations. Although ships could have been clad with iron earlier, they would have been clumsy to sail. The earlier invention of the steam engine had made this practical.

Ironclad ships created a military crisis for Great Britain's ocean military supremacy, shown by its earlier defeat of Napoleon's French navy (Bastable, 1992). In 1860, a new Bonaparte, Napoleon II, was on the French throne, and he was having steam engines installed in France's fleet, launching its first ironclad frigate (a midsized warship), named *La Gloire*. The British admiralty made a mock-up target of *La Gloire's* sides—wood clad with four-inch iron plate. They fired on the target with existing naval cannons, and the cannon balls only bounced off the mock target. Cannons then were made of cast iron, smooth bore, loaded at the mouth, and fired round iron balls. The military point was obvious to them. No ship in the British fleet could sink *La Gloire*. If the French chose to invade England's island with a fleet of ironclads, the British fleet could do nothing to stop them. The British army's prestige rose at the expense of the navy, for the army would then become the country's last resort. The prime minister, Lord Palmerston, appointed a royal commission on the defense of the United Kingdom to design a series of coastal fortifications with guns large enough to sink ironclads—monster guns. A British inventor, Sir William Armstrong, would supply these guns.

Armstrong had pioneered the development of large guns made of layers of wrought iron rather than cast iron. Armstrong added rifling to the layered iron gun, and breechlocks with which to load the gun from the rear rather than down the muzzle. Some of Armstrong's ideas were not original but borrowed from recent technical advances in shoulder-held guns (which we now call *rifles*). The technical problem that Armstrong had to solve was how to make the rifled barrels stronger for cannon and how to devise strong breechlocks for cannon.

His invention was to use a coiled arrangement of strips of wrought iron wound into a barrel and welded into a solid tube. The wound strips provided rifling for the cannon, and layers of wound strips built up around the smallest core provided strength for the explosions in the large cannon to shove out the shell. Armstrong's revolutionary large guns had already demonstrated military superiority over existing cast-iron smooth-bore cannon. He produced these for the British government, and such guns firing 40-pound conical shells demonstrated their long-range accuracy and deadly efficiency in the British participation in the Anglo-French invasion of China in 1860. But a 40-pound shell could not penetrate 4 inches of iron plate at 1200 yards, which is what would be required to sink *La Gloire*. Armstrong calculated that it would take at least a 300-pound explosive shell to penetrate 4-inch plate iron from that distance.

In 1862, Armstrong completed his first 300-pounder early and demonstrated it at the government firing range at Shoeburyness, at the mouth of the Thames. The next day, the *Times* published an enthusiastic account: "With an indescribable crash that mingled fearfully with the report of the gun, the shot struck upon a comparatively uninjured plate, shattering the iron mass before it into little crumbs of metal, splintering the teak (behind that) into fibers literally as small as pins" (Bastable, 1992, p. 232).

Armstrong supplied not only the British military but also sold monster guns to the world market. These provided a major turning point in military technology. Soon wrought iron was replaced by steel. The German armament maker Krupp perfected the steel cannon, which played an important role in the German defeat of the French in 1975 in the Franco-Prussian War. In 1914, in World War I, the world would begin to be subjected to the terrible destruction that this new artillery was to produce in the twentieth century.

The deliberate cooperation between an inventor such as Armstrong and government opened a new era in the management of military technology. As the historian Bastable (1992) commented: "Armstrong's first breechloaders and his monster guns set in motion technological, political, bureaucratic, and industrial forces that established a more intimate link between industry and government. This new relationship reflected the fact that the ability to wage war and build empires had come to depend more than ever before on the industrial capacity of an economy and continual technological innovation by its engineers" (p. 246).

APPLICATION SYSTEM

Technological progress in commercially innovated high-tech products and services that provide new functional capabilities to customers results in functional value-added products and services. A complementary idea to a technology system that is embodied in a high-tech product or service is the idea of the system of application in which the customer uses the high-tech product or service.

A customer application system is an ordered set of useful tasks having functional value to the customer.

Accordingly, it is important to distinguish between three kinds of systems relevant to technological progress:

1. Technology systems
2. High-tech product systems
3. Customer application systems

The economic value of a technological innovation lies in its application system. The purpose to which a technology is put is called its *application*. The ability to do something for that application is called its *functional capability*. How good that ability is in accomplishing the application is called its *performance*. Until the performance of a new technology is sufficient for a customer application, the technology is not yet ripe. The critical performance measure of a technology is the minimum performance necessary for the technology to perform in a customer application.

An innovative leader should think about technological progress in terms of function, customer, product, application, and critical performance.

As technologies occur in various system configurations, the selection of the particular system configuration to be used depends on how the technology is to be used—its application. An application is also a system (transforming inputs into desired outputs that accomplish a purpose). An application system consists of:

1. A major device system and all the technologies embodied in the device (e.g., a battleship in naval warfare)
2. Key peripheral systems and all the technologies embodied in the peripherals (e.g., naval cannon)
3. Strategies, tactics and control technologies for using the major device system and peripheral systems in the application

The concept of a technology system focuses on the techniques necessary to attain a functional capability. The concept of an application system focuses on how a functional capability as used by a customer. The concept of a major device system of an application focuses on the primary technological artifice used in an application.

Sometimes the term *technology* is used to cover several meanings. Some use the term to indicate the inventive idea of a functional capability; others use it to indicate the idea of the system of the capability, the technology system; some use it to indicate the context of the technical application, the application system; and still others use it to indicate key artifices used in the application, product systems. However, let us distinguish these different uses of *technology:*

1. *As invention:* the *generic concept* of technology
2. *As system:* the *various configurations* of technology
3. *As application:* the *various applications* contexts of technology
4. *As artifice:* the *various devices* embodying technology systems that are used in an application

For example, flight is a generic concept of a technological capability of transport through air. Fight as a generic technology depends on applying power to provide lift in air. Airplanes and helicopters are different technology systems for flight. Applications of military or commercial transportation systems use airplanes and helicopters as artifices (major device systems) within the transportation system.

USES OF TECHNOLOGY IN BUSINESS

The applications systems for technologies used in a business are the ways a business uses technologies to produce economic value for customers. A business produces products to sell to a customer, and in so doing, buys supplies from vendors. Within the business, products are produced from such, and other operations are performed that are necessary to design and sell products. Information must be obtained, generated, and communicated about operations and production within the business and between the business and vendors and customers. In addition, revenue must be received from customers for sales and paid to vendors for supplies Accordingly, technologies in a business system are used immediately as:

1. *Product technologies:* Technologies used in the design of product/services
2. *Production technologies:* Technologies used in manufacturing/service delivery
3. *Operations technologies:* Technologies used in operations
4. *Information technologies:* Technologies used in informing business activity
5. *Communication technologies:* Technologies used in communications
6. *Capital technologies:* Technologies used in financial activities

CASE STUDY:

Innovation of the Internet

Application systems are used by customers purchased as products and services from industry. But as technologies are systems and applications are systems, so is the complete functional capability provided by a new basic technology. The customer uses new technology in its applications but obtains this new technology in the products and services of a new high-tech economic functional system.

Earlier we noted that the Internet was an example of a new functional capability in society in the last decade of the twentieth century. We looked at the economic impact of the innovation of the Internet, and now let us review its actual invention.

The origin of the Internet goes back to an earlier computer network called ARPAnet, whose creation can be traced back to J. C. R. Licklider. Licklider served in 1962 in a government agency, the Advanced Research Projects Agency (ARPA), which funded advanced military research projects for the U.S. Department of Defense. He headed research in ARPA on how to use computers for military command and control (Hauben, 1993). Licklider began funding projects on networking computers in a newly created ARPA research program on information processing techniques. He wrote a series of memos on his thoughts about networking computers that were to influence the entire computer science research community.

About the same time, a key idea in computer networking derived from work of Leonard Kleinrock. He had been doing research on the idea of sending information in packaged groups—packet switching. He published a first paper on packet switching in 1962 and a second in 1964. Then in 1965, Lawrence Roberts at the Massachusetts Institute of Technology (MIT) connected a computer at MIT to one in California through a telephone line. This was the first prototype of computer communications. In 1966, Roberts submitted a proposal to ARPA to develop a computer network for U.S. military communications under nuclear attack. This was called the Advanced Research Projects Administration Network, ARPAnet.

By this time, Robert W. Taylor had replaced Licklider as program officer of ARPA's Information Processing Techniques Office. Taylor had read Licklider's memos and was thinking along the same lines of the importance of computer networks. He funded Robert's ARPAnet project. Earlier, Taylor had been a systems engineer at the Martin Company and then a research manager in the National Aeronautics and Space Administration (NASA), funding advances in computer knowledge. Next he went to ARPA and became interested in the possibility of communications between computers. In his office there were three terminals, time-sharing computers used in three different research programs that ARPA was supporting. He watched how communities of people built up around each time-sharing computer. Taylor was also struck by the fact that each time-sharing computer system had its own commands.

In 1965, Taylor proposed to Charlie Herzfeld, head of ARPA, the idea for a communications computer network using standard protocols. In 1967, a meeting was held by ARPA to discuss and reach consensus on the technical specifications for a standard protocol for sending messages between computers, and this was called the *interface messaging processor* (IMP). Using IMP to design messaging software, the first node on the new ARPAnet was installed on a computer on the campus of the University of California at Los Angeles. The second node was installed at the Stanford Research Institute, and the third and fourth at Stanford and the University of Utah. Then ARPAnet began to grow from one research setting to another. By 1969, ARPAnet was up and running, and Taylor left ARPA to work at Xerox's Palo Alto Research Center.

As the ARPAnet grew, there was a need for control of the system. It was decided to control it through another protocol, called *network control protocol*

(NCP); and this was begun in December 1970 by a committee of researchers called the *network working group*. The ARPAnet grew as an interconnected independent sets of smaller networks. In 1972, a new program officer at ARPA, Robert Kahn, then proposed to advance protocols for communication as an *open architecture,* accessible to anyone. The open architecture would emphasize that (1) each network would stand on its own, needing no change; (2) communication would be on best effort, so that if a packet could not be transmitted, its source would resend it; and (3) the overall system would have no global control. These ideas were formulated as the *transmission control protocol/Internet protocol* (TCP/IP). ARPAnet was converted to the TCP/IP protocol in 1983, and this protocol was to become the open standard on which the worldwide Internet would later be based.

While the ARPAnet was being expanded, other computer networks were being constructed by other government agencies and universities. In 1981, the National Science Foundation (NSF) established a supercomputer centers program, which aimed to have researchers throughout the United States able to connect to the five NSF-funded supercomputer centers. NSF and ARPA began sharing communication between the networks, and the possibility of a truly national Internet became envisioned. In 1988, a committee of the National Research Council was formed to explore the idea of an open, commercialized Internet. They sponsored a series of public conferences at Harvard's Kennedy School of Government on the "Commercialization and Privatization of the Internet."

In 1989, the first Internet search program, called *Archie,* was developed, soon followed by a simpler search software called *Gofer,* developed by the University of Minnesota. Gofer was used in the first commercial online service, established by Delphi in 1992, which allowed individuals to connect to the Internet. Both Archie and Gofer used Unix commands to search. In 1991, a new protocol for the Internet was developed at CERN, based on the concept of hypertext. The idea was to embed links inside text to link to other text, and the concept was called the *World Wide Web*. A browser called *Mosaic* was developed to use hypertext, and the World Wide Web (WWW) exploded in usage. Then in April 1995, NSF stopped supporting its own NSFnet "backbone" of leased communication lines, and the Internet was privatized. The Web sites expanded from about 130 in 1993 to 10,000 in 1995, to 100,000 in 1996, to 650,000 in 1997, and so on.

On October 24, 1995, the Federal Network Council defined the *Internet* as:

- Logically linked together by a globally unique address space based on the Internet protocol (IP)
- Able to support communications using the transmission control protocol/ Internet protocol (TCP/IP) standards

In summary, one can see in this case that innovation for the Internet was first motivated by military applications and next by science applications—researchers seeking ways to have computers communicate with each other. This was a new kind of functional capability in computation. The invention of computer networks required several new technical ideas. The first technical idea was that computer messages should be transmitted in small, brief bursts of electronic digital signals rather than the continuous connection used in the human voice telephone system. Computers could "talk" to each other in bursts of digital bits, different from how humans talked to each other in continuous streams of analog sounds. Thus the physical basis of computer communication, packet switching, differed from the technology of the human phone system, continuous connection. The second technical idea was that formatting of digital messages between computers needed to be standard in format to send message packets, and these open standards became the Internet's TCP/IP standards. The third technical idea was that a universal "address" repository needed to be created so that computers could know where to send messages. This became the Internet directory (when all addresses would end in, for example, .edu for universities, .com for businesses, .org for other organizations).

The architecture of the Internet system that evolved is illustrated in part in Figure 4.3, where two kinds of contexts are shown connected to the Internet. There is a home context, connected through modems and telephone lines (or cable or DSL connections) to an Internet portal server [such as America Online (AOL)], whose server and router are connected to the Internet. There is also a business context, with an internal local area network (LAN) connected to the Internet through its own server and router. Also shown are two of the important technical ideas of the Internet system, message standards and a registered address system. The Internet standards (TCP/IP) provide the formats for all messages exchanged on the Internet. Each router accesses a universal directory for registered addresses of all sites. It is also interesting to note that in the Internet system, only three

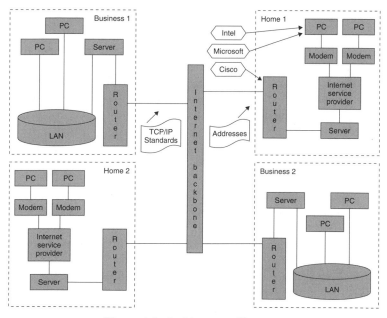

Figure 4.3. Architecture of Internet.

information technology companies built early dominant commercial positions: Intel in personal computer (PC) central processing unit chips, Microsoft in PC operating system software, and Cisco in routers.

The Internet is both an inventive idea and an implementation of a technology. As an idea, the Internet consists of the technical knowledge of how to connect computers into networks—information technology. As an implementation, the Internet is a commercialized system of hardware, connections, and software that enable attached computers of different organizations to communicate with people in different locations. The Internet is a commercialized new functional system in a society.

ECONOMIC FUNCTIONAL SYSTEMS

The Internet is an example of a class of systems that we will call *economic functional systems,* providing an economically functional capability in society. The Internet extended the functional capability of communications to computer-to-computer communications. As a technology, the Internet was implemented into the world's economies as a functional system of telephone communications and computers wired to telephone communication systems.

The Internet is a subsystem of a larger set of communication systems in an economy. In the U.S. society, the oldest communication system was the U.S. Postal Service, established early in the new nation's life. Following on the postal system, other innovations in communications were the telegraph, telephone, radio, and television. Computer networks and the Internet are the latest innovations in the economic function of communications. The innovation of the Internet resulted from (1) research to develop the *technical ideas* on which computer networks are based, and (2) *commercial introductions* of products and services for which it was hoped to establish their capability in the U.S economy.

New economic functional systems provide a new functional capability in a society and are implemented as innovative and connectable products and services—an economic functional system. Examples of other historically important innovations upon which major economic functional systems were based are listed in Table 4.1. As we can see in this partial list of some of the world's most significant inventions, innovation has played a major role in the development of human civilization.

An economic functional system is an interconnectable set of products and/or services that together provide a useful functional capability in a society.

The profitable opportunities for new businesses and industries in new economic functional systems occurs in providing products and services to establish and run the functional capability system. For example, the innovation of the Internet provided business opportunities for new information technology companies to provide goods and services for the Internet. Examples of new Internet information companies are (1) Cisco, which provided routers to connect the Internet; (2) AOL, which

TABLE 4.1 Some Important Technological Innovations since the Fourteenth Century

Approx. Dates	Material Technologies	Biological Technologies	Information Technologies	Energy Technologies
1300s				Gun
1400s			Printing	
1700s	Coke-fueled steel Powered machinery			Steam engine
1820s				Battery
1830s				Railroad
1850s	Aniline dye		Telegraph	Steamship
1860s			Photography	Electric lighting
1880s			Telephone	Electrical power Internal combustion engine
1890s		Vaccines		Automobile
1900s		Anesthetics	X-ray photography Cinema	Airplane
1910s	Plastic		Electron vacuum tubes	
1920s			Radio	
1930s		Penicillin	Radar	
1940s			Electronic computer Transistor	Controlled rockets
1950s			Satellite communications Laser	Nuclear power
1960s			Integrated-circuit chip Fiber-optic communications	
1970s	Computer-controlled machines	Recombinant DNA	Computer networks Computer-aided design	

provided portal services to connect to the Internet; and (3) Netscape, which provided browser software to explore the Internet.

SUMMARY AND PRACTICAL IDEAS

In case studies on

- Sikorsky's helicopter
- Systems control in helicopters
- Monster guns of the nineteenth century
- Innovation of the Internet

we have examined the following key ideas:

- Visions of a technology system
- Systems concept
- Technological transformation
- Technology system
- Technological invention
- Loci of change in a technology system
- Scales of innovation
- Technology control subsystem
- Application system
- Types of technology in business
- Economic functional systems

The practical use of these ideas provides a view of any technology as a system of manipulation. Invention of a technology requires the creation of a functional scheme mapped to sequences of states of a natural phenomenon. Technological progress in a technology requires improvements in the system of the technology. Technology systems are innovated as the basic system ideas in the design of product systems, production systems, operations systems, information systems, communication systems, capital systems, and customer application systems.

FOR REFLECTION

Identify a basic invention that established a new technology. When was it invented, and by whom? Describe the history of the circumstances leading up to the invention. What was the first application of the invention? How well did it perform?

5

PRODUCT SYSTEM

INTRODUCTION

We have noted that a new economic functional system is constructed from the new high-tech products of the new technology. We have also seen that technologies are a kind of system and emphasized that *invention* of the technology system is the first part of innovation. The second part of innovation, *commercialization,* occurs by designing the new technology into new products (or services or operations). Now we look at high-tech products, and we will see how products, too, are systems. As the idea of a *technology system* is fundamental to innovation theory, so is the idea of a *product system.* But there is an important difference between them. A technology system is *knowledge,* and a product system is a *utility.* The knowledge of a technology is embodied in the design of a useful product. We will review the idea of a product as a system, including the basic ideas of: *product architecture* and *complexity* and the differing roles of technology in *product, service, production,* and *operations systems.*

As an example of a product system, we use the case of the computer. It makes an interesting illustration, since the technology in a computer is complex in both physical morphology and schematic logic. First we look at the invention of the electronic computer and then we look at the design of an early computer product, the PDP-8 minicomputer.

CASE STUDY:

Invention of the Computer

Like the ancient invention of iron, the computer was one of those great inventions that change human civilization—an invention of mythic scale—a *Promethean* invention. Prometheus was a titan in ancient Greek mythology to whom the Greeks ascribed the invention of the technology of fire as a gift to humans. The ancient Greeks thought the gift of fire so valuable—and so godlike in power—that they imagined that Prometheus was punished for his gift. The angry gods chained

Prometheus to a mountain, to be exposed eternally to the harshness of weather, blasts of lightning, and the flesh-tearing beaks of vultures. Fascination with the power of technology has long been a part of human culture. The computer changed the mythic power of humanity to acquire and process information. Central to its invention were five people: von Neumann, Goedel, Turing, Mauchly, and Eckert (Heppenheimer, 1990), each of whom came to a tragic end.

John von Neumann was born in Hungary in 1903, the son of a Budapest banker. He was precocious; at the age of 6 he could divide eight-digit numbers in his head and talk with his father in ancient Greek. At 8 years of age, he began learning calculus. He had a photographic memory; he could take a page of the Budapest phone directory, read it, and recite it back from memory. When it was time for university training, he went to study in Germany under a great mathematician, David Hilbert. Hilbert believed that all the diverse topics in mathematics could be established on self-consistent and self-contained intellectual foundations. In a famous address in 1900, Hilbert expressed his position: "Every mathematical problem can be solved. We are all convinced of that. After all, one of the things that attracts us most when we apply ourselves to a mathematical problem is precisely, that within us we always hear the call: here is the problem, search for the solution; one can find it by pure thought, for in mathematics there is no ignorabimus (we will not know)" (Heppenheimer, 1990, p. 8).

As a graduate student, von Neumann worked on the problem of mathematical foundations. But in 1931, Kurt Goedel's famous papers were published, arguing that no foundation of mathematics could be constructed wholly self-contained. If one tried to provide a self-contained foundation, one could always devise mathematical statements that were formally undecidable within that foundation (incapable of being proved or disproved purely within the foundational framework). This disturbed von Neumann, as it did all other mathematicians of the time. But Goedel also introduced an interesting notation in which any series of mathematical statements or equations could be encoded as numbers. This notation would later turn out to be a central idea for von Neumann's vision of a stored program computer. All mathematical statements, logical expressions as well as data, could be expressed as numerically encoded instructions.

However, the first person to take up this idea was not von Neumann but Alan Turing. Turing was a 25-year-old graduate student at Cambridge University when he published his seminal paper, "On Computable Numbers," in 1937. He had used Goedel's idea for expressing a series of mathematical statements in sequential numbering, Turning proposed an idealized machine that could do mathematical computations. A series of mathematical steps could be expressed in the form of coded instructions on a long paper tape. The machine would execute these instructions in sequence as the paper tape was read. His idea was later to be called a *Turing machine*. Turing had described the key idea for what would later become a general-purpose programmable computer. Although many people had thought of and devised calculating devices, these had to be instructed externally or could solve only a specific type of problem. A machine that could be instructed generally to solve any kind of mathematical problem had not yet been built.

Back to von Neumann. After finishing his graduate studies in Germany, von Neumann emigrated to the United States and joined the faculty of Princeton University in 1930. There Turing's and von Neumann's paths crossed temporarily when, in 1936, Turing came to Princeton to do his graduate work. He was thinking about the problem of his idealized machine (which he would soon publish), and he worked with von Neumann, exposing von Neumann to his ideas. Von Neumann offered Turing a position as an assistant after Turing received his doctorate. But Turing chose to return to Cambridge where in the following year, he published his famous paper.

So Turing's ideas were in the back of von Neumann's mind. But meanwhile, war was to intervene, beginning with the German invasion of Poland in 1939. During that war, Turing went into England's secret code-breaking project and helped construct a large electronic computer to break enemy codes. Called the *Colossus,* it began operating in 1943, but it was a single-purpose computational machine (Zorpette, 1987).

At the same time in the United States, von Neumann was involved in the Manhattan Project, to create the atomic bomb. Earlier at Princeton, von Neumann had been exploring the mathematics of problems in fluid flow. In the design of one of the atomic bombs, it was proposed to place a sphere of dynamite around pie-shaped wedges of plutonium. When the dynamite exploded, the wedges were intended to be blown together into a small sphere (slightly smaller than a soccer ball). This plutonium sphere would then undergo the violent nuclear chain reaction of an atomic explosion. Von Neumann designed the dynamite trigger for this bomb, called the "Fat Man" version of the atomic bomb.

The trigger problem was: To precisely what thickness should the dynamite around the wedges of plutonium be formed so that they would all be blown in at the same time? To design this, one had to calculate the physical form of the shock waves from the dynamite explosion that would push the plutonium wedges inward. It was a flow type of mathematical problem, and von Neumann would be just the one to calculate it. But it was a tough problem because of the accuracy required in describing the shock waves. To do it, von Neumann and a colleague at Los Alamos, Stanislaw Ulam, devised a kind of human computing system. They had one of their colleagues, Stanley Frankel, devise a lengthy sequence of computational steps that could be carried out on mechanical calculating machines made by IBM. Frankel then had a large number of Army enlistees running these steps on the calculating machines. It was a slow kind of human computer, but it worked. von Neumann got the solutions he needed to design the explosives for the Fat Man. The Fat Man bomb was the second atomic bomb to be exploded. It was dropped on Nagasaki, killing more than 100,000 Japanese. This technical challenge taught von Neumann an important lesson, the need for a general-purpose computer. (Afterward, one of von Neumann's favorite events was observing atom bomb tests, and exposure to radiation may have been the cause of von Neumann's fatal tumor.)

One day in August 1944 (before the atomic bombs were dropped), von Neumann had gone to the Army's Aberdeen Proving Ground in Maryland on a consulting assignment. Afterward, he waited at the station for a train. On the same platform happened to be Lt. Herman Goldstine (who before the war had taught mathematics at the University of Michigan). He recognized von Neumann, already a world-famous mathematician. Goldstine introduced himself. Personally, von Neumann was a warm, pleasant man. He chatted amiably with Goldstine, asking him about his work. Later Goldstine said of the meeting: " 'When it became clear to von Neumann that I was concerned with the development of an electronic computer capable of 333 multiplications per second, the whole atmosphere changed from one of relaxed good humor to one more like the oral examination for a doctor's degree in mathematics.' " (Heppenheimer, 1990, p. 13). It turned out that Goldstine was one of Mauchly and Eckert's team building the ENIAC, the world's first electronic computer; and it was in this meeting that von Neumann first heard of the ENIAC project, a project conceived by John W. Mauchly. The story of the computer now switches to Mauchly.

Mauchly had been born in Cincinnati, Ohio in 1907. In 1932, he had received a Ph.D. in physics from Johns Hopkins University. Teaching physics at Ursinus College from 1933 to 1941, Mauchly had begun to experiment with electronic counters while doing research on meteorology. In 1941, he attended a summer course in electronics at the Moore School of Electrical Engineering at the University of Pennsylvania. He was then invited to join the faculty. In 1941, Mauchly heard of some computational work by a physicist, John Atanasoff at Iowa State University. He visited him, and Atanasoff showed Mauchly an experimental electronic adder which Atanasoff had completed in 1940. That gave Mauchly an idea. In the fall of 1942, he wrote a memorandum on the idea of using electron vacuum tubes for a calculating machine (Brittain, 1984). The key idea in Mauchly's proposal was to use Atanasoff's flip-flop circuit as a basic logic circuit for computers, since it could record a binary state of either 0 or 1. A signal applied to the grid of a tube could turn the tube on, and thereafter it would stay on and conducting (a state of 1). Another signal might then be applied to the tube, and it would turn off (a state of 0). In either state, the flip-flop circuit would be stable until a new signal arrived to flip it. This circuit used a pair of tubes hooked together. Mauchly outlined in the memorandum how one could use a set of flip-flop circuits to express a number system (binary numbers) and so construct a reconfigurable calculating machine.

Mauchly submitted his proposal to the Army Ordinance Department: Colonel Paul N. Gillon, Colonel Leslie E. Simon, and Major H. H. Goldstine. They approved the idea and gave an R&D contract to the Moore School to build the machine, which Mauchly would call the ENIAC. For this project, Eckert was the chief engineer and Mauchly the research engineer. J. G. Brainerd was project supervisor; and several others made up the team, including Arthur Burks, Joseph Chedaker, Chuan Chu, James Cummings Leland Cunningham, John Davis, Harry Gail, Robert Michael, Frank Mural, and Robert Shaw. It was a large undertaking and required a large R&D team (Brainerd and Sharpless, 1984).

When constructed, the ENIAC took up an entire large air-conditioned room, whose four walls were covered by cabinets containing electron-tube circuits

Figure 5.1. ENIAC, 1946. Eckert is in the foreground and Mauchly is in the center.

(Figure 5.1). Altogether the circuit cabinets weighted 30 tons and drew 174 kilo-watts of power. The major problem was tube failure. Mauchly and Eckert had calculated that of the 17,468 tubes they were using in ENIAC, they were likely to have on average one failure every 8 minutes. This would have made ENIAC useless, not able to compute anything that took 8 minutes or longer. Cleverly, they decided to run the tubes at less than one-half of their rated voltage and one-fourth of their rated current, which reduced the failure rate to one tube about every two days. Still, the ENIAC was not easily programmable. To set up a new problem for calculation, one had to connect circuits physically using patch cords between jacks, and the cabling ran up to a total of 80 feet in length. The patching task itself could take at least two days.

It all came together in the summer of 1944 after Goldstine met von Neumann. The ENIAC was already working, and Eckert and Mauchly were thinking about an improved successor, which they intended to call EDVAC (electronic discrete variable automatic computer). At Goldstine's urging, the Army was considering awarding the Moore School $105,600 to build the EDVAC: "Into this stimulating environment stepped von Neumann. He joined the ENIAC group as a consultant, with special interest in ideas for EDVAC. He helped secure the EDVAC contract and spent long hours in discussions with Mauchly and Eckert" (Heppenheimer, 1990, p. 13). The contract was let in October 1944, and in June 1945 von Neumann completed his famous paper "First Draft of a Report on the EDVAC." This was one of the most influential papers in what was to become computer science.

Goldstine circulated the draft with only von Neumann's name on the title page: "In a later patent dispute, von Neumann declined to share credit for his ideas with Mauchly, Eckert, or anyone else. So the "First Draft" spawned the legend that von Neumann invented the stored-program computer. He did not, though he made contributions of great importance" (Heppenheimer, 1990, p. 13).

However, neither von Neumann nor Eckert nor Mauchly built the EDVAC. The ENIAC group broke up. The University of Pennsylvania hired a new director of research, Irvin Travis, who quarreled with Eckert and Mauchly over patent rights. Eckert and Mauchly had a letter from the university's president stating that they *could* hold the patents on ENIAC. But Travis told them that they must sign patent releases to the university. Mauchly and Eckert refused and resigned from the university in 1946. They formed the Electronic Control Company, which became the Eckert–Mauchly Computer Corporation in 1947. It built the first commercial general-purpose electronic computer, the UNIVAC, selling its first product to the U.S. Census Bureau. But running out of funds to build that first machine, Eckert and Mauchly sold out to Remington Rand, which in turn became Sperry Rand in 1955. In 1959, Mauchly left Sperry Rand to form a consulting firm.

After the war, von Neumann returned to the Institute for Advanced Study in Princeton. He decided to build a computer, and Goldstine joined him, along with Julian Bigelow (who had worked on radar-guided antiaircraft gun control during the war). Goldstine and Bigelow built a 2300-vacuum-tube computer for von Neumann in the boiler room of the main building at the Institute. This was the first fully automatic stored-program computer.

A patent on this basic invention would have been valuable. Somewhere between Turing, von Neuman, Mauchly, and Eckert the inventive idea arose. But von Neuman published the idea before filing for a patent. In patent law, this invalidated anyone from filing for a patent. So it happened that there was no basic patent on the stored-program computer.

In summary, John von Neumann created the world's first stored-program electronic computer. He was influenced by the earlier ideas of Kurt Goedel and Allen Turing. He also drew upon the contemporary ideas and work of John Mauchly and J. Presper Eckert. Mauchly and Eckert created one of the world's first all electron-vacuum-tube special-purpose computers (about the same time, other special-purpose computers were being created independently in England by Turing and in Germany by Konrad Zuse). Mauchly had also borrowed earlier ideas from John Atanasoff. Mauchly and Eckert were in turn influenced by von Neuman's ideas. Mauchly and Eckert did create the world's first commercially successful stored-program electronic computer, the UNIVAC.

Mauchly was very bitter that von Neumann never acknowledged him as a coinventor of the stored-program concept. Adding injury to insult, von Neumann testified against Mauchly in an important trial that challenged Mauchly and Eckert's patent claim from the ENIAC. The judge ruled that Mauchly and Eckert were not the true inventors of the electronic computer. This ruling contributed to Mauchly's final penury: " 'Lawyers keep making money,' said Mauchly toward the end. 'We've got down to the point where maybe we can buy some hot dogs with our Social Security.' " (Heppenheimer, 1990, p. 16).

And the gods seemed to add to Mauchly's woes. In 1980 he died from a genetic disease that had terribly disfigured him. Terrible also was the gods' fate for von Neumann: "For von Neumann, it was even worse. In the summer of 1955, he was diagnosed with bone cancer, which soon brought on excruciating pain. In the words of his friend Edward Teller, 'I think that von Neumann suffered more when his mind would no longer function than I have ever seen any human being suffer.' Toward the end there was panic and uncontrolled terror" (Heppenheimer, 1990, p. 16).

Things also went hard for Allen Turing, the person who had first conceptualized general computation as programmable operations: "Convicted in England of soliciting sexual favors from a teen-age boy, he was given a choice of prison or hormone treatments. He chose the hormones and soon found his breasts growing. Driven to despair, he made up a batch of cyanide in a home laboratory and died an apparent suicide" (Heppenheimer, 1990, p. 16).

Even the famous logician, Kurt Goedel, whose work influenced the computational ideas of both Turing and von Neumann, was personally tormented: "It was his own personal demons that would drive him to death. . . . After his wife underwent surgery and was placed in a nursing home, in 1977, Goedel refused to take any food. He starved himself to death" (Heppenheimer, 1990, p. 16).

Apparently, Promethean inventions make gods angry.

COMPUTER ARCHITECTURE

Von Neumann's famous report had described the logical scheme of a computer operational architecture: single-instruction single-processing (SISP) architecture. The key advance was the inclusion of stored-program instructions, which made the machine a general-purpose computational device. First a program and data must be input into computer memory. A central processing unit (CPU) calls up the program instructions in sequence from the stored program and executes each instruction on the data. After the data have been transformed according to each instruction, the result calculated is stored temporarily and then transformed further by the next instruction in the program. After all instructions in the program have been executed, the final results are both stored in the memory and outputted from the computer.

In this computational scheme, the program expresses the logic of the mathematical transformation. The operation of the computer is to map each instruction of this logic onto a specific configuration of the electronic circuits within the computer. This is a mapping of logic to physical electronic processes within the computer, and it is a high-level program logic mapped to physical structures of the computer sequentially for calculations.

At a lower logical level of computer technology, a binary number system has been mapped onto physical circuits of the computer. The instructions of the program and the data upon which calculations are performed are expressed in *binary numbers,* which constitute the lowest level of logic mapped into the physical structure of the computer.

A binary number system is a way of writing integer numbers with only two symbols (0 and 1). In a binary system:

zero	is symbolized as	0
one	is symbolized as	1
two	is symbolized as	10
three	is symbolized as	11
four	is symbolized as	100
five	is symbolized as	101
six	is symbolized as	110
seven	is symbolized as	111
eight	is symbolized as	1000
nine	is symbolized as	1001
ten	is symbolized as	1010 and so on

In a decimal system, an addition of numbers such as $1 + 2 = 3$ would be in the binary system $1 + 10 = 11$. In the von Neumann architecture, one could build a calculating machine using only binary-state electron-vacuum tubes and all calculations in the machine would be performed in the binary number system. This is an example of a mapping of a logic onto a physical structure (the logic of numbers to binary physical states of electron-tube flip-flop circuits) In the mathematical operations, computer calculations are performed as a coordination of physical processes to logical operations. Arithmetical operations are coordinated to as a series of physical processes, such as turning some tubes in circuits off and others on.

Von Neumann used Turing's idea for a mathematical algorithm to be performed as an ordered sequence of operations, beginning with the initial data and transforming them to calculated results. This is the concept of the *stored-program computer:* to store both the data and the program algorithm in the active memory of the computer architecture. A von Neumann computer architecture is then run, timed to an internal clock, performing the following calculational cycle:

1. Initiate running the program.
2. Fetch the first instruction from main memory to the program register.
3. Read the instruction and set the appropriate control signals for the various internal units of the computer to execute the instruction.
4. Fetch the data to be operated on by the instruction from main memory to the data register.
5. Execute the first instruction on the data and store the results in a storage register.
6. Fetch the second instruction from the main memory of the program register.
7. Read the instruction and set the appropriate control signals for the various internal units of the computer to execute the instruction.
8. Execute the second instruction on the recently processed data, whose result is in the storage register, and store the new result in the storage register.

9. Proceed to fetch, read, set, and execute the sequence of program instructions, storing the most recent result in the storage register until the complete program has been executed.

10. Transfer the final result calculated from the storage register to the main memory and/or to the output of the computer.

The von Neuman stored-program computer worked like a Turing machine. Any mathematical calculation can be expressed as a sequence of ordered algorithmic steps that transform initial data into calculated results (with both the data and program stored in the main memory of the computer).

CASE STUDY:

PDP-8 Computer

The idea behind computer architecture just discussed represents *technical knowledge,* knowledge of a technology system. To actually make a computer that uses this knowledge, the ideas must be translated into the *design* of a computer product. We see how the ideas of a technology are translated into the design of a product by looking at the design of an early minicomputer in 1970, Digital Equipments PDP-8.

After the invention of the computer, technical progress in computer development came first as improved components for memory and logic, ferrite core memories, and transistorized circuits. Next, the invention of the semiconductor integrated-circuit (IC) chip provided the parts for both memory and logic circuits. By the late 1960s, technological progress in IC chips made possible a new product line, minicomputers, lower in performance and price than mainframe computers. The PDP-8 was innovated by a new company, Digital Equipment Corporation (DEC), founded by Ben Olsen.

As a student at MIT, Ben Olsen had worked on the world's second general-purpose computer, developed by Jay Forrester at MIT (after Forrester had seen Mauchly and Eckert's computer). Later, Olsen founded DEC to produce transistorized circuits boards for computer applications. By the late 1960s, Olsen saw that he could make a relatively simple transistorized computer that although lower powered than the expensive mainframe computers of the time, could be afforded by a single scientist rather than only by a research laboratory. This was the commercial focus of the PDP-8, and over 100,000 of these units were built and sold to scientists. It launched the minicomputer industry as a market niche beneath the higher-priced mainframe industry.

The PDP-8 was configured on a von Neumann architecture. Its central processing unit was an arithmetic–logic unit (ALU) to perform both arithmetic (e.g., addition, multiplication, integration) and logical operations (e.g., and, or, if–then). It had a memory unit storing 4096 words, with each word of 12-bit length. It had a data bus (internal electrical connections to move data around inside the computer from input to memory, from memory to ALU, from ALU back to memory, and from memory to output). It had an input/output (I/O) and control panel with switches with a lighted display register for inputting, displaying output, and over-all control. In the simplest form, data were entered manually by the graduate student (or technician). Most scientists also added other devices to the PDP-8, for automatic data entry and output (such as punched tape drives and readers).

The units noted above constituted the central components of the technology system of the stored-program computer. The PDP-8 added several other units. The first addition was an accumulator (AC) register, which stored the intermediate computational results from the ALU as it stepped through a computation. It also had a storage register, called LINK, to store temporarily arithmetic overflow bits in the calculations (e.g., to add the units and carry the overflow to the tens). It also had a memory address (MA) to indicate where in memory to find the data (or instruction) to read and where to store the computed data result from the ALU. There was also a memory buffer (MB), where data read from memory was stored temporarily to be ready at the exact time the ALU needed it. The ALU ran on a very strict and rapid sequence, according to stored-program instructions. There was also a program counter (PC) to specify the location in memory where the instruction could be found that was to be executed next by the ALU. There was also an instruction register (IR) into which had been read from memory the instruction that the ALU was currently executing.

The difference between the PCP-8's principal units (arithmetic–logic unit, memory, input–output) and its secondary units (arithmetic overflow, memory address, memory buffer, program counter, instruction register) is an illustration of one of the important differences between a *generic technology* and an *actual product* that implements the technology. Additional features are necessary to make the technology perform adequately for an application.

In addition to adding features to a generic technology, a product may also arrange a hierarchy of technology subsystems. The concept on which the stored-program computer was based used the idea that a binary number system could be mapped directly to the two physical states of a flip-flop electronic circuit. This circuit was one in which one tube was constantly conducting electricity (on) until a signal arrived to turn it off. Then it remained off (nonconducting) until another signal arrived to turn in on again. The logical concept of the binary unit 1 could be interpreted as on (or conducting) state, and the binary unit 0 could be interpreted as the off (or nonconducting) state of the electric circuit. In this way the functional concept of numbers (written on a binary base) could be mapped directly into the physical states (morphology) of a set of electronic flip-flop circuits. The number of these circuits together would express how large a number could be represented. This is what is meant by *word length* in a digital computer. The PDP-8 was designed with a 12-bit word length, which required 12 transistorized flip-flop circuits to be constructed wherever a word was to be expressed morphologically in the computer. Binary numbers must encode not only data but also the instructions that perform the computational operations on the data. So a word length of 12 bits had the important constraint of limiting the complexity of the instruction. (One of the points of progress in computer technology has been how long a word length the computer can use. For example, in microcomputers (computers on a chip), the word length began at 8 bits in 1975, increased to 16 bits in 1980, and increased to 32 bits in 1988. Each increase in word bit length lead to faster processing speeds and faster memory access.)

The design of the PDP-8's technology system used a hierarchy of subsystems. The lowest level of the computer morphology mapped to logic was the mapping of a set of bistable flip-flop circuits to a binary number system. At the highest logic level, von Neumann's stored-program concept was a clocked cycle of fetching and performing computational instructions on data. At an intermediate logic level between the binary number logic and the program logic of the computer was an intermediate level of logic in arithmetic operations. Thus, the PDP-8 computer had several hierarchical levels of schematic logics mapped to physical morphologies:

1. Binary numbers mapped to bistable electronic circuits
2. Boolean logic operations mapped to basic electronic circuits
3. Mathematical basic operations mapped (through Boolean constructions) to electronic circuits
4. Computer architecture mapping algorithms into temporally sequenced electronic circuits

TECHNOLOGY SYSTEM AND PRODUCT SYSTEM

The configuration of a technology system indicates only the principles necessary for the functional transformation of the technology. But the configuration of a product system embodying the technology must add secondary steps and processes needed *to fill in the details and fully connect* the principal logic steps and *other secondary technology systems to complete the product.*

> **The differences between a technology system and a product system are (1) in the details and fully connected steps of the core technology in the product system, and (2) in completing the product system with the necessary secondary technologies.**

For example, consider the familiar automobile as a product system. The functional transformation of the automobile provides land transportation for moving passengers and goods from one location on land to another. In logical order, an automobile must be fueled, entered, started, directed, and stopped. To accomplish this, the physical structure of the automobile has several subsystems which together configure the automobile system:

- A body subsystem
- A fuel subsystem

- A starting and ignition subsystem
- An engine and transmission and drive train subsystem
- A wheels and suspension subsystem
- A breaking subsystem
- A steering subsystem

In addition, there are subsystems for passenger comfort and safety:

- Seating
- Windows and climate control
- Seat belts and airbags

Knowing how a technology works is not sufficient to know how to design a product of that technology. Knowing how to design a product provides a source of technical differences between competitors.

A product system is a completed and connected transformational technology system used by a customer.

A product system such as a computer involves not just one simple logic in its technology system but many complex logics. The schematic logic of the computer technology requires several hierarchical logic levels from transistors to circuits to binary numbers to Boolean algebra to mathematical operations, and above that, a Turning machine procedure. Although information technologies (such as the computer) have complex logics, not all technologies have complex logic. Most physical technologies have relatively simple schematic logic but have complex physical morphologies in their product systems. For example, the relatively simple schematic logic of a helicopter technology is to provide lift by rotating propellers and to provide stability by controlling the tilt of the propeller blades and a counter-rotational tail propeller.

A complex technological system may be constructed of parallel subsystems, or of a hierarchy of subsystems, or of both parallel and hierarchical subsystems. Parallel subsystems are component systems of a system which operate simultaneously but independently in the system's transformation function. Hierarchical subsystems are functionally lower-level systems whose operations determine the operation of the system at a synthetic upper level.

As we saw, the logic of the computer system is constructed of hierarchical subsystems: beginning with transistorized bit word lengths, scaling up to basic logic operations and up to basic arithmetic operations and up to programmed mathematical and logical operations. Information products perform logically complex transformations.

TECHNICAL PROGRESS IN PRODUCT SYSTEMS

Technical progress in a product can occur from technological inventions to improve the product design or to improve its production process. To see the need for technical progress in products, one should ask the following kinds of questions of an existing product system:

1. What level of product performance is minimally acceptable for an application, and what increments in performance would be clearly noticeable by a customer?
2. What features of the product are used in which application?
3. What peripheral devices are essential to an application?
4. What aspects of the product have glitches or breakdowns or require frequent maintenance in an application?
5. What aspects of the product operating in various environments create safety or pollution risks?
6. How does the current cost of a product limit the number of applications?
7. What factors in the application system determine the product's replacement rate by customers?
8. What factors in the product facilitate (or inhibit) brand loyalty in replacement purchases?

TECHNOLOGY IN PRODUCT, SERVICE, PRODUCTION, OR OPERATIONS

Finally, we note that business technology systems are embodied not only in product systems but also in production systems, service systems, and operations systems:

- A *production system* is a completed and connected set of transformational technology systems used in producing a product.
- A *service system* is a completed and connected set of transformational technology systems used in communicating and transacting operations within and between producer/customer/supplier networks.
- An *operations system* is a completed and connected set of transformational technology systems used in communicating within and operating a business enterprise.

SUMMARY AND PRACTICAL IDEAS

In case studies on

- Invention of the computer
- PDP-8 computer

we have examined the following key ideas:

- Computer architecture
- Product system complexity

- Technology system and product system
- Technical progress in product systems

These theoretical ideas can be used practically to understand how technological invention occurs and how technology becomes embedded into products, production, services, or operations that are high-tech—embodiments of the latest in technological progress.

FOR REFLECTION

Identify a key invention and read a history of how it came to be invented. What was the functional logic in the invention? Into which products, processes, or services was the technology embedded?

6

PRODUCT DEVELOPMENT PROCESS

INTRODUCTION

Earlier we emphasized that innovation is a business process. Now having seen how products are systems, we look next at the business process for developing new product systems—the *product development process*. We will look at how innovative products are created within organized activities of design and development and production—seeing new product creation as *cycles* and *processes* of development that determine the *quality* of products.

Commodore 64 Personal Computer

As an illustration of the product development process, we look at the development of a new computer product, the Commodore 64, back in the nascent days of the new personal computer industry in the early 1980s. For a very brief time, Commodore affected the very new and rapidly growing personal computer market. Commodore no longer exists. Often, fame and fortune are brief and fleeting, particularly in the new days of a high-tech industry. So it was with almost all new personal computer companies in the beginning. This is typical of the many new firms created at the beginning of a Kondratief economic long wave—some brilliant short-term successes and then many business flame-outs—as the competition in a Kondratieff long wave heats up.

In 1981–1982, when Jack Tramiel was president of Commodore, the company designed and produced a new personal computer called the Commodore 64. The "64" indicated that it was the first personal computer to use the then new 64K memory chip. Although now long obsolete, it was then a hot, competitively priced product for the emerging home computer industry. In fact, it pushed the prestigious Texas Instruments Company (TI) out of the home computer market and

99

contributed to the death of a then high-flying company, Atari, whose founder had marketed the first computer game, called "Pong."

The strategic technology underlying the personal computer was the semiconductor integrated-circuit (IC) chip, which although not invented until 1959, had by 1980 already passed through three generations of rapid technological change: small-scale integration (SSI), medium-scale integration (MSI), and large-scale integration (LSI). LSI was a set of production processes enabling the fabrication of IC chips with *thousands* of transistors on a single chip. That transistor density had enabled computer designers to fabricate on a single chip the heart of a computer, its the central processing unit (then called a microprocessor). The microprocessor chip made possible the new personal computer industry, and the first relatively complete personal computer to be marketed was the Apple in 1976. By 1981, personal computer companies were emerging as a new industry. Then the personal computer market served two distinct and new markets, a business market for word processing, spreadsheets, and accounting, and a home market for games. Both markets were served by the same computers. Apple was the leader, closely followed by Tandy, Texas Instruments, Atari, and Commodore. The computers were priced in the range $1500 to $2500.

The home market was price sensitive to the $1000 barrier. Tramiel at Commodore first produced a personal computer called PET, which competed poorly against the then leader, Apple. Tramiel also watched the rapid sales of a new personal computer product from England, the Sinclair, then selling for $100. But the Sinclair did not even have a real keyboard. So at Commodore, Tramiel stopped producing his PET computer and introduced a cheaper model, called the VIC-20, priced around $500 ($1000 cheaper than the Apple. This was Commodore's first really big seller, but the VIC-20 was limited in speed and memory compared to its competitors, the Apple II and the Atari 800. All computers then contained 8-bit-word-length microprocessors (except the TI 99/4a, which did use a 16-bit microprocessor, but its designers had not utilized it to show a performance advantage over competitor's products).

Back in 1981, it was the video game market that was the hot and rapidly increasing market for personal computers. There were a lot of kids who loved the new video games, and a lot of parents willing to spend $100 on a child, but few parents then were willing to spend more. To get performance up and price down in the game market, the product required increased speed for graphics and sound. The Atari 800 had pioneered a technical solution to speeding up graphics performance by designing a special video display chip to assist the microprocessor.

This was the background, and the Commodore 64 story began in 1981 at a small chip manufacturing company, MOS Technology, then located in West Chester, Pennsylvania. In January, a group of semiconductor engineers decided to design two special chips for graphics and for sound. Albert J. Charpentier led the group, and he was head of MOS's LSI group. The use of these new chips would be for whoever then wanted to make the "world's best video game" (Perry and Wallich, 1985, p. 48). What Charpentier's group wanted to do was to follow up on Atari's technical solution by designing a newer and more advanced set of graphic and sound chips to provide the same fast game capability as the chips in the Atari 800. Charpentier proposed this project to his boss, Charles Winterble, who was then director of worldwide engineering for Commodore. Winterble approved the project.

It is interesting to note why Commodore owned MOS Technology. MOS Technology had been a new startup company in the early 1970s when chip technology advanced from a density of hundreds of transistors on a chip in MSI technology to thousands of transistors on a chip in LSI processing technology. MOS Technology had designed and produced the then popular microprocessor 6502, which was the 8 bit-microprocessor that powered both the original Apple computers and Commodore's PET computer. Yet by the middle of the 1970s, several Japanese firms (i.e., Hitachi, Toshiba, and others) mastered LSI technology and produced a 16K chip, which was then the most dense memory chip available. Memory chips sold as the largest volume of the IC chip market. In 1975, Japanese chip producers grabbed a significant world market share (20%) of the 16K market. In addition, they aggressively lowered the price of memory chips and raised production quality. That hurt many of the new small chip producers, like MOS, many of which had been started on the new LSI technology. When MOS was only a few years old, the entry of larger and better-financed Japanese competitors into its market dealt it a death blow. (This is just one of many examples of the action of Kondratieff long economic waves throughout the nineteenth and twentieth centuries.) So in 1976, MOS was on the financial ropes; and Commodore, one of its customers for its 6502 microprocessor chip, picked up the company "at 10 cents on the dollar" (Perry and Wallich, 1985, p. 48). That was how Charpentier of MOS came to work for Winterble of Commodore (thanks to the changing conditions of international competition in strategic technologies). This also illustrates what happens to many small high-tech firms as radical technological innovation in a new Kondratieff long wave allows new small firms to start up but then later, many established competitors from other industries enter and drive down prices and profitability.

KNOWLEDGE-TO-VALUE TRANSFORMATION

Let us look again at an S&T infrastructure, but this time sketching it in a diagram for a knowedge-to-value translation. It is in high-tech products that new knowledge is transformed into new economic value, as a knowledge-to-value translation. Just how a particular high-tech business and its new products fit into an S&T infrastructure as innovation process logic can be envisioned as in the Venn diagram of Figure 6.1. A Venn diagram is a symbolic way of depicting the overlapping areas of logical sets of things to show logical interrelatedness. What this Venn diagram emphasizes is that in innovation there are lots of interactions between technology and business and industry and university and government and between technology and product, customer, and application.

Research for scientific and technological progress connects industry, university, and governmental sectors to both a high-tech business and to a technology. A high-tech business uses research and technology in the design and in production of its

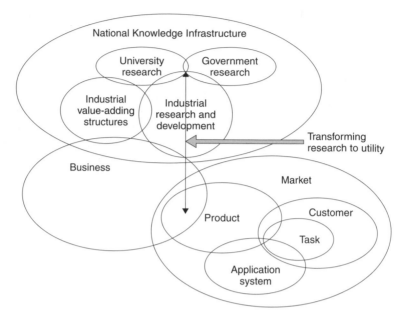

Figure 6.1. Transforming knowledge to value.

high-tech products. *It is important to note that the business and customer are connected through the products of the business.* The customer experiences directly the product and the application of the product, yet a business does not experience the customer's application directly, *although it does so indirectly through its product.*

These connections are important for several reasons. First, the technological sophistication of a high-tech business is bounded by the *research capability* of the industrial, university, and governmental R&D infrastructure in which a business exists. Second, the research and technological capability of a business is known to a customer *only through the business's products.*

> **The research and technical capability of a high-tech business that does not contribute directly to product performance, quality, or price is not valuable to the business because it is not seen by a customer.**

Therefore businesses must be very selective about what research and technologies engage their interests and investments. Thirdly, since the satisfaction of a customer with a product depends upon its performance in an application (and since a business does not directly experience the application), it is the *application* which is the greatest source of uncertainty about commercial success in the design of a product.

> **The direct connection between a business and a customer is through the product or service; but products and services use scientific and engineering knowledge created not only in the business but also the universities, in governmental research, and in the industrial sector.**

Ultimately, all the technology used by a business is seen, directly or indirectly, by a customer through the product or service, yet the engineering and marketing personnel of a business may not fully anticipate the customer's application. This is a common reason why product designs fail to satisfy a customer—when a high-tech business's engineering and marketing personnel do not fully recognize the customer's needs. In Figure 6.2, note the line drawn across the Venn diagram that indicates the *price-to-value divide*. This line indicates three economic ideas about value:

- *Utility:* between the customer and application systems
- *Value:* between the customer, application, and product systems
- *Price:* between the business, product, and customer systems

Figure 6.2 indicates where the economic concepts of utility, value, and price differ as perceptions shift from application to customer to business. This perceptual shift emphasizes *where the difficulty lies* in successful commercialization of a new high-tech product or service. The dark line that shows the divide between the business and application systems demonstrates that the *direct experience* of the business System with the application system often does not occur because the business and application systems do not overlap. This creates a kind of divide in perception between the business systems perception of the value of a product as the product price and the customer systems perception of the product as product value in its contribution to utility in the application system. This perceptual divide in the experiences

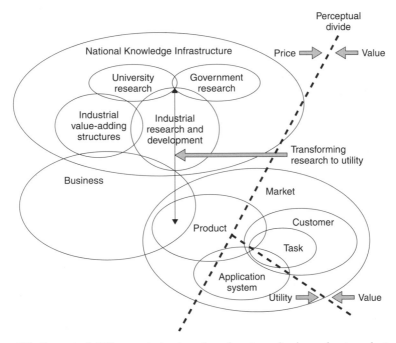

Figure 6.2. Perceptual difference in business's and customer's views about product value.

of product price and product value is inherent as an experience missing from the business system.

> **The perceptual divide between price and value in a new high-tech product is the basic reason that marketing forecasts fail regarding the potential of a basic new high-tech product, when such forecasts are derived from only a market analysis of existing markets.**

The marketing problem for the commercial success of a new high-tech product or service arises from a lack of detailed understanding and knowledge regarding the matching of its price and value. Hence, marketing of new high-tech products is always a commercial gamble. The gamble is won by the firm that best guesses how to match superior product performance to customer value in a new market.

For a business to design a commercially successful product or service to sell to customers, the business needs to have specific knowledge of the needs of the customer—the "face" of the customer (in terms of the kinds of applications, tools, and tasks regarding which the customer will use the product). The technical risks in designing a product or service for this "face" arises from knowing these needs exactly and precisely. The technical risks arise from uncertainties about *what* the product or service can really do for a customer in applications, how well it does it, how many resources it will consume, how dependable it is, how easy it is to maintain and repair, and how safe it is:

- Functionality
- Performance
- Efficiency
- Dependability
- Maintainability and repairability
- Safety and environment

The *functionality* of a product, process, or service indicates for what kinds of purposes it can be used. For example, different industries are often classified by purpose: food, transportation, clothing, energy, health, education, recreation, and so on. The goods and services within these industries satisfy the various purposes. Furthermore, within a purpose are usually a range of applications. For example, in transportation, there are applications for business, vacation, and personal travel.

The *performance* of a good or service indicats the degree of fulfillment of a purpose that is achieved. For example, different food groups provide different kinds and levels of nutritional requirements.

The *efficiency* of a good or service for a level of performance indicates the amounts of resources consumed to provide a unit level of performance. For example, different automobiles attain different fuel efficiencies at the same speed.

Dependability, maintainability, and *repairability* indicate how frequently a product or service will perform when required and how easily it can be serviced for maintenance and repair.

Safety has both immediate and long-term requirements: safety in performance and safety from after-effects over time. The environmental impact of a product or service includes the impacts on the environment from production, use, and disposal.

In contrast to technical risks in trying to design products for the face of the customer, there also are commercial risks, from uncertainties about just who the customers are, how they use a product, what specifications are necessary for use, how they gain information and access to a product, and what price they are willing to pay:

- Customer type
- Application
- Specifications
- Distribution and advertising
- Price

The *customer, application,* and *specifications* together define the market niche of a product, process, or service. The *distribution and advertising* together define the marketing of the product, process, or service. The *price* set for a new product, process, or service needs to be acceptable to the market and include a large enough gross margin to provide an adequate return on the investment required to innovate the new product, process, or service.

Resolving the technical variables correctly is both necessary and costly (they constitute the research and development costs of developing and designing a product). Even if accomplished successfully, the commercial variables must be correctly apportioned (they constitute the production and marketing costs). Despite how much a business may learn of a customer or a set of customers during market analysis, the variability of the desired balance of the technical and commercial factors will vary across a market and even across a market niche. A range of technical variables is always possible in the design of a new product, process, or service. Which variables will turn out to map correctly to the set of commercial variables required in the future is never clear initially, only in retrospect. We see this in continuing the case of the Commodore 64.

CASE STUDY CONTINUED:

Commodore 64 Personal Computer

When Charpentier's group decided to make graphics and sound chips, their first step in the design process was to find out what the current high-quality chip products could do. This kind of practice in design is called *competitive benchmarking.* They spent the first two weeks finding and looking at comparable graphics and sound chips industrywide. They also looked at the graphics capabilities of the currently most advanced game and personal computers: " 'We looked heavily into the Mattel Intellivision,' recalls Winterble. 'We also examined the Texas Instruments 99/4a and the Atari 800. We tried to get a feel for what these companies could do in the future by extrapolating from their current technology. That made it clear what the graphics capabilities of our machine had to be' " (Perry and Wallich, 1985, p. 49).

Competitive benchmarking for new product design requires not only finding and looking at the best and most advanced competing products, but also *extrapolating* the future technical capability of competitors. One should design new products not only to beat the *current* products of competitors but also to beat the *future* products of competitors.

> **The design of products to beat future products of competitors is the fundamental reason why technology *forecasting* is vital to new product development.**

Next, the design team at MOS borrowed all the ideas from competitors' products that they liked. One can borrow a competitor's ideas as long as they are not patented, copyrighted, or obtained illegally as trade secrets. In designing new products, technology should be approached as *a cumulative as well as an anticipatory state*. This means that in the design and production of new products, the technical goal should be to include at least all the best current practices and highest current performances and then add to them.

Yet it is not enough just to try to include all the best features in a product because that might make the product too expensive for a target market.

> **Innovative product design should balance best practice with a competitive advantage in the proper ratio of performance to cost; and this is the fundamental reason why technology *strategy* is vital to new product development.**

Next, the design team looked at *constraints* that would limit how much of the best practice and how much of the highest performance and improvements they could squeeze into the product design. In the case of the design of new chips, the most important constraint was that the line width of etching structures into the silicon for making the transistors on the silicon was then limited to no smaller than 5 micrometers (5 millionths of a meter) in size. This constraint limited the numbers of transistors and parts, hence the numbers and complexity of circuits, and finally, the number of features and speed of performance. The fundamental challenges in new product design are always the trade-offs required between desirable performance and technical and cost constraints.

> **The fundamental reason that technology *implementation* is important to new product development is to provide improved product performance at a lower price.**

The design team then had two things: a *wish list* of features and performance they would like to see in the chips, and the *limit* to the numbers of circuits and features they could put on one chip with 5-micrometer technology. Then came the next step in the design process: " 'Then we prioritized the wish-list from what

must be in there to what ought to be in there to what we'd like to have' . . ."
(Perry and Wallich, 1985, p. 49).

This kind of innovation logic in product design is basic. First, there should be
a decision as to what the customer needs that is better than what competitors now
supply or could (by extrapolation) supply in the future. Second, there should be
a decision to prioritize desirable features into "must," "ought," "like to have." If
a design cannot accomplish both the "must" and "ought" features, it will probably be a technical failure. If the design cannot also accomplish many of the like-
to-have features, it probably won't be a commercial success. Accordingly, we can
see that a new technology must be anticipated correctly, and the proper trade-off
between performance and cost must be made in the initial phases of product
design.

> **Commercial success by a firm in the early part of an economic long wave
> depends critically on appropriate design decisions made between corpo-
> rate research and product development.**

INNOVATION IN THE PRODUCT DEVELOPMENT CYCLE

As we saw in the Commodore case, new technology is introduced into a product
during a product development process. New product design is not a single event but
repetitious design activities taking place within the ongoing activities of the firm.
Current products are being produced, new product models are being designed, production improvement is being planned, and implemented, and new generations of
product lines and/or new product lines are being researched, planned, and designed.
Innovation in product development is a cyclic innovation process focused on incremental innovations.

In this case, we are seeing a new product generation being developed, not directly from science but from advanced industrial technology. Technical progress
in chip design and in chip memory capacity was advancing at an industrial level.
Charpentier's group was designing a next-generation product model with improved
graphics and memory. The innovation in the product development came from acquiring technology from without the firm. Also, the innovation was not basic but
incremental. So a second kind of relationship of innovation to product development
was being used in this case, which has been called the *product development cycle*
or *circular innovation*.

As depicted in Figure 6.3, the logic of technological innovation within a product development for incremental innovation involves four concepts:

- Technology anticipation
- Technology acquisition
- Technology implementation
- Technology stimulation

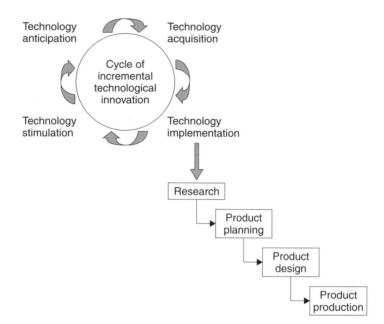

Figure 6.3. Product innovation cycle for incremental technology progress.

It is necessary first to *anticipate* technological change so that one has time to prepare and take advantage of it. Next it is necessary to *acquire* new technology for use by the firm. Then new technology can be *embedded* into new products, processes, or services. The next logical step is to *implement* the new technology commercially through new product introductions and aggressive pricing or quality strategies. Customer's subsequent experiences with the new products suggests new needs and stimulates requirements for more new technology. The cyclic innovation process is driven by periodic, incremental improvements to technology to improve the *value* of products to the customer. For firms, these logical steps form a kind of circular pattern in the technical and business activities of the firm. We will call this the *cyclic incremental innovation process.*

CASE STUDY CONTINUED:

Commodore 64 Personal Computer

Returning to the Commodore case study, the design was a prioritized wish list of required and desirable features that Charpentier's group laid out in the circuits and transistors for the computer's chips. Several more product design decisions had to be made (e.g., sophistication versus simplicity, time to completion versus complexity): "Because design time rather than silicon was at a premium, the chips were laid out simply rather than compactly. 'We did it in a very modular fashion' . . ." (Perry and Wallich, 1985, p. 49). The actual design process consists of many details and decisions and trade-offs. Scientific and generic engineering principles can help guide some of these. Other decisions are based on practical engineering experience, others on practical product and production experience. Some simply are judgments about what the customer should want and some about technical pride. Design decisions are a mixture.

Design is the first step in a new product development. Next comes prototype production and testing. Fortunately, MOS had a chip-fabrication line that could be interrupted to produce samples of the new design. This was necessary for debugging and refining a design until it worked: "David A. Ziembicke, then a production engineer at Commodore, recalls that typical fabrication times were a few weeks and that in an emergency the captive fabrication facility could turn designs around in as little as four days" (Perry and Wallich, 1985, p. 49).

The cooperation between production and product design here facilitated the exceptionally fast design to product time of nine months. Thus, by November 1981, the chips were completed and samples produced and tested. All designs require debugging and refining; and reducing redesign time is one of the major keys to competitiveness. This is another important point about competitiveness that we will explore later in managing the product development cycle.

> **For economic success, the links between technology strategy and product development must include production strategy.**

Next came a very significant change in this case, a changed vision of the product. In November 1981, Charpentier and Winterble reported the success of the project to the CEO of Commodore, Jack Tramiel. While listening to their intended sales of the chips, Tramiel had in the back of his mind the recent success of Commodore's new personal computer product, the VIC-20.

So when Charpentier and Winterble showed Tramiel the new chips, Tramiel imagined a new personal computer product for Commodore that would use the competitive advantage of the new technology and provide a product successor to the popular VIC-20. Tramiel saw that the new chips could be used in a new product as good as the best on the market but at a fraction of the price. He was also anticipating other new technological products soon to be introduced, new 64K random access memory (RAM) chips: " 'Jack [Tramiel] made the bet that by the time we were ready to produce a product, 64K RAM's would be cheap enough for us to use,' Charpentier said" (Perry and Wallich, 1985, p. 51).

> **This is an example of the *integration* of technology strategy with business strategy; and it is that *integration* that drives economic long waves and industrial dynamics.**

Tramiel decided to use Charpentier's graphic and sound chips not to sell to competitors but to produce a new product, called the Commodore 64 personal computer, with a major competitive advantage in price: "When the design of the

Commodore 64 began, the overriding goals were simplicity and low cost. The initial cost of the Commodore 64 was targeted at $130; it turned out to be $135" (Perry and Wallich, 1985, p. 51).

The competitive advantage of the Commodore 64 was its price; the product was designed and produced specifically for low cost. At the time it did have roughly comparable or superior performance to competing products, but at one-fourth the cost of production. Tramiel had set performance and price goals for a new product, the Commodore 64, which would use Charpentier's new chips. But the next problem was to design that computer. This new design was relatively straightforward since it was a redesign of existing personal computer designs. The system architect for the design was Robert Yannes.

PRODUCT DEVELOPMENT PROCESS

As we noted, product development is a cyclic process in a business. In a new product develop lies the periodic opportunity for innovation in product models. There is always a timely window of opportunity only when innovating new technology in a product that can gain a positive competitive advantage. A speedy product development process is necessary to use technological innovation as a competitive advantage. Moreover, once in that window, the product must have been designed correctly for the customer in terms of proper application focus, adequate performance and features, and reasonable pricing, safety, and dependability.

> **Speed and correctness are the criteria for a good product development process and are essential when innovating new technology.**

Why are speed and correctness not always easy to achieve in the product development cycle? The reasons relate to the uncertainties and variation and changes that must be managed in the cycle:

1. There are *uncertainties about timing* in the anticipation of new technology—when it will be ready for embodiment in new products.
2. There are *uncertainties about performance* of new technologies—how much actual achievement the technology will provide in a new product.
3. There are *uncertainties in customer requirements* in the new product—how the new technology will alter the customer's use of the product.
4. There are *uncertainties in the trade-off between performance and cost* that the product designer can provide.
5. There are *uncertainties in production* of the product from the inherent variability in the physical production processes.

Jim Solberg emphasized that these kinds of uncertainties taken together are the sources of delays and errors in the overall product design and production system (Solberg, 1992). He suggested that each of these provide feedbacks of uncertainty in the prior decisions of parts of the overall system, as depicted in Figure 6.4. These uncertainties results in changes and in demands for new work by preceding units that introduce delays into new product innovation. For example, production variation can create feedback loops in the product development process that create delays. A new product designed for production may turn out not to be producible in terms of projected quality and cost without improving the production process, and production may therefore be delayed until research creates the needed production improvement. Cost and quality problems may require changes in product plans or redesign of a product.

The performance–cost trade-offs in product design may also create delays that derive from further product development from research or alteration of product plans and schedules for product introduction. Changes in customers or in customer requirements can also produce delays in planning new products by requiring more research for the creation of the products. Finally, the technical risks in research may result in research projects being delayed or in not achieving a desired performance level, which would delay the entire downstream product development process.

Solberg (and others) have suggested that the development of procedures and software tools to promote concurrency and virtual and rapid prototyping in the product development cycle will speed up and aim toward correctness activities in

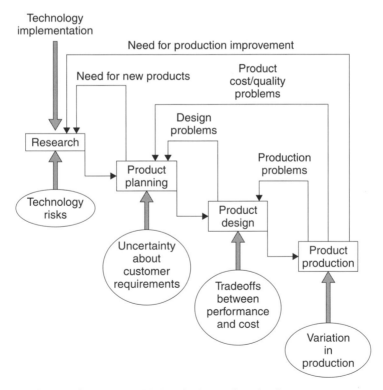

Figure 6.4. Sources of delays in the product development process.

the product development cycle. These include product development procedures and tools for managing innovation in the product development cycle.

Design and development time is an important factor in the commercial success of new products. Being too late into a market after competitors have entered is a disadvantage unless one comes in with a superior product or substantially lower price. We will later discuss that critical to fast product development times is the use of multifunctional product development teams. For example, Zirger and Hartley (1996) studied the product development times of several electronics companies. They found that the fast product developers had cross-functional development teams that included explicit goals about fast time to market and overlapped their development activities in concurrent engineering practices.

As another example, Funk (1993) studied product development at Mitsubishi's Semiconductor Equipment Department and at Yokogawa Electric. He found that studies about the product development practices of Japanese firms identified several strengths in product development, including (1) the use of multifunctional teams in problem solving, (2) close relationships with customers and suppliers, (3) a focus on incremental improvements to product and production, and (4) learning how to improve the product development process.

While product development time is important, it is, of course, not the only factor. For example, Meyer and Utterback (1995) examined whether increased product development time alone guaranteed commercial success and, not surprisingly, found that it did not. Commercial success for new products depends on many factors, of which time to market is only one. There are many factors that may increase development time in innovation, such as the complexity and number of new technologies that have to be integrated into the product. The minimum condition for a good chance at success is to get the product performance right at the price for the right application for a customer. The sooner one can do this, the better. In software development, the complexity of the logic and the production process particularly interact, creating special challenges for software development processes. For example, Rowen (1990) emphasized that technical difficulty in large software projects arises both from the complexity of the application and the process of a large team of programmers. Accordingly, as we noted earlier, the criteria for judging the fitness of a design process are *speed* and *correctness*.

There have been many other studies about the importance of the product development cycle to commercial success. For example, interviewing senior managers, Cooper (1990) studied a sample of 203 new product projects in 125 firms and identified several factors that financially successful product development projects had in common:

1. Superior products that delivered unique benefits to users were more often commercially successful than "me too" products.
2. Well-defined product specifications developed a clearly focused product development.
3. The quality of the execution of the technical activities in development, testing, and pilot production affected the quality of the product.

4. Technological synergy between firms' technical and production capabilities contributed to successful projects.

5. Marketing synergy between technical personnel and a firm's sales force facilitated the development of successful products.

6. The quality of execution of marketing activities was also important to product success.

7. Products that were targeted for attractive markets in terms of inherent profitability added to success.

We note that in these factors, the first and last have to do with the relationship of the product design to the market—a superior product aimed at a financially attractive market provides both competitiveness and profitability. Factors 5 and 6 have to do with marketing activity. Sales efforts are helped by products that fit well into an existing distribution system and also by proper management of marketing activities that test and adapt a product to a market. Factors 2, 3, and 4 have to do with the quality of the technical development process. Good management of the technical activities in product development help produce good products, and good project management includes having a clearly focused product definition, managing well the various phases of the product development process, and having the proper technical skills to execute the project.

The success of a new product introduction requires not only a good product design but also good management of both the product development process and the marketing process. The product development cycle should be managed to provide (1) innovation in product performance, (2) innovation in productivity, (3) responsiveness to market changes, and (4) competitive advantages.

CASE STUDY CONTINUED:

Commodore 64 Personal Computer

Yet in the design of a new product (particularly in every new high-tech product) there are always bugs in design and in production. Unfortunately, even small bugs can have a major competitive impact. This is why the process of innovation is not a one-shot affair. Commercially successful innovation requires a series of subsequent innovations.

Despite the cleverness of the new Commodore 64 and its very low price, the product quickly gained a reputation for poor quality which eventually hurt Commodore's business reputation. For example, one of the bugs was in Charpentier's graphics chip design, which required a patch in order to allow the computer to display both black-and-white and color information without conflict. This problem was fixed quickly by Commodore and corrected later in a improved model of the graphics chip.

Another bug occurred in the new design as a result of using a read-only memory (ROM) chip from the earlier VIC-20. Used in the new product system of the Commodore 64, it proved sensitive to spurious voltage spikes that occurred when control over the bus changed. Unfortunately, the result was to produce a visible "sparkle," small spots of light dancing randomly on the display screen. This defect was found after three weeks of hard work and was corrected.

There were other problems that came both from design choices and from production. For example, cheap components were purchased that degraded performance. Assembly errors were made that also affected performance. Finally, there was also a very major problem with the disk drive, which operated too slowly. This problem came from a marketing-influenced decision for compatibility with the VIC-20 and from a deliberate decision by Commodore to scimp on software development. In the end, all the bugs, problems, and poor market and production decisions added up to a public perception of a poor-quality product: "The one major flaw of the C-64 is not in the machine itself, but in its disk drive. With a reasonably fast disk drive and an adequate disk-operating system (DOS), the C-64 could compete in the business market with the Apple. . . . With the present disk drive, though, it is hard pressed to lose its image as a toy" (Perry and Wallich, 1985, p. 51).

One sees in this example that the design of a product is only part of the competitive problem. Bugs will always occur and wrong decisions will always get made. It is important to correct problems rapidly, improve production quality, and produce a new model correcting poor decisions. Otherwise, the competitive advantage of the new technology can rapidly be lost for other-than-technical reasons.

> **In the product development cycle, rapid product improvement is fundamental to maintaining market share after a successful innovation.**

With its relatively low price, the Commodore was a commercial hit for a time. The Commodore 64 was singlehandedly responsible for driving Texas Instruments (TI) out of the home personal computer market and a major contributor to the bankruptcy of Atari.

> **Even early in a new technology industry, price still counts as an important competitive factor.**

In summary, the price war in home personal computers began as early as 1980, in the infancy of the new industry. Tramiel had produced the Commodore PET computer in 1979 as a competitor to Apple and Tandy, but it did only so-so. During 1980, Tramiel had watched a very cheap personal computer made by Sinclair in England, with only a toy keyboard but priced at $100, sell well. In contrast, personal computers with usable keyboards and peripherals then sold from $1000 to $2000. Responding to this, early in 1981 Tramiel had introduced the VIC-20, priced at $250, and watched it sell well. It sold despite the fact that it had limited performance compared to the Apple, Tandy, Atari, and TI machines. Texas Instruments then responded to Tramiel's aggressively priced VIC-20 by cutting the price of its 99/4A to an even lower $199. It also sold well, and TI launched a crash program for a new, cheaper product, the TI/8.

Thus, by early January 1982, when TI personnel went to Las Vegas for the Winter Consumer Electronics Show, they felt pretty good because they did not see anything yet with a 16-bit processor. But at that show they did see the new Commodore 64, although they did not then appreciate its potential price significance. However, the Atari people caught the significance: " 'All we saw at our [Commodore's] booth were Atari people with their mouths dropping open, saying, 'How can you do that for $595?' " (Perry and Wallich, 1985, p. 51).

After that show, Commodore turned on the competitive heat. Tramiel cut the price of the VIC 20 to $125. TI had to match that price, but their profit margin on the 99/4A fell to zero. In April, Tramiel struck again, lowering the VIC-20 price to $99, at which Commodore could still make a little money. TI again had to match their price, and their 99/4A continued to lose money. All through 1982 and 1983, TI's home computer division lost money—much money. Atari also lost money.

Texas Instruments was putting their hopes on a new model, the TI 99/2. Finally, in 1983, TI took the new product to the summer consumer electronics trade show, a full year after the introduction of the Commodore 64. Of course, by then, Tramiel had already aggressively lowered the price of the Commodore 64 (from its initial price of $595 of the preceding year). TI then saw that their new product could still not match the rapidly declining price of the Commodore 64. After the show, Fred Bucy of TI called a meeting of the TI group and said: " 'I don't think this product can make any money. Does anyone disagree?' " (Nocera, 1984, p. 65).

No one disagreed. The new product model of the TI 99/8 was dead before production, and with it died TI's home computer business. Texas Instruments then announced their withdrawal from the home computer market. This is the reason that TI stayed a chip-producing company and never became a computer manufacturer. The great irony, of course, is that the technical and scientific capability of Texas Instruments was then, and still is, far greater than the capability of Commodore. Another game computer competitor at the time, Atari, could not meet Commodore's price and their corporate owner closed Atari, selling off its inventory to Tramiel.

> **Commercially successful technology strategy is not the same as technical capability; it requires a good product development process.**

The case of the Commodore 64 product innovation shows the major commercial impact that an appropriately timed innovation in the product development process of one competitor can have upon competitors whose product development cycles lag seriously in timeliness and effectiveness. Thus, we see how the product development cycle at that point of time at Commodore created an enormous competitive success. But business competition continues, and although Commodore was very successful at that time, it subsequently lost its competitive advantage.

In summary, there are several questions to discuss about this case:

1. How did Charpentier anticipate the possible technical change?
2. How did Charpentier plan and create technical change?

3. How did Tramiel, Winterble, and Charpentier plan a product to use the new technology?

4. How was the product/technology strategy integrated with Commodore's business strategy?

5. Why did the product, after such dramatic initial success, eventually fail to build the business position of Commodore?

Charpentier had two different sources of information in mind when his group decided to make a new generation of graphics and sound chips. The first source was a market need, as the home electronics game market was exploding in 1980 and 1981. The second source was the technical advances in chip density in 1980–1981, which was moving toward very large scale integration (VLSI) and which Charpentier understood as providing the capability of creating a chip with 5-micrometer-line-width technology. Together these provided a *market* and a *technical basis* for anticipating new technology. These two kinds of information sources for anticipating new technology are called *market pull* and *technology pull*.

Implementation of technical change occurs within discrete technical projects that are planned, budgeted, and managed. Regarding the second question, we note that Charpentier first obtained approval from Winterble, Commodore's director of engineering, for the project with an agreed-upon project budget and deadlines. Technical goals, market goals, and budget and schedule goals were set. Technology strategy must be translatable into research and development projects.

What kind of product by which to implement new technology requires involves business criteria as well as technical criteria. Regarding the third question, we note that the final product use of the new technology involved not just engineering personnel (Charpentier and Winterble) but the CEO of Commodore, Tramiel. The product use of the technology was redefined by Tramiel, from the selling of the new chip to the internal use of the chip for a new personal computer model. Tramiel's business strategy focused on the opportunities of the personal computer market as providing a larger market and higher profitability than those available to a chip supplier.

The integration of technology and business strategy requires a focus on both the competitive advantages of a new technology product and a marketing and pricing strategy that exploits these advantages. About the fourth question, we note that Tramiel's business judgment focused on the market size and price sensitivity of the home computer market. Tramiel judged correctly that the price was more immediately important than the technical capability. If the Commodore could deliver the same or slightly better technical performance than that of competing home personal computers but at a fraction of the cost, it could knock out competition. And it did.

Product performance must include not only technical performance but also quality. Any new product introduction by itself will never ensure the continuing business success of any company. Product improvement must be continual. For the last question, we note that Tramiel did not apparently appreciate the impact that customer perception of quality would have on the long-term competitive position of Commodore. Poor decisions were made in design, production, procurement, and software development, which together created a public perception of Commodore's poor quality and disappointing performance. Quality was never emphasized, nor was the product improved in a timely manner for better performance. The C-64 did not ensure Commodore's future, as it quickly became obsolete. Yet every product provides an opportunity for impressing customers about the quality of the company.

In the long term, a company survives through a perception of quality, particularly as technological change slows and competition forces prices down toward commodity-type pricing.

PRODUCT QUALITY

The term *quality* is an important term in business, but it is used with a variety of meanings (Godfrey and Kolssar, 1988):

1. *Performance quality:* how well a product technically performs in its central function—a high-tech product of superior performance
2. *Design quality:* as focused on a particular application—a high-tech product focus on application
3. *Production quality:* process that reproduces quantities of a product rapidly, without defects, and at low cost—a high-tech production process
4. *Service quality:* as to durability and maintenance—a high-tech serviceable product

A technological innovation may affect one or more of the various kinds of quality: performance, design, production, service. Products can be high-tech products (differentiable in technical performance) or commodity products (not differentiable in technical performance). High-tech products can compete through performance, but commodity products must compete through cost, production quality, and/or fashion. Porter (1989) once summarized the technology contribution to competitive strategy as either product differentiation or low-cost, high-quality production.

SUMMARY AND PRACTICAL IDEAS

In the case study on

• Commodore 64 personal computer

we have examined the following key ideas:

- Transforming knowledge to utility
- Product development cycle
- Product development process
- Product quality

These theoretical ideas can be used practically in understanding how to manage innovation in new product design for high-tech products.

FOR REFLECTION

Look up the industrial history of a product system (such as automobiles and airplanes) and identify what innovations appeared over time in the product system and when and by which company.

7
INDUSTRIAL RESEARCH
AND DEVELOPMENT

INTRODUCTION

We have seen that long-term economic development occurs from technological progress, but how, precisely, is such development organized? New industries are established to implement a new economic functional system enabled by a new technology. Next we review the idea of *industrial structures* as the organization of an economic functional system. Furthermore, it is in the structure of industries that research and development activities are performed to advance the technological progress of an established industry. We review the basic ideas of industry (1) as *structures* of industrial sectors with differing foci on *R&D* and (2) how technological change can force *industrial reorganization*.

CASE STUDY:

Innovation in the Steel Industry

We can see how change occurs in industrial organization from technological innovation by looking at the historical case of innovation in the steel industry. Bela Gold and co-workers (1984) described technical progress in steel production in the industrial value chain of iron ore and coal acquisition through the refining and shaping of steel products. In the technology of steel production, iron ore is mixed with metal alloys and scrap iron and steel to produce iron and steel. In the refining of the ore to steel, fuels (such as coke or electricity) and fluxes (chemicals) and oxygen are used. Steel thus produced is shaped into ingots and steel products such as rails or rolls of flat sheets of steel.

Since the iron age, the ancient technology of iron production used charcoal from partially burned wood to melt iron ore to obtain iron. Charcoal had to be used instead of wood for heating iron ore because iron obtained from ore by heating with a wood fire was too soft and brittle to use. Why a charcoal fire was

119

better than a wood fire for melting iron ore was not understood by ancient iron-makers. But they knew this from practice. Next, iron was hammered into desired shapes by reheating iron in smoky braziers and hammering the hot iron on anvils. The reheating in the smoke of the braziers could produce steel surfaces on the hammered iron. Again the ancients knew how but not why. The difference be-tween iron (which is relatively soft and brittle) and steel (which is hard and tough) is a few percent of carbon atoms mixed in with the iron atoms—this we now know from the science of chemistry.

We recall that the first industrial revolution in the world, which began in England in the late eighteenth century, was dependent on an innovation in the new steelmaking process, the Bessemer process. Bessemer enabled steel production fueled by partially burned coal, coke, to melt the iron ore. By the middle of the eighteenth century, the new science of chemistry was evolving, and Bessemer then understood that it was chemical impurities that made iron brittle. Bessemer un-derstood that partially burned wood, charcoal, rid the fire of the impure chemi-cals that might be absorbed by the melted iron to make it brittle. Bessemer imagined that partially burned coal, coke, would similarly be free of impure chem-icals to make iron properly. He was right. Also, with the new science of chem-istry, Bessemer understood that to produce steel from iron, one needed to add in with the iron atoms a proper proportion of carbon atoms. The Bessemer innova-tion make possible a gigantic leap in the iron and steel production, providing steel in the great quantities needed for the industrial age.

INDUSTRIAL VALUE CHAINS

The general form of organization of industrial activities for commericalization of the products and services of an economic functional system is called an *industrial value chain*. The idea of a chain of value-adding industrial sectors was popularized in the 1980s by Porter (1985) and later elaborated by Steele (1989).

> **An industrial value chain is a connected set of industries required to ac-quire natural resources and to transform these to a final customer product or service.**

The general form of an industrial value chain is sketched in Figure 7.1. Indus-trial value chains begin with some kind of resource-acquisition industrial sector. The next sector is usually a materials-processing sector. Resource-acquisition and materials-processing sectors are usually organized separately because their tech-nologies and investments and operations differ markedly. Materials-processing sec-tors usually supply materials to parts-producing sectors. For example, the automobile industry is supplied by many different kinds of parts-producing firms, such as tires, wheels, brakes, battery, and electronics.

The next industrial sector is that of major device fabrication. For example, in trans-portation applications, key major devices are automobiles for use on land, airplanes for use in the air, and ships use on the seas. In addition, within a major-device

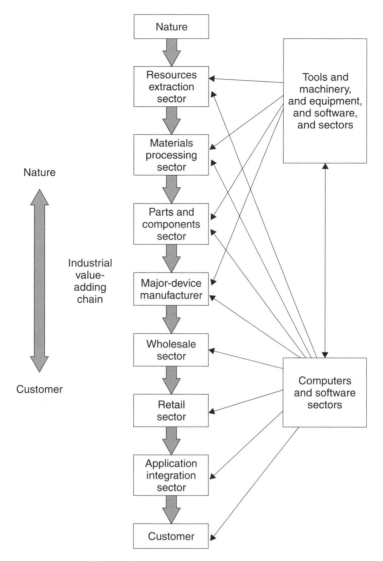

Figure 7.1. Industrial value-adding chain.

sector, subindustrial sectors are frequently organized. For example, the automobile industry is divided into cars, trucks, tractors, motorcycles, tanks, and others. The airplane industry is divided into military and commercial sectors. Sometimes major devices are retailed directly to customers (such as cars) and sometimes they are integrated into a service-providing industry (such as airlines). In either case, major device industries or service industries are often retailed through a distribution system. For example, in the case of cars, there is an automobile dealer industry for sales and service. In the case of airlines, tickets may be sold directly by airlines or through the travel agent industry.

Finally, the customer is served by products and services created in the industrial value chain. These customers may be consumers or business firms (themselves involved in different industrial value chains) or government. After purchase of a major device, the customer will continue to purchase supplies for the device and maintenance and repair services. All along the chain, sectors of tool and equipment manufacturers and computer and communications manufacturers provide equipment to the various businesses in the various sectors of an industrial value chain.

> **The industrial structures of an economy consists of all the transforming sets**
> **of industrial value chains that connect nature to customer.**

When a new technology creates a new functionality, an industrial structure will grow to supply products to the customer embodying the new functionality. The concept of industrial structure provides a way to analyze the economic leveraging of the new technology. Not only are new firms established to produce the new technology, but new firms are established to provide supplies and parts to that firm. New firms may also be established to distribute and service the new product. Thus, the total market creation of a new-functionality technology goes far beyond the market itself for the final product. Also, an industrial value chain can be used to identify all the kinds of technologies necessary to an industrial value chain to satisfy customer needs from nature. Moreover, the science bases relevant to industrial value chains can then be traced from these technologies.

CASE STUDY CONTINUED:

Innovation in the Steel Industry

The industrial value chain for steel products begins with the mining of two natural resources, iron ore and coal. After mining, coal must first be coked (partially burned) to remove chemical impurities that impede iron refining. Iron ore is mined or imported and transported to refineries. Iron and steel scrap is recovered from the production process and also reclaimed from junked products that use large amounts of steel (such as automobiles). Oxygen, flux chemicals, and alloy metals are also used in iron and steel production; and these are produced by various industrial sectors in chemicals and metals.

Mining Technology for Coal

We look first at the innovations in one of the supply industries to the steel value chain—the technologies of coal mining. Coal mining technologies divide basically into underground mining and surface strip mining. Underground mining technology was historically an early process, and underground mining of any ore was restricted to depths no deeper than the water table. In the eighteenth century, the innovation of the steam-engine-driven water pump by Newcomen in the eighteenth century allowed mining to go deeper than the water table. In the nineteenth century, the innovation of dynamite by Nobel allowed miners to fragment rock and ore more efficiently.

From then until early in the twentieth century, the sociotechnical system of coal mining required coal miners to descend on foot (or in lifts) into the depths of the Earth and blast their way in tunnels to coal seams, then pick and blast coal into chunks for loading into cars on rails, which were pulled to the surface by men (or mules or electrical trains). The hazards to people working in the coal mines included mine cave-ins and mine explosions, machinery accidents, and partial deafness from machine noise and black lung disease from the coal dust. Innovations in the coal mining technology included drills to fragment the ore, hoists and conveyors to lift the ore, and machines to rip the face of the ore.

Later innovations in other industries (notably the construction machinery industry) made possible the alternative technology of open-strip mining which involves the removal of surface soil with large tractors and shovels and the systematic removal of ore. This process reshapes the landscape, leveling and terracing hills and digging large holes in the Earth. After strip mining the landscape requires extensive reconstruction for environmental recovery. By 1963, the ratio of ore output to labor was a factor of over four times higher for open-pit mining than for underground mining (Gold et al., 1984, p. 339).

Coking and Ore Benefication Technologies

In modern steel processes, both coal and iron ore receive a processing step before going directly into iron refining. For coal, this is *coking,* a partial burning to rid coal of such impurities as sulfur. For iron ore, this is *benification* (a process for fragmenting ore to a smaller particle size, concentrating its iron content, and agglomerating the concentrated particles into ore pellets).

The early coking process, called *beehive coking,* was carried out at the minehead. In 1887, an innovation, the *Koppers oven,* was used to capture the waste gases. Koppers ovens were boxlike structures in which coal was partially burned to coke. Vent pipes captured the coal gas and delivered it to recovery plants for electrical and chemical production. Later, petroleum and natural gas were substituted for coal gas as chemical feedstocks.

> **Sometimes a stimulus for technological innovation in one part of an industrial value chain occurs when other industrial value chains find uses for the by-products of the first value chain.**

Iron ore benefication made possible the use of ores from different sources in the same refining process and reduced the cost of transporting iron ore from the source to the refiner. Innovations in concentration included washing, magnetic concentration, gravity, and froth methods.

> **Sometimes innovations in both the technologies of processing natural resources and in using natural resources are required for a new form of a natural resource to find economic utility.**

Ore Transportation Technology

In the making of iron and steel, transportation of iron ore from the mines to the furnaces required loading the ore into rail cars and then into ore ships. Incremental innovations in the transporting ships then focused on loading and unloading procedures. Unloading equipment was built into the ship, as hoppers emptying onto conveyor belts. This technology was particularly useful for loading and unloading beneficiated and pelletized iron ore.

In the acquisition and transportation of natural resources, the quality and location of the resources provide technical challenges and stimulate technical change. Once a technical solution has been found (and until better solutions are invented) the technological system will be stable. As the quality and quantity of natural resources decline, technological innovation will again be required.

R&D IN THE RESOURCE EXTRACTION INDUSTRIAL SECTOR

We pause in this case to look at innovation in any resource extraction industry. The industrial value chain idea allows one to see what kinds of technological progess are desirable in an industrial structure. First we look at desirable innovation in the resource extraction industrial sector of an industrial value chain. In this sector, sources of raw materials are located and collected, mined, or extracted from nature. (For example, the asset value of oil production firms can be measured in the estimated barrels of oil in their reserves and on the proprietary knowledge they have in the field of oil exploration techniques.) One competes by finding and owning the rights to natural resources and by producing raw materials efficiently from these natural resources. Extraction industry firms provide raw materials to their customers, and the availability, quality, and cost of these materials influence sales to customers.

For resource extraction industries, the most important assets are (1) access to sources of raw materials, and (2) technologies for extraction of raw materials.

The competitive factors that especially discriminate among firms in this sector include:

1. The effectiveness of resource discovery techniques
2. The magnitude and quality of the resources discovered
3. The efficiency of the extraction technologies
4. The capacity for extraction and utilization of capacity
5. The cost and efficiency of transportation for moving resources from extraction to refinement

We recall that research for technological innovation in industry is called industrial research and development (R&D). Innovative firms in the resource extraction sector focus their R&D on improving:

- Techniques for resource deposit discovery
- Techniques for resource extraction
- Techniques for transportation or modifying resource for transportation

Innovation in the Steel Industry

We return to the case of the steel value chain and look at innovation in the iron- and steelmaking firms of the value chain. Steelmaking is a two-step process. First, iron is refined from iron ore and/or scrap, and then steel is made from the iron. In both processes the core technology is the furnace. In ironmaking, the blast furnace provides the principal transformation of iron ore to pig iron, and consists of:

1. Raw materials handling and storage facilities
2. Equipment for moving materials into the furnace
3. The blast furnace
4. Facilities for heating and blowing air into the furnace
5. Equipment for drawing the molten iron and slag from the furnace
6. Equipment for removing waste gas from the furnace top and reducing emissions to the atmosphere
7. Cooling equipment requiring large quantities of water

Although the blast furnace as a basic technology was unchanged for nearly 200 years, the details of the blast furnace plant process were improved continuously by incremental innovations. From 1890 to 1960, output per person-hour increased sixfold. Performance improvements came from the incremental innovations and from economies of scale in production capacity. Many of the incremental innovations in the blast furnace system centered upon improving control of the operation. Control of the furnace system is very important for both productivity and safety. For example, a stoppage of the air blast to the furnace stops the operation immediately. When the furnace is shut down, it will be out of production for a long period. If there is a delay in drawing off molten metal or slag, the materials may pile up beyond the melting zone of the furnace, seriously damaging the furnace.

Innovations were also made in other parts of the furnace technology system. One important example involves the mechanization of materials handling and charging of the furnace with materials. At first, raw materials were shoveled by manual labor into wheelbarrows and taken to the top of a platform and dumped manually into a hopper (which then discharged into the furnace). The innovation of a skip-hoist conveyed materials directly from the dock to the furnace. This mechanization of operations standardized the composition of the charge into the furnace, permitting better quality control.

Another part of the furnace technology system involves the means of removing the iron from the furnace. Hot iron is tapped and removed from the top of the furnace, and innovations in drilling through the clay plug in the tap hole (an automatic high-speed drill) and in removal of final product also improved performance. These innovations in the mechanization of loading and removing material from the furnace permitted enlargement of the size of the furnace for economies of scale. Scaling up the size of furnaces provided major gains in profitability by improving energy efficiency and decreasing labor requirements. Furnace diameters increased from the 13 feet in 1900 to over 30 feet by 1935. From 1900 through 1960, the number of blast furnaces in the United States declined by a factor of 5 while the output increased by a factor of 3. During the 1930s, efficiency was also increased by enriching the air with oxygen blasted directly into the furnace. Furnace lining materials were improved to withstand higher temperatures. Waste products were recovered and reused. The physical output of iron product in blast furnaces grew from 1899 to about 1950 and then leveled off.

Improvements in many parts of a technology system may also make possible economic benefits from economies of scale.

After reducing iron from iron ore, it must be transformed into steel. The carbon content of iron must be adjusted into the right range to become the stronger structure of steel. Hot iron from the blast furnaces, along with cold iron from scrap and alloying metals and fluxes, are put into a steelmaking furnace and the furnace transforms these into steel. The quality of the steel varies according to the proportions of materials put into the furnace and the details of the process of transformation within the furnace. In steelmaking from 1798 to 1970, there were three different basic technologies: the Bessemer furnace, the open hearth furnace, and the basic oxygen process furnace.

Prior to the middle of the nineteenth century, steel had always been made in crucibles; then a new process was invented by Sir Henry Bessemer. The basic equipment was a large pear-shaped vessel that could be tilted along a horizontal axis. Hot iron was placed into the vessel and air was blown into it through tuyeres in the bottom of the vessel. This blowing of air into the iron was the inventive idea that distinguished the new technology from the older crucible method. It provided the first technique for the mass production of steel to specific metallurgical standards.

Yet although the Bessemer furnace was long used, after about 100 years it began to be displaced by another furnace, the open hearth. The open hearth was first proposed by C. W. Siemens in 1861. Pierre and Emil Martin obtained a license and began production in 1864. The open hearth used heat in the waste gases from the furnace to preheat air and gas fuels and to increase the temperature. This enabled it to use inputs of scrap and other cold metal in addition to molten iron. The apparatus consisted of a rectangular container and two sets of brick chambers through which the waste gases were alternately blown in order to heat them. The air and fuels were blown into the hearth through the preheated chambers.

In contrast, the Bessemer process depended entirely on heat generated by the reaction of metalloids with the blast of air. This gave this process several technical limitations, including an inability to process inputs of various mixes. Also, it was difficult to control the transformation in the Bessemer technology because the conversion was short and violent. This resulted in ingots of poor quality, with included gas bubbles. It also resulted in the rapid and thus costly deterioration of the furnace lining. It was due to these technical limitations that the open hearth process replaced the Bessemer process. Steels produced in the open hearth produced uniformly stronger, higher-quality steel, and maintenance was easier.

There also came a third competing basic furnace, the electric furnace. The electric furnace was actually thought of as early as 1889, but the high cost of electricity then made it not commercial. Later, the electric furnace was used for the production of calcium carbide, aluminum, and precious metals. During and after World War II it began to be used for steelmaking, taking advantage of the availability of steel scrap.

As to the utilization of the various processes from 1900 to 1960, there was a steady decline in Bessemer capacity from 1910 through 1960 and an increase in open hearth capacity over the same period. After 1935, electric furnace capacity began to grew rapidly. Technological progress in the open hearth process was incremental, focused around the use of oil as an alternative fuel and improvement of the furnace lining (doubling the lining life from about 200 heatings in the 1920s to as many as 500 heatings in the 1940s). The war production of steel in the early 1940s stimulated further progress in the use of thinner linings and in building furnace walls higher. New spectrographic methods for measuring temperature led to improved process control.

In the 1920s, a technological innovation in another industry provided increased supplies of oxygen and decreased its price: the Linde–Frankl process. Injecting oxygen with air into the open hearth improved the steelmaking process. However, it was not until the mid-1950s that the new process began to replace the older processes. Experimentation to learn to control the process began in Europe in the late 1940s. Problems had to be solved that involved rapid destruction of furnace linings. It was found that blowing the oxygen onto the center surface of the hot metal bath avoided contact of the oxygen with the furnace linings.

The oxygen converter consisted of a pear-shaped vessel that could be tilted in two directions. The vessel is charged by introducing hot metal and scrap when it is in a horizontal position and then turned upright. Next, a water-cooled lance is introduced through an opening at the top through which oxygen is blown off the center surface of the charge. The heating times vary from a half hour to one hour, then the vessel is tipped and the steel poured off. The first plant in the United States using this process was built in 1954. Then its attractiveness was lower cost for equipment and operating savings over other processes.

> **In an industrial value chain, technical progress can occur not only incrementally in basic processes in parts of the chain but also as new, discontinuous basic processes in other parts of the value chain.**

We see that four different basic inventions were used in the steel furnace technologies. The Bessemer process was eventually replaced completely by the open hearth process, due to the latter's improved technical performance. The basic oxygen process was beginning to replace the open hearth process, also due to improved technical performance. The electric furnace provided a competing alternate process to the basic oxygen process, depending on the availability of scrap steel and the relative costs of electricity versus that of other fuels and on the price of oxygen.

> **In competing technologies for the same functional transformation, the basic invention with inherently better technical performance will eventually replace the others, but the rate of replacement may be slowed by economic factors.**

Finally, in the steel industry, steel material is shaped into steel products through the mechanical shaping of steel. After molten iron has been converted to molten steel, the next transformation step is to form the steel into its product forms, which include girders, rail, pipe, bars, wire, and sheet. There are alternative processing paths for raw steel into steel product. First, a blooming mill or slabbing mill shapes the steel into blooms, billets, or slabs. From these, a hot-rolling process produces the final shapes:

- From blooms are produced structural shapes, and rails and tube rounds are formed (with the tube rounds subsequently shaped into seamless pipe).
- From billets are produced bars and wire rods, which may be shaped further into cold-drawn rods (by cold drawing) or wire (in a wire mill).
- From slabs are produced sheet and strip-steel or plates or skelp (which processed further in a pipe mill may become welded pipe or tubes).

The basic technologies for shaping of produced steel are rolling processes, wire drawing processes, or die and tube extrusion processes. These basic technologies have not changed, but many incremental improvements to them have provided significant progress. These improvements are in the areas of equipment size, power sources, control, and sensing. The total market growth for steel products in the United States stopped growing after 1950, due to the substitution of other materials, such as nonferrous metals, plastics, and composites.

R&D IN THE MATERIALS REFINING, SYNTHESIS, AND SHAPING INDUSTRIAL SECTOR

In the materials refining sector of an industrial value chain, material products are produced from raw materials. The material products are processed and shaped into material products that can be used in parts manufacture or fabrication. For this sector the most important assets are plant and synthesis/processing technologies and new application technologies. Competitive factors for firms in this sector include:

1. Patents and proprietary knowledge about the creation of material products
2. Patents and proprietary knowledge about the processes and control in production processes
3. Quality of materials produced
4. Cost of materials produced
5. Timely delivery of materials produced
6. Assisting customers in developing applications for material products

Here firms focus their R&D on:

- The invention of new material products
- The modification and variation of material products
- The invention of new unit production processes and processes for producing material products
- The improvement and integration of unit production processes
- The development of new applications for material products

CASE STUDY:

Samsung Electronics

The depiction of the iron and steel industrial value chain ends with the materials refining and shaping sector. Accordingly, for an illustration of technological innovation of the parts and components industrial sectors, we look to a different industrial value chain, computers, and to a different kind of company, Samsung Electronics. Laptop computers and their parts supply provide a nice illustration of the technical progress in parts and in the industrial value chain.

The decreasing size of computer components and the increasing computational power made possible several product lines of computers, the smallest being the laptop computer. To make laptop computers possible, technical progress in computer parts required: (1) liquid-crystal displays, (2) miniaturization of disk drives, and (3) low-powered IC chips.

The first laptop computer was marketed by Tandy as the model 100 and designed and produced by Kyocera, without a disk drive and weighing 3 pounds. In 1982, a new company, Grid, produced a laptop with disk drives weighing 10 pounds. In 1986, Zenith won a government contract to supply 15,000 laptop computers to the Internal Revenue Service. In the same year, IBM introduced a 12-pound laptop (which was unsuccessful and was withdrawn in 1989), and Toshiba introduced a 10-pound laptop (which was soon redesigned into a 7-pound laptop that became successful). By 1990, the laptop market had grown to 832,000 units annually, over $2\frac{1}{2}$ million by 1993, and continuing. The U.S. firms Intel and Microsoft dominated personal computers in the the CPU chip and operating software areas. Japanese firms attained about 40% of the total world market in laptop computers. Next came new competitive parts suppliers from Korea and China.

In 2001, the Korean company Samsung Electronics made significant progress in becoming a global parts supplier and in moving up the personal computer industrial value chain: "Last summer [2001] Chin Dae Je, the head of Samsung Electronic's digital media division, sent a laptop to Michael Dell. The two men had never met, but Samsung had recently signed a multi-year $16 billion deal with Dell computer to provide components to the Texas PC maker. The laptop— thinner and lighter than a Sony Vario—made an impression on Dell. . . . Now Samsung is making a similar model that will soon be sold in the U.S. under the

Dell label. And Dell himself travels with a Samsung-made laptop" (Holstein, 2002, p. 89). The ambition of Samsung Electronics was to become a global laptop competitor, not merely a parts supplier. Earlier, Samsung had bought chips from their competitors and were not getting the cutting edge of technology. Chin commented: 'Five years ago we had to buy chips from Sony or Matsushita, so we were always behind. Now we can be number one. There's no doubt in my mind' " (Holstein, 2002, p. 89).

The challenge of firms in the parts supply industry is to be both technically competitive and profitable. In 1970, Samsung Electronics had begun making inexpensive small black-and-white TV sets sold by the Japanese firm Sanyo. Over the years Samsung developed research capability. In 2001, Samsung Electronics ranked fifth in the world in patents, and Samsung became profitable that year, selling $24.7 billion in products, with a profit of $2.2 billion. It manufactured products in 14 countries and generated 70% of its revenues on foreign sales. The case of Samsung Electronics illustrates a vertical integration upward from a parts supplier toward a major device producer.

R&D IN THE PARTS, COMPONENTS, AND SUBSYSTEMS INDUSTRIAL SECTOR

In the parts, components, and subsystems industry, the parts and components or subsystem assemblies are manufactured and fabricated for the producers of major device systems. The parts and components industries constitute the greatest numbers of firms in the manufacturing sectors of industry and the greatest diversity. Generally, they divide into electronics and mechanical production. Electronic parts suppliers include IC chip and other electronic part manufacturers, printed-circuit-board manufacturers and assemblers, and electronic subsystem designers and assemblers. Mechanical parts suppliers include mold and die makers, fabricators, and subsystem designer and assemblers. For the parts, components, and subsystems industry, the most important assets are (1) proprietary knowledge of part design, (2) production control, and (3) unique equipment for part production.

Competitive factors for these firms include:

1. Patents and proprietary knowledge about the design of products and components
2. Patents and proprietary knowledge about the processes and control in production processes
3. Quality of materials produced
4. Cost of materials produced
5. Timely delivery of materials produced
6. Assisting customers in developing applications for material products
7. Concurrent design capability with customers

Here firms focus their R&D on:

- The invention and design of improved parts, components, and subsystems
- The modification, variation, and redesign of lines and generations of parts, components, and subsystems
- The invention of new unit production processes and processes for producing the parts, components, and subsystems
- The improvement and integration of unit production processes

R&D IN THE MAJOR DEVICE INDUSTRIAL SECTOR

Firms in the major device systems integrator industry include automobile manufacturers, airplane manufacturers, shipbuilders, building construction industries, weapon systems manufacturers, and computer manufacturers. They are basically product system integrators, utilizing parts and components and subsystems from the prior industry. These firms can vertically integrate backward into parts and materials to gain cost and competitive advantages. On the other hand, they can sometimes deverticalize to gain cost and competitive advantages. In the major device industry, the most important assets are (1) proprietary knowledge of design and production control, (2) production facilities, and (3) brand name and access to market distribution channels.

Brands and distribution distinguish the asset value of the major device industry from the parts supplier industry. Competition early in the technology life cycle of an industry usually depends on proprietary technology, but as the industry matures, firms compete predominantly by brand recognition, production quality, price, luxury features, fashion, customer service, and cost to the customer of device replacement. The device systems industry principally supplies "tools," not solutions, to customers since major devices are usually used by customers for many different applications.

Competitive factors for these firms include:

1. Patents and proprietary knowledge about the design of the major device system and about key competitive components and subsystems
2. Patents and proprietary knowledge about the processes and control in production processes
3. Performance and features of the major device system
4. Costs of purchase and maintenance of the major device system
5. Dependability and cost of repair of the major device system
6. Availability and cost of distribution channels and timely delivery
7. Availability and nature of peripherals to complete a major device system as an applications system
8. Assisting customers in developing applications systems around the major device system

Here firms focus their R&D on:

- The modification, variation, and redesign of lines and generations of major device systems
- The modification, variation, and redesign of key competitive components and subsystems of the major device system
- The invention of new unit production processes and for producing parts, components, and subsystems
- The improvement and integration of unit production processes
- Assisting customers in developing applications systems around a major device system

R&D IN THE EQUIPMENT AND INFORMATION INDUSTRIAL SECTORS

In the equipment and information industrial sectors, equipment and computers and software are produced that are essential for the production capability necessary for the other producing sectors of an industrial value chain. Competition between firms in the equipment and information industries focuses on performance capability and equipment capacity, price, and production system integratibility.

Competitive factors for these firms include:

1. Patents and proprietary knowledge about the design of the production equipment and tools
2. Proprietary knowledge about control in production processes
3. Performance and features of equipment and tools
4. Costs of purchase and maintenance of equipment and tools
5. Dependability and cost of repair of equipment and tools
6. Availability and cost of distribution channels and timely delivery
7. Availability and nature of peripherals to complete production equipment and tools into a production process system
8. Assisting customers in automating and controlling production processes

Here firms focus their R&D on:

- The modification, variation, and redesign of lines and generations of production equipment and tools
- The understanding and control of industrial processes for which a piece of equipment is central
- The development of specialized equipment customized to a particular customer's production process

- The development of software for planning, scheduling, and controlling production processes
- The development of software for generic product design and integration with production control

The equipment and information industries supply production equipment and control systems for all producing sectors of an industrial system economic value chain.

R&D IN THE SERVICE SECTORS

Firms in the service industries divide principally into service delivery firms and retail firms. Service delivery firms include airlines, bus and railroad lines, transportation firms, telephone and communication firms, and others. Information and control integration firms include engineering design firms, information integration firms, and others. The major professional services firms are medical and legal businesses and engineering consultant businesses. The application system for a customer is designed and/or acquired and controlled for the system developed.

Competitive factors for firms in the applications systems integrator industries include:

1. Proprietary tools for systems design
2. Patents and proprietary interconnect hardware and software
3. Patents and proprietary sensing and control hardware and software
4. Professional expertise

Here firms focus their R&D on:

- Tools to improve applications system design capability
- Interconnect hardware and software
- Sensing and control hardware and software
- Tools to improve consulting services capability
- Tools to improve logistics and scheduling

Firms in the wholesale and retail distribution and service Industry include appliance, food, apparel, household furnishings, automobile dealerships, and others. Generally, each sector specializes around a functional group of products (e.g., food, apparel, automobiles). There is some grouping of products in general retail establishments, such as department stores and large grocery stores. These firms have traditionally done little R&D of their own but have been affected strongly by service technologies. A few (such as Sears) do product specification and testing. Advances in service technologies of logistics, computerized information, and communications have a significant effect on their competitive capability.

Competitive factors for firms in the retail service industry include:

1. Location
2. Brand franchises
3. Inventory control and logistics capability
4. Price-sensitive advertising
5. After-sales service
6. Point-of-sale information systems
7. Computerized communication with customers

Competition between retail and wholesale firms focuses on location, prices, brand product lines, and customer service.
Here firms focus R&D on:

- The development of customer service applications
- The development of logistics and scheduling technologies

INNOVATION AND INDUSTRIAL COMPETITION

As we see, an industrial structure for commericalizing an economic functional system is organized into different sectors with different kinds of R&D. This industrial structure provides the focus for innovation and competition. We recall that there are different levels of innovation, macro and micro. The macro levels are at the national/international and industrial levels; the micro level is at the business level. The industrial level sits between the national/international and business levels. It is technological progress at the industrial level that defines the technical conditions for competition of any business in that industry.

Competition occurs with an industry as its businesses provide competing products and services to customers of the industrial market. Technology itself does not have a positive competitive value for any business within an industry unless that business's technology is superior to other businesses within the industry.

> **If one competitor changes to a superior technology, all other competitors are temporarily at peril—until they catch up in technology—provided that they survive long enough.**

It is the change in technology in an industry—the technological progress in an industry—that provides a competitive advantages. Accordingly, the industrial context of a business is important to an understanding of whether and how any technology strategy can provide a competitive advantage. For example, in the 1980s, Porter (1985) emphasized the importance of looking at the competitive situation in an industrial sector to understand what provides competitive advantages, and he called this situation the *Five Forces Model of Competition* (Figure 7.2):

1. Business selling a product/service
2. Competitors selling similar products/services
3. Customers as buyers of the product/service
4. Suppliers of resources to the business and its competitors
5. Potential substitute products/services
6. Potential new entrant sellers into the industry

The smaller dashed box in Figure 7.2 delineates the boundary of the competitive situation for the business. Porter argued that the struggle of the participants for power determined relative competitive advantages. Any strategy for a competitive advantage needs to consider the traditional situation of the five participants (competitive forces) within an industry. But now this description of a competitive situation that bounds a business in an industry is itself bounded by the state of knowledge in the industry. Thus one needs to add to Porter's model a larger *contextual bound of the state of industrial knowledge,* shown in Figure 7.2 as a larger dashed box encompassing the competitive situation.

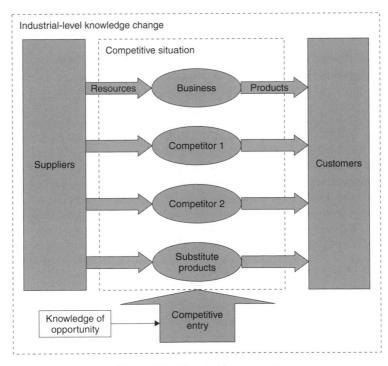

Figure 7.2. Competitive context.

For example, in the middle 1990s, a new competitor, Amazon, entered the book retailing business through online use of the Internet. The founder of Amazon, Jeff Bezos, used his knowledge of the new Internet system to envision a new business opportunity. The change in knowledge at the industrial level of the new information technologies and Internet system provided Bezos with the *competitive entry opportunity* of becoming a new book-selling competitor through his *knowledge of the Internet opportunity*. Thus, the competitive situation was changed for older book retailers (such as Barnes & Noble), and they had to respond by setting up their own Web businesses.

Change in industrial technical knowledge alters the competitive situation. In a modern competitive model, one must indicate explicitly that the boundary of a business competitive situation is encompassed in a larger boundary of change in industrial-level knowledge. New knowledge can alter competitive situations by providing new strategic business opportunities to those who have new knowledge at the industrial level and can envision business opportunities in the new knowledge.

When a technology discontinuity occurs in a core technology of an industry, the entire organization of an industrial value chain can be altered. An example of this was the replacement of electronic vacuum tubes by transistors and then by IC chips in the 1950s and 1960s. In these reorganizations, the U.S. consumer electronic industry was lost to Japanese firms.

When one can forecast dramatic technical change, one can envision its impact on an existing industrial structure as possibly:

- Altering vertical integration in the chain
- Creating new product line variations in segments of the chain
- Making some product lines in segments of the chain obsolete
- Providing substituting-technology products in segments of the chain
- Fusing two different industrial value chains together or making an entire industrial value chain obsolete with a substitute value chain

Within an industrial value chain, the upstream sectors are viewed as suppliers and the downstream sectors are viewed as customers. But from the perspective of any given industrial sector, the technologies of the upstream sectors are viewed as industrial technology systems, whereas the technologies of downstream sectors are viewed as industrial application systems. Upstream technical change can affect downstream applications systems in several ways:

1. Can lower cost to move applications into lower-priced market niches
2. Can improve quality to improve substitution into current applications systems
3. Can improve performance to increase substitutions
4. Can improve cost, quality, and performance simultaneously to increase substitutability and move into new market niches

5. Can lower cost to provide multiple-copy ownership of products in a market niche

6. Can improve performance to adapt technology to new application systems (and new markets)

For example, in the late 1990s, one pervasive technological impact of the Internet on business was in changing the way to manage an industrial value chain. Value chains were altered by how vendors and customers communicated. They can exchange models of desired parts and components (computer-aided designs); inform each other about required tolerances of parts; think about innovations in parts and materials; decide on the appropriate balance in purchasing parts between price, quality, and capacity of suppliers; communicate in bidding for supply contracts; and control the scheduling and delivery of parts and materials.

SUMMARY AND PRACTICAL IDEAS

In case studies on

- Innovation in the steel industry
- Samsung Electronics

we have examined the following key ideas:

- Industrial value chains
- R&D in industrial sectors
- Competitive factors in industrial sectors
- Industrial structure reorganization

The practical use of these ideas provides an understanding of how industrial structures depend on technology and change through research and development.

FOR REFLECTION

Choose a major consumer product or service (e.g., automobiles, airplanes, medicine) and describe the industrial value chain connecting nature to consumers for that product or service.

8

TECHNOLOGY FORECASTING

INTRODUCTION

We have examined industrial R&D as a source of continuing technological progress. But the next question is whether or not technological possible is always possible. Is technical progress infinitely possible, or only finite? Will there be no end to technical progress? Or might there come a time of stable technology, a time when no more significant technological progress occurs? This may not be a future fantasy. Before science, technological progress was infrequent. For ages, technologies were stable. But even now, with science, is it possible that technological progress could still be finite? That is the issue we examine next: if technological progress is a finite concept. We recall that technology enables functional transformations of nature as a functional logic of a sequence of natural states—manipulation of nature. If nature is finite, the manipulation of nature can be finite, and technological progress using that nature is finite. The evidence for this view comes from a topic called *technology forecasting*. We will examine the *finiteness* of technological change as the basic ideas of (1) *rates* of technological progress, (2) *underlying* natural phenomena and logic, and (3) *translating* technological forecasting into planning technological change.

CASE STUDY:

Illumination

As an example of progress in a physical technology over time, progress in illumination technology is plotted in Figure 8.1. This shows the improvements in energy efficiency (lumens per watt) of the incandescent lamp technology. One notes that the shape looks like a lazy kind of "S." In the early years of the technical progress from 1889 to 1909, the rate of improvement in technical performance was increasing exponentially with time. From 1909 to 1920, further progress was linear with time. Finally, the rate of technical progress turned over into an asymptotic region, approaching a finite limit to progress, during the period 1920 to 1960.

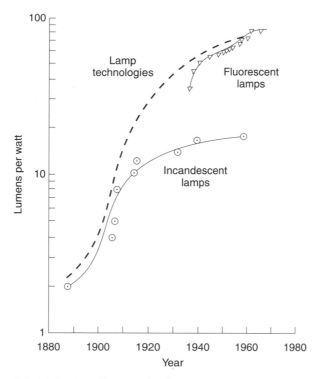

Figure 8.1. Technology S-curves for fluorescent and incandescent lamps.

The electric light bulb was innovated independently by an American, Thomas Edison, and an Englishman, John Swann. Both used carbon as the filament in their incandescent bulbs. Thus the first point in Figure 8.1, at 1 lumen per watt in 1889, was the efficiency of either Edison's or Swann's bulb. Commercially, however, Edison was more successful and came to dominate the new electrical illumination industry. However, to sell his light bulb, Edison also had to innovate the supply of electricity for the light. He established the first coal-fired power plant business, which in 1882 was serving the Wall Street district of New York City.

Later, the Edison Electric Light Company became the General Electric Company (GE). In 1900, GE was still producing the carbon-filament type of lamp that Edison had invented. However, GE was then facing new technological threats to its incandescent lamp (Wise, 1980). One threat was a "glower" lamp, invented by a German chemist, Walther Nernst, which used a ceramic filament. This produced a higher efficiency, 2 lumens per watt, due to the higher temperature that the ceramic filament could bear. Second, there was the new mercury vapor lamp, developed by an American inventor, Peter Cooper Hewitt. George Westinghouse, a major competitor to GE, supported Hewitt's work and obtained the American rights to Nernst's lamp.

Employed in GE at that time was an outstanding engineer, Charles Steinmetz. He had visited Hewitt's lab and recognized the commercial threat to GE of

Hewitt's mercury vapor lamp. Steinmetz strongly urged GE to perform basic research to meet such threats and to ensure its long-term future. In Figure 8.1, which compares the progress in the technical performance of fluorescent and incandescent lamps, one can see the basis of this threat, the intrinsically higher efficiency of fluorescent lamps over incandescent lamps. The reason is that fluorescent lamps are based on different physical phenomena than those for incandescent lamps. Fluorescent lamps utilize the natural processes of the passage of electrons through gases in which electron and atomic collisions excite gas atoms to higher electronic states that relax by fluorescence, that is, giving off photons. In contrast, incandescent lamps utilize the natural processes of the passage of electrons through solid materials, in which electron/atom collisions yield energy as photons from the electrons' altering course.

The intrinsic difference in efficiency of conversion of electricity to light arises from the physical differences of these two processes. Here we see two different processes of nature, electrons traversing gases and electrons traversing solids. Both technologies, incandescence and fluorescence, provide light but at intrinsically different efficiencies.

TECHNOLOGY PERFORMANCE PARAMETERS

We have seen that any technology provides a functional transformation that transforms purposely defined input states to purposely defined output states. The measure of performance of this transformation is the efficiency of the transformation.

> **A technology performance parameter is a measure of the efficiency of the transformation.**

The appropriate performance measure for a technology is one that relates technical progress directly to customer utility. The choice of the technology performance with which to characterize the utility of technical progress is therefore a critical choice and may not be obvious.

In the preceding case, the appropriate technology performance parameter was the efficiency of illumination (lumens per watt). This was the measure of the energy consumed per unit of illumination by a bulb. Another example is semiconductor integrated-circuit (IC) chip technology, for which the performance parameter is the density of transistors on a chip. This transistor density parameter has been useful for technology forecasting of progress in IC chip technology because increased numbers of transistors per chip allow more circuits per chip which in turn allows increased functionality (or utility) to the chip's users. Whereas the lamp parameter measured energy efficiency, the IC chip parameter measured transistor density. For the IC chip, energy efficiency was a less important parameter than density, although as a secondary parameter, chip efficiency was very important to the notebook personal computer market. As another example, for commercial airplane manufacturers, two performance parameters have been important: cost per passenger-mile and transit time.

> **The choice of a technology performance parameter to forecast technical progress for a technology system is a means of communicating the value of technical progress to the customer of the technology.**

PHYSICAL TECHNOLOGY S-CURVE

In general, for any physical technology, one can plot the rate of change of a performance parameter over time, and one will probably see a kind of S-shaped curve, such as in Figure 8.2. Historically, this common pattern has occurred for physical technologies and is called the *technology S-curve*. It has been used as a basis for extrapolative forecasts of technology change (Twiss, 1968; Foster, 1982; Martino, 1983). In a technology S-curve, there are three periods of the rate of progress of technology: an early period of new invention, exponential in temporal form; a middle period of technology improvement, linear in temporal form; and a late period of technology maturity, asymptotic in temporal form.

Exponential Portion

The exponential portion of the physical technology S-curve occurs in the beginning of technology development, just after the basic invention. Then when some people

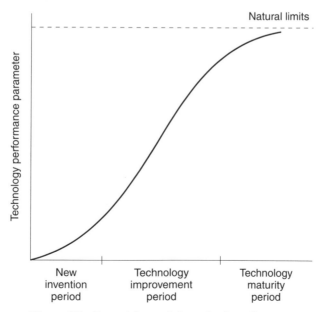

Figure 8.2. General form of the technology S-curve.

see the new technology, they immediately have ideas how to improve it. For example, after the invention of the airplane, many people began improving on the invention. The Wright brothers had used cables to twist the wingtips for left–right control of the roll of the airplane. Curtis invented wing flaps at the back sides of the wings to provide improved control of roll. The Wright brothers had used wheels for takeoff and landing on land. Curtis replaced them with pontoons for water take-off and landing. Sikorsky had the idea to replace the wings with rotors and invented the helicopter.

In any physical technology, the first invention is crude, unsafe, and limited in performance. Almost immediately, many other inventors have ideas to improve the new technology. This is the beginning exponential portion of the physical technology S-curve. This temporal part of the S-curve is mathematically an exponential form, because such initial growth consists of events that have a growth rate proportional to the state currently existing.

> **Expressing the beginning of a technology S-curve mathematically suggests that new ideas for improving a new technology will be proportional to an existing set of ideas in the beginning of the new technology.**

Since the creation of new ideas are not necessarily proportional to the quantity of preexisting ideas, one must take the technology S-curve formula not as a model of technological progress but as a historical *analogy*. But it is true that for a brand-new technology, when ideas are new, they stimulate new thinking and thus new ideas. This analogy of the technology S-curve states that there will probably be an explosion of new thinking, which can roughly be described as exponential. When the ideas of a radically new technology are so exciting and intriguing, many new people start to think about how to improve it.

Linear Portion

The linear portion of the technology S-curve occurs when incremental improvements in technology are being made. Historically, many incremental improvements gradually improve the performance of the technology. For example, although in the case of winged flight in 1904, so many improvements occurred in that first decade that during World War I the aeroplane became a dependable and deadly front-line observer and was used in fierce plane-to-plane combat. Yet progress in aviation was linear throughout the 1920s and 1930s until the performance limit of propeller-driven aircraft was reached in the early 1940s in World War II (in terms of speed, range, height, and payload).

Asymptotic Portion

Since all physical technology is based on a natural phenomenon, in which a sequence of natural states is selected in manipulating nature, the ultimate limit of the technology's performance will be determined by nature in terms of the highest-performing natural state reachable. This is the natural limit of any physical technology. For example, earlier we saw that oil is created by nature from fossil remains and is found, extracted, and refined into petroleum products by means of technologies. Yet the steps of changing initial states of nature (fossil oil) into technology-final states of nature (petroleum products) are all states of nature. Nature is discovered in natural states and manipulated into artificial states. Artificial states are natural states of nature brought about through human intervention in nature.

Nature as used by technology (in natural states or in artificial states) is still nature.

We saw in the example of technologies of illumination that the technical performance parameters of the different technologies of incandescent and florescent illumination were fixed by the natural states of electrons passing through metals and through gases. This is an important point because it points to a limit to technology. Technology neither creates or destroys nature but only alters nature.

The potential of any technology is ultimately limited by nature.

In flight, the speed of propeller-driven airplanes increased between 1904 and 1946 from about 70 miles an hour to about 250 miles an hour for level flight. But in World War II, German aircraft designers innovated a jet-propelled fighter reaching a speed of nearly 500 miles per hour. (Fortunately for the Allied forces of that war, Germany did not produce many of the new fighters.) The jet engine passed the physical limitation on speed of the propeller-driven aircraft.

In using a technology S-curve to forecast the rate of technological progress, the most important parameter to try to estimate is the natural limit, and the ability to estimate that limits lies in an understanding of the natural phenomenon of a technology—the science base of the technology. Science provides detail through quantitative modeling of the physical processes of the phenomena that a technology uses. A scientific model allows prediction of behavior of a phenomenon. To obtain forecasts, a manager of technology needs to consult the research community that is studying the science base of a technology.

CASE STUDY:

Organophosphate and Pyrethroid Insecticides

Robert Becker, director of the Agricultural Research Division of American Cyanamid Company, and Laurine Speltz, staff assistant, decided to evaluate the S-curve technique and chose to conduct a retrospective study of a class of insecticides, organophosphates. First they had to chose a measure of technological progress, and they decided to use the cumulative number of organophosphates introduced by industry "based on the assumption that each introduction, i.e. a new chemical moiety, represented some improvement in terms of cost, spectrum, market niche, or the like, over those products already in the marketplace" (Becker and Speltz, 1983, p. 31).

Figure 8.3 displays their results, forming an S-curve. In 1974, American Cyanamid discontinued research on organophosphates: "The analysis suggests that research should have been discontinued in about 1968. In fact, we might better have spent the associated man-years on a new technology" (Becker and Speltz, 1983, p. 31).

Encouraged by this example, Becker and Speltz next tried the technique on one of their current research efforts on another class of insecticides, the pyrethroid insecticides: "At the time this [pyrethroid] analysis was made in 1982, American Cyanamid had established a commercial position in the pyrethroid market. A pyrethroid, 'PANACTO' cypothrin, had been introduced in 1979 and a second 'PAYOFF' flucythrinate, in 1982" (Becker and Speltz, 1986, p. 21). Figure 8.4 displays their results. Since this class of insecticides was then relatively new, they had only points from 1975 through 1983 to graph, which showed clearly that the technology was still in an early exponential growth-rate phase: "The S-curve for the pyrethroids . . . its shape is not obvious" (Becker and Speltz, 1986, p. 21).

Therefore, the important analytic problem was to establish the level of natural limits to the technology. They next identified five properties that they thought

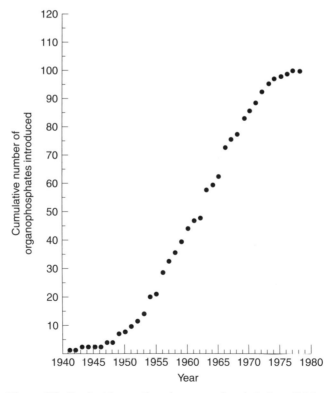

Figure 8.3. Product innovations in organophosphate insecticides.

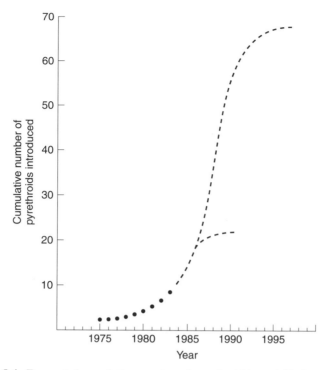

Figure 8.4. Forecasted cumulative number of pyrethroid insecticide innovations.

critical to significant technical improvement: (1) spectrum of activity, (2) increased soil activity, (3) systemic activity, (4) reduced fish toxicity, and (5) reduced cost. The *spectrum of activity* is the range of parasites against which the insecticide is affective. A major weakness of the class of pyrethroid insecticides is marginal performance against mites: "It was believed that this weakness could lead to the discovery and development of several niche products but the size of the market opportunity of such a product was questionable" (Becker and Speltz, 1986, p. 23).

The property *increased soil activity* indicated if the nature of the soil impaired the insecticide performance. They knew that pyrethroids were deactivated by soil binding to them. Their own research gave them little hope that there was much chance of overcoming that deactivation. The property of *systemic activity* indicated if the molecular structure system offered opportunities for technical improvement. They also thought that due to the hydrophobic nature of pyrethroids, there was also little hope that systemic activity could be much improved. The property *reduced fish toxicity* indicated an important safety feature of the insecticide in its impact on the environment. But they felt that improvement here would not have much impact on commercial sales. On the final criterion of cost, they also thought that not much improvement could be made over the cost of existing pyrethroids.

Accordingly, their forecast was that the lower of the curves on Figure 8.4 would prove to be the eventual outcome of the rate of growth of the pyrethroid insecticide technology: "It is recognized that this analysis is only a forecast, and that it has the uncertainty and risks associated with any forecast. Additionally, while we believe the development of a pyrethroid which truly represents a technical advance is unlikely, we do not believe that the development of a 'me too' pyrethroid which could be commercially important is precluded. Furthermore, every R&D manager must deal with the difficult decision to change the direction of a research program in the face of uncertainty about the outcome, and any decision to change a research program invariably involves some type of forecast" (Becker and Speltz, 1986, p. 23).

CHARACTERISTICS OF TECHNOLOGY S-CURVES

Let us summarize the preceding concepts that we have been illustrating.

Definition

A technology S-curve is a common pattern of progress in a technology's principal performance parameter over time, with an initial exponential growth, intermediate linear growth, and an eventual asymptotic limit.

Properties

- A technology S-curve is dependent on an underlying generic physical process.
- The natural limit of progress on a technology S-curve arises from intrinsic factors in the underlying physical process.
- Discontinuities in technical progress for a technology occur when alternate physical processes can be used in inventions for the technology.
- Such discontinuities express themselves as different and disconnected S-curves.

A word of caution about using the mathematics for technology S-curve. The difference between a model of a phenomenon and an analogy of the phenomenon is that the model both describes and explains the phenomenon. An analogy may describe a phenomenon but does not explain it.

> **The physical technology S-curve is not a model of the process of physical technological change but an analogous pattern frequently seen in the histories of technical progress.**

It is not a model because it does not explain the rate of change of a technology's progress. Such an explanation must be derived from an understanding the physical morphology of the technology, and how progress in the technology depends on changes in the physical morphology. Accordingly, the technology S-curve is merely a *descriptive analogy*. But as an analogy, it can be used for an intelligent "guess" at the rate of change, that is, anticipation and forecasting, but cannot be used to "predict" how change may occur.

Technique

We may now summarize the technique of fitting technology S-curves to technical data on the rate of progress of a technology in the following manner:

1. Identify a key technical performance parameter.
2. Collect existing historical data on technical performance since the date of innovation of the technology and plot them on a time graph.
3. Identify intrinsic factors in the underlying physical processes that will ultimately limit technical progress for the technology.
4. Estimate the magnitude of the natural limit on the performance parameter and plot this asymptote on the graph of the historical data.
5. Estimate the time of two inflection points between the historical data and the asymptotic natural limit (first inflection from exponential to linear rate of progress and second inflection from linear to asymptotic region).
6. The reliability of expert forecasting of the exact times of inflection will probably be more unreliable than their anticipation of the research issues required to be addressed for inflection points to be reached.

Underlying Phenomenon

Underlying any physical technology is a physical phenomenon. Underlying the Edison lamp bulb was the physical phenomenon of electron conduction through metals (which generate heat and light from collisions with the metal atoms).

> **The intrinsic limit to technical performance of a physical technology is determined by the natural physical phenomenon of the technology.**

Technical progress in a physical technology based on one physical process will show the continuity of form along an S-curve shape. However, technical progress that occurs when a different generic physical process is used will jump from the original S-curve to a new and different S-curve.

> **Every technology S-curve is expressive of only one underlying physical phenomenon.**

CASE STUDY:

Twentieth-Century Progress in Semiconducting Chips

This case illustrates the first instance in the history of technical progress whereby the rate of progress occurred as a sequence of exponential phases, because technical progress occurred in IC chips through next-generation technologies. The historical context was the period from 1960 to 1990, during the early period of technical progress in production of IC chips. New manufacturing processes were devised to achieve increasing densities of transistors on a chip.

The technology performance parameter is the number of transistors that could be fabricated on a chip. Since transistors worked as electronic valves and at least one valve is usually needed in a circuit for one logic step, increases in the numbers of transistorized valves on a chip correlates with increases in complexity of functionality for which a chip can be used. Figure 8.5 shows the density of transistors achievable on an IC chip over time. The graph is plotted on a semilog scale, and one notes that *this is not the form of the technology S-curve*. In the figure, the vertical axis of technical performance is on a log scale, so that the entire chart is in semilog format. A straight line of a semilog chart indicates exponential increase on the vertical scale. [Mathematically, a chart plotting a function as linear on a semilog scale indicates that the function is exponential on a linear scale. We can see this by taking the logarithm of an exponential function $f(t) = e^{kt}$, or $\ln f(t) = \ln e^{kt} = kt$. Thus, $\ln f(t) = kt$. Plotting $\ln f(t)$ thus scribes a straight line in the temporal variable t with a slope of k.]

The different slopes of the lines in Figure 8.5 show different slopes k of technical progress in producing chips of higher transistor density. Although the rates of progress from 1959 to 1972 and from 1972 to 1984 were both exponential, the rates of exponential increase differed in the two periods. Therefore, region A is one exponential region of increase in chip density, region B is a second exponential region, and region C is also an exponential region. Initially, from the invention of the chip in 1960 until 1972, the growth rate was such that on the average

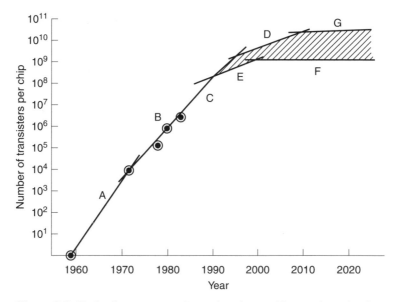

Figure 8.5. Technology progress in semiconductor chip transistor density.

the density of transistors on a chip were *doubled each year*. However, in the second period from 1972 to 1984, while the growth was still exponential, the rate was such as to achieve an average *doubling each 1.5 years*.

The five data points at 1960, 1972, 1979, 1980, and 1981 are actual historical data and show that technical progress in IC chips occurred through next-generation technology innovations rather than incremental innovations. In 1970, the feature size of transistors was about 12,000 nanometers. In 1980, feature size was down to 3500 nanometers, in 1990, to 800 nanometers, in 1997, to 300 nanometers. (One nanometer is one billionth of a meter; and for comparison, a human hair has a width of about 100,000 nanometers.) This reduction in feature size allowed the production of chips with greater densities of transistors.

For example, by 1995, the transistor density in Intel CPU chips had the scale shown in Table 8.1.

The era when hundreds of transistors could be placed on a chip by 1970 was the time of medium-scale integration (MSI). After 1972, densities of transistors on a chip leapt to thousands of transistors on a chip, large-scale integration (LSI). By 1981, densities of transistors increased to tens of thousands of transistors on a chip, very large scale integration (VLSI). During the 1980s, chip densities increased to hundreds of thousands. As the 1990s began, chip densities were in the millions of transistors on a chip, ultralarge-scale integration (ULSI).

What would be the natural limits of silicon chip technology? In the 1983 meeting of the International Electron Devices, James Meindl, then of Stanford University, predicted: "It could happen by the turn of the century" (Robinson, 1984, p. 267). At that meeting there was the anticipation that the density limits could be in the billion of transistors per chip. The actual limit would be in the tiny, tiny size of the fabricated features on a chip. IC chips through the 1980s were produced with light exposure to define features on the chip. Smaller sizes could be reached with x-ray exposure. But the shorter the wavelength of the x-ray, the more accidental damage to the features will occur. The ultimate limit to transistor

TABLE 8.1   Progress in Chip Density

Year	Intel Chip	Number of Transistors	Word Length (bits)
1971	4004	2,300	4
1974	8080	6,000	8
1978	8086	28,000	16
1982	80286	134,000	16
1985	80386	275,000	16
1989	80486	1,200,000	16
1993	Pentium	3,100,000	32
1995	Pentium pro	5,500,000	32
2001	Pentium IV	40,000,000	32
2007	(anticipated)	1,000,000,000	64

Source: Intel Corporation.

density on silicon chips will occur from technical trade-off between fineness of feature definition and material damage from the exposing radiation.

What, then, was seen as a natural limit to transistor size? In the year in which this forecast was made (1983), researchers estimated that the natural limits to chip density would eventually restrict the number of transistors per chip in the range 10^9 to 10^{10}, or a natural limit on the order of 1 billion to 10 billion transistors per chip. That estimate was presented by James Meindl (at that time director of Stanford University's Center for Integrated Systems) at a research meeting composed of expert researchers in chip technology (the third annual International Electron Devices meeting, which was held in Washington, DC, in December 1983): "Projection of the future course of semiconductors has become a liturgical requirement of integrated circuit meetings since Gordon Moore of Intel Corporation . . . in the mid-1970s . . . observed that the number of transistors on a chip had been roughly doubling each year since Texas Instruments and Fairchild Semiconductor independently developed the integrated circuit in 1959" (Robinson, 1984, p. 267).

At that time, Meindl also expressed the consensus of experts who anticipated an eventual limit to the growth: "Meindl's message . . . is that further moderation of the growth curve is in store. . . . The reason for the anticipated decline in rate is that engineers are approaching a number of theoretical and practical limits on the minimum size of transistors. . . . Meindl calls this future era (of natural limits) ULSI, for ultra-large-scale integration, as opposed to the current VLSI or very-large-scale integration epoch" (Robinson, 1984, p. 267).

In the middle of the 1990s, research laboratories began experimenting with trying to create a single-electron transistor using only one electron to move in and out of a storage area for a binary-state transistor. This would provide an ultimate natural limit on transistor size: "Scientists around the globe are making a furious assault on the last frontier of electronics. . . . They are striving to create transistors that work by virtue of the movement of a single electron . . ." (Broad, 1997, p. C1).

Then one prototype of a single-electron transistor was a kind of metal-on-oxide field-effect transistor (MOSFET) technology. In this technology, there is a flow of electrons from source to drain electrodes through a channel controlled by a gate electrode. The idea for a single-electron MOSFET is to make a small island in the channel that would allow only one electron at time to pass through the channel. Such small islands, also called *quantum dots,* operate according the laws of quantum mechanics (which specify that only one electron can occupy one quantum state, i.e., Fermi statistics). For example, in 1996, research teams at NTT's Basic Research Laboratory in Atsugi, Japan and at the University of Minnesota in Minneapolis demonstrated such transistors with islands only 5 to 10 nanometers in size, small enough to allow only a few electrons at time through the channel (Service, 1997).

> **The linear portion of the technology S-curve for IC chips was absent, as technical progress in processing IC chips to achieve increasing densities occurred in successive generations of chip technologies (produced with different physical processes).**

A historical note: The technical progress of chips at an exponential rate was first noted by Gordon Moore of Intel. The popular press called this *Moore's law*. It is an example of progress by a sequence of discontinuous technology progress (changes in the physical bases of producing chips). There was no "law" here but a very clever sequence of inventions by chip producers to continually change the physical phenomena of the chip production processes.

Early chip production used photographic images of a circuit design projected by a lamp on a silicon chip that had first been coated with a photo resist. The lamp imaged the circuit on the photo coating on the chip, and then the image was developed chemically (much like a camera's plastic film is developed). Then an acid etched away portions of the silicon surface not projected by the developed image of the photo coating. These steps were repeated until the transistors and components of the circuit were etched into the silicon material of the chip and then aluminum connecting wires were evaporated onto the chip. Finally, the chip was covered by a plastic (or ceramic) covering to protect the chip. Progress in chip densities were made by improving the imaging capabilities, the etching capabilities, and new circuit design software to use all the hundreds, then thousands, and then millions of transistors that could be etched onto a chip.

TECHNOLOGICAL DISCONTINUITIES

We recall that technologies are composed of inventive logics to manipulate states of natural phenomena. Therefore, the phenomenal base of a technology is the nature manipulated by the logic of the technology. Technologies can be altered by changes in either the logic schema or phenomenal bases.

Every technology S-curve depends on a phenomenal base, and when new phenomena is substituted, a new technology S-curve begins.

When progress in innovation has occurred through substitution of a new underlying physical phenomena for the technology, a chart of the performance parameter's rate of change will jump onto a new S-curve. Then the overall technology S-curve is a composite of the underlying technology S-curves. In Figure 8.1 we depicted the composite technology S-curve for lamp technologies, including both incandescent and fluorescent technologies. A composite technology S-curve is formed from two technologies of the same function but of different phenomenal bases of their physical morphologies.

The envelope of a series of discontinuous S-curves is not simply a kind of "sum" of S-curves. Each curve represents a particular phenomenal base to a technology. When phenomenal bases change, the S-curves change, and it is the phenomenal bases that provide the "natural limits" to a technology. Thus using an "envelope" as a simple S-curve can hide the most important idea that an S-curve can really present, which is the notion of a natural limit to a given phenomenally based technology.

The discontinuous technology S-curves for changes in the phenomenal bases of technology are some of the most important information that technology forecasting can provide for business and competitive strategies.

Historically, most technologically based failures of businesses have occurred at such technology discontinuities; conversely, most successful new high-tech businesses have been started at such technology discontinuities.

Therefore, there are two kinds of technical progress. The first is *incremental progress* along an S-curve as improvements are made in a technology utilizing one physical process. The second is *discontinuous progress,* jumping from an old S-curve to a new S-curve as improvements in a technology use a new and different physical process.

When progress in innovation has occurred through substitution of a new underlying physical phenomenon for the technology, a chart of the performance parameter's rate of change will jump onto a new S-curve. A technological discontinuity occurs as progress in the form of *next generations of technology.*

Incremental technological progress in physical technologies always involves progress using the same underlying physical phenomenon. Discontinuous technological progress in physical technologies involves progress though replacement of one underlying physical phenomenon with another one. The commercial implication of incremental and discontinuous technological progress is in how these affect differently the competitive structure of an industry.

Incremental technological progress reinforces an industrial competitive structure, but discontinuous technological progress disrupts an industrial competitive structure (competitive discontinuities).

METHODOLOGY OF FORECASTING

We conclude by discussing the methodological issues about forecasting in general. Any forecast is an attempt to anticipate the course of future events. But this attempt to anticipate can proceed with different levels of knowledge about the nature of the events being forecast and can result in deepening levels of forecasts: (1) extrapolation, (2) generic patterns, (3) structural factors, and (4) planning agenda.

When a forecaster has almost no knowledge about events except historical data on past occurrence, the forecaster can do little more than *extrapolate* the direction of future events from past events. Extrapolation forecasting consists of fitting a trend line to historical data. When a forecaster has some knowledge about the general pattern of a class of events but little knowledge about the specific exemplar of that class at hand, the forecaster can use the *generic pattern to fit the extrapolation* of the specific exemplar case. The technology S-curve of this chapter is just such a generic pattern for the class of innovations. Fitting generic pattern to an extrapolation has more knowledge than mere extrapolation because one knows beforehand the form of the curve to be extrapolated.

In addition to knowing the generic pattern of an event, knowing something about the kinds of factors that influence the directions and pace of the events provides the basis for even better anticipation. Extrapolations from past data always assume that the *structure* of future events is similar to the structure of past events. Changes in structural factors will render extrapolation meaningless and create the most fundamental errors of forecasting. The deepest level of forecasting requires understanding not only the generic pattern of the class of events to be anticipated but also understanding the structure of the events and then proceeds to intervene in the future by *planning* to bring about a desired future event. A research agenda provides an anticipatory document required to bring about a technological future.

Until both a scientific basis has been discovered for a basic invention and until a basic invention for a technology does occur, there is no reasonable basis on which to predict technical progress. Even after a basic innovation, some experts may be too pessimistic about progress because they are very familiar with the problems and difficulties of new technology. Thus one does require more objective and systematic procedures for technology forecasting than simply the subjective judgments of individuals (no matter how brilliant and successful).

Yet (and despite this caveat about experts and prediction) there do exist experts, such as researchers in science and engineering, who are attempting to create knowledge for inventing and improving technologies. One technique that can be used for anticipating technological progress is to use such experts to predict the future. In the 1930s, one of the first technology forecasting efforts was commissioned by the National Research Council in the United States and was carried out by the National Resources Committee, chaired by a sociologist, William Osburn (Osburn, 1937). Ayres (1989) commented on this report: "This ambitious study used technological trend curves almost exclusively to illuminate the past rather than to forecast the future" (p. 52).

After World War II, the military began supporting formal methods of forecasting technology. Theo von Kármán chaired a group to produce a technology forecast for the U.S. Air Force. The Rand Corporation was then created to continue and extend this work: "It [Rand] carried out many such forecasts and pioneered in developing technological forecasting methodologies such as the 'Delphi' method" (Ayres, 1989, p. 52).

Experts can be brought face to face to produce a report, such as the von Kármán report cited above. However, in developing what they called the *Delphi method,* Dulkey and Helmer (1963) also sometimes saw a usefulness in querying experts by questionnaire and recirculating questions until a consensus was obtained among the experts. Martino (1983) has succinctly summarized the Delphi technique: "Delphi has three characteristics that distinguish it from conventional face-to-face group interaction: (1) anonymity, (2) iteration with controlled feedback, and (3) statistical group response" (p. 16).

During a Delphi technique, the members of the group do not meet and therefore may not know who else is included in the group. Group interaction is carried out by means of questionnaires circulated to members of the group. Each group member is informed only of the current status of the collective opinion. Reiterations are continued until some consensus is reached. Used for forecasting, such Delphi groups have attempted to predict what technologies will occur and when. However, technology-predicting forecasts by experts have often turned out to be notoriously wrong, particularly about dates. Rowe and co-workers (19) reviewed the studies that have been performed on the efficiency of the Delphi methods versus any method of consulting experts. They concluded that the technique is probably no more useful than any technique for gaining group consensus among experts: "We conclude that inadequacies in the nature of feedback typically supplied in applications of Delphi tend to ensure that any small gains in the resolution of 'process loss' are offset by the removal of any opportunity for group 'process gain' " (p. 235).

By *process loss* they meant factors in group decision activities such as that in some circumstances the group's collective judgment may be inferior to the judgment of the group's best member, premature closing of group judgment results in poor results, or individual jockeying for power may skew results. The Delphi technique focuses on reducing group process interaction, not on eliciting structural information.

> **Any refinement of techniques for forecasting that does not focus on eliciting information about the underlying structures of forecasted events is not likely to improve the forecast.**

All forms of forecasting, such as: Delphi consensus, extrapolation, leading indicator, etc., depend on expert opinion. Therefore, it is important to understand what opinions experts are really experts about. Are experts expert in predicting the future? Probably not. What research experts really know are (1) a technological need, (2) scientific theory underlying the technology, and (3) imaginative research approaches to attempt to improve a technology. Therefore, the proper use of experts in technological anticipation and creation is to:

1. Identify future technological needs
2. Identify technology areas that require improved understanding in order to advance technology
3. Propose research directions to study these areas
4. Use detailed information about the research directions to forecast the direction and pace of technological progress
5. Use the research directions to structure research programs

FROM FORECASTING TO PLANNING

In any forecast, four general classes of structural features will underpin events:

1. Structures of current technological capabilities
2. Structures of economic activities and markets
3. Structures of nature and natural potential
4. Structures of demographics and cultures

Whenever underlying structures alter, forecasts based principally on extrapolation will be in error.

We saw in the case of progress in IC chip technology in the twentieth century a proper use of forecasting and experts who used the forecast as *a basis for motivating future research*. Forecasting technological change is an effort to foresee invention before it occurs. But on the way to invention, forecasting is not sufficient. One also needs to plan a new technology and perform the research necessary to invent it. The germanium transistor, silicon transistor, and IC chip are all examples of planned research and invention. Planning technological change provides the basis for supporting efforts of invention. The combination of anticipating and planning technological change is usually called *technology strategy*.

All together, the anticipation, planning, and research for invention is the business function of research. When a business does all these together—anticipation, planning, invention, and implementation—we have what we call *technological innovation*. These terms are used in technology management in the following way:

- Forecasting   anticipating
- Invention   creating
- Strategy   anticipating, planning
- R&D   anticipating, planning, creating
- Innovation   anticipating, planning, creating, implementing

We should note that the technology S-curve summarizes the *rate* of change for a technology but does not indicate the *direction* of change—how a technology system evolves. Although the patterns of the rate of change of technologies are similar for different technologies, the direction of change in a technology will differ for different technologies. To plan the directions of change for a specific technology, one needs another concept, that of morphological analysis (which we discuss in Chapter 20).

SUMMARY AND PRACTICAL IDEAS

In case studies on

- Illumination
- Insecticides
- Semiconducting chips

we have examined the following key ideas:

- Rates of change of technologies
- Natural phenomena underlying a technology
- Changing the inventive logic of a technology

- Extrapolating a technology S-curve
- Anticipating the natural limits to a technology
- Moving from forecasting to planning

These theoretical ideas can be used practically as a technique for extrapolating a rate of technology and understanding the kinds of underlying factors that could invalidate such extrapolations.

FOR REFLECTION

Identify several technologies and their key technical performance parameters. Why were those particular performance parameters used? Plot the historical technology S-curves for the technologies. Which have show one or more technology discontinuities? What were the physical sources for these discontinuities?

9

INDUSTRIAL DYNAMICS

INTRODUCTION

Since technological progress in any technology is finite, what effect does this have on industry and competition? We have seen that industrial structures can be created from new basic technologies, but can industries mature and even die? The answer is yes, when their technologies mature or become obsolete. Technological finiteness affects industry in creating a dynamical path in an industry from high-tech toward a commodity status. This has been called an *industrial life cycle*. We now look into the dynamics of change and competition in industry due to the finiteness of technological progress. We examine this impact in the basic ideas of *core technologies, industrial life cycles, high-tech versus commodity industries,* and *industrial technical obsolescence.*

U.S. Auto Industry

We look at the historical case of the U.S. automobile industry, which began with the innovation of putting an engine onto bicycle technology. The bicycle had been invented in the middle of the nineteenth century, taking technical advantage of new high-strength, lightweight, low-cost steels produced in quantity by the then new steel industry using Bessemer's invention. Toward the end of that century, some bicycle manufacturers were further innovating by adding engines onto a four-wheeled bicycle frame. Abernathy (1978) wrote: "Alan Nevins observes that the [automobile] industry was born from the consumer's desire for a light personal transportation vehicle, a desire stimulated by the bicycle boom of the 1890s. . . . Men with experience in the bicycle industry were the first to see the possibilities of the automobile as a means of personal transportation. Their technological orientation led them to improve the automobile's performance through lightweight designs, high-strength materials, and low-friction ball bearings rather than increased motor power" (p. 12).

The year 1896 marked the beginning of the U.S. automobile industry because in that year more than one auto was produced from the same plan: J. Frank Duryea made and sold 13 identical cars in Springfield, Massachusetts. During the next few years, many new automobile firms were founded and a variety of auto configurations were offered (Abernathy, 1978). Races were held between the three principal configurations of automobiles in steam, electric, or gasoline power. In 1902, a gasoline-powered car defeated electric and steam cars at a racetrack in Chicago, establishing the dominance of the gasoline engine. Thereafter, this engine was to become the core technology for the automobile. Also in 1902, the Olds Motor Works constructed and sold 2500 small two-cylinder gasoline cars priced at $650. The next six years in the United States saw the growth of many small automobile firms selling different versions of the gasoline engine auto.

Other configurational choices were made. In 1903, Buick relocated the engine from the rear to the front of the car (as in most current designs). The bicyclelike drive chain was replaced by a direct drive shaft, connecting the front-placed engine to rear wheels. In 1904, Packard patented the four-position transmission with positions in the shape of an H, which subsequently became the standard for manual transmissions (with later alternative choices of three or five speeds). Ford redesigned the earlier one-piece block-and-head engines in two separate pieces for ease of casting and machining. So by 1907, the automobile system began to look more like modern designs than like the early carriage/bicycle with rear-mounted engine.

The next key event in the history of the U.S. auto industry was Henry Ford's introduction of the model T. Ford was producing automobiles and racing them to establish a reputation for performance for his automobiles. His cars were expensive, as were all other cars, which were principally for the well-to-do. But Ford had in mind a large untapped market, a car for people living on farms. Around 1900, half of Americans lived on the farm. Ford wanted to build a practical high-quality automobile priced at $450. His commercial strategy was price and his technical strategy was durability. The rural application required an inexpensive, reliable, and durable car, which had a high clearance for dirt roads and easy maintainability by mechanically minded farmers. The key to his technical innovation would be in the weight and strength of the chassis of the automobile structure.

Materials in the early automobile formed a very large part of its cost. If Ford could reduce the weight of the model T by at least one-half that of competing designs, the technology would produce an enormous competitive advantage for his grand strategy of a "car for the people." Ford's innovation for decreasing the weight of the automobile would be to use high-strength steel for the chassis, made of the element vanadium as an alloy. Henry Ford learned of this new steel when attending an automobile race. In one of the unfortunate accidents that occurred that day, a racing car imported from France was wrecked: "In 1905 [Ford] saw a French automobile wrecked in a smash-up. Looking over the wreck, he picked up a valve stem, very light and tough. . . . it proved to be a French steel with vanadium alloy. Ford found that none [in the United States] could duplicate the metal. [Ford] found a small steel company in Canton, Ohio [and] offered to guarantee them against loss. The first heat was a failure . . . the second time the steel came through. Until then [he] had been forced to be satisfied with steel running between 60,000 and 70,000 pounds tensile strength. With vanadium steel, the strength went to 170,000 pounds" (Nevins and Hill, 1954, p. 349).

Making the chassis of this steel meant that he could reduce the weight of the chassis by nearly a third and get the same strength. It was a technological breakthrough that allowed Ford to imagine an innovative new product design: " 'Charlie' (Ford said to Charles Sorensen, who helped him design the Model T), 'this means entirely new design requirements and we can get a better, lighter and cheaper car as a result of it.' " (Sorensen, 1956, p. 98).

Ford used the new vanadium steel to fabricate the chassis of the automobile, which reduced the overall weight of the model T to about half that of then-existing automobiles. In addition, Ford also innovated in the design by mounting the motor to the chassis with a three-point suspension. The prior practice had been to bolt the engine directly to the frame, and often the cylinder blocks of the engines had been twisted in half by the enormous strain that occurred when the automobile bounced over a hole or rut.

Ford also designed the model T to be a "best of breed." He used the other best ideas in other contemporary automobiles. For example, he replaced the then traditional dry-cell batteries for starting the car with a magnet-powered ignition (one cranked the model T to start it). He also designed the car with a high road clearance for rural roads. The model T became a design standard for automobile technology and remained so for a long time: "For eighteen years the design of the model T chassis was not significantly changed. During this period the industry's production of passenger cars increased nearly sixtyfold, from 64,500 cars annually to 3,700,000. . . . Ford maintained about a 50 percent market share through 1924" (Abernathy, 1978, p. 18).

Ford's model T was the right product at the right time for the right market at the right price. Performance, timing, market, price—the four factors for commercial success in innovation. Ford captured the auto market from 1908 through 1923, selling the majority of automobiles in the United States in those years.

CORE TECHNOLOGIES OF AN INDUSTRY

There are four basic technologies characteristic of all industries:

- Core technology   necessary and unique
- Supportive technology   necessary but not unique
- Strategic technology   rapidly changing technology
- Pacing technology   rapidly changing technology that is defining
                             competition

The technologies that provide the basic functional transformations for products and production processes are called *core technologies* for an industry. In the automobile business, the core technologies for the engine technologies and the assembly line technologies. For the earlier case of helicopters, core technologies were the gasoline turbine engine, rotors, and controls.

> **The core technologies of an industry provide the essential functional transformations in the industry's products and production.**

A core technology is essential to an industry, as the industry could not provide its product lines before a core technology was invented. Improvement in the technical performance of the industry depends on progress in its core technologies. Thus the coke-fueled steel industry could not exist before the invention of the Bessemer furnace. Nor could the automobile, airplane, and helicopter industries exist before the invention of the gasoline engine.

All industries use several technologies in their products, production, and services. Those technologies that are uniquely necessary to the product or production or service systems of the industry are the core technologies for the industry. Other technologies that an industry uses are also important to the features of the product but may not be uniquely necessary to the basic functional capability of the product. These are called the *supportive technologies* for the industry.

> **Lagging in technical progress in a core technology will put a business at a serious competitive disadvantage; leading in technical progress in a core technology will defend against competitor's encroachments.**

Some core or supportive technologies will usually be changing at a much faster rate than others, and these may be called the *strategic technologies* for the industry. The strategic technologies that are also core technologies for the industry can be called *pacing technologies,* for competition in the industrial structure will be technically paced by these technologies.

The core technologies in the subsystems of a business enterprise that are developed and held in-house for competitiveness are proprietary core technologies of the firm. As such, proprietary core technologies can be in products, production, distribution, or information. If a core technology is acquired from outside the firm, the core technology cannot provide a competitive advantage to the firm.

INDUSTRIAL LIFE CYCLE

The rate of technical progess of the core technologies of an industry affects the dynamics of the industry. William Abernathy (1971) first suggested the dynamical idea of an industrial life cycle due to technical progress. Later, Ford and Ryan (1981) proposed that a chart of market volume over time for an industry would reflect the underlying maturation of the core technologies of its product. They called this chart an *industrial life cycle*. Figure 9.1 shows the general pattern of growth for a new industry begun on a basic technological innovation. Market volume does not begin to grow until the application launch phase of the new core technology begins with

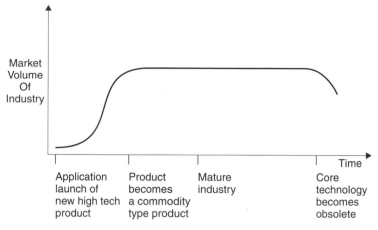

Figure 9.1. Industrial life cycle.

an innovative product. Figure 9.2 shows the industrial life-cycle chart for domestic production of automobiles in the United States.

The first technological phase of an industry will be one of rapid development of the new product during the applications growth phase. (For the automobile this lasted from 1989 to 1902, as experiments in steam-, electric-, and gasoline-powered cars were tried.) When a standard design for the product occurs, rapid growth of the market continues. (For the automobile this occurred with Ford's introduction of the model T design.) Industrial standards ensure minimal performance, system compatibility, safety, repairability, and so on. Sometimes these standards are set through an industrial consortium and/or government assistance (e.g., safety standards). But usually in a new technology, a performance standard emerges from a market leader.

Abernathy and Utterback pointed out that the pattern of early innovations in a new technology–based industry will usually be product innovations (improving the performance and safety of the product) and that later, innovation shifts to improving production process make the product cheaper and with better production quality (Abernathy, 1978; Utterback, 1978). Figure 9.3 plots the rate of product innovations over time and rate of process innovations. One sees that the rate of product innovations peaks about the time of the introduction of a design standard for the new technology product. Thereafter, the rate of innovations to improve the product declines, and the rate of innovations to improve production increases. This occurs because until the product design has been standardized, a manufacturer cannot focus on improving the production processes that will produce such a design.

After the key technologies of the industry mature, the market for the industry will eventually saturate. This level of market for the industry will continue unless the key technologies for the industry become obsolete by technology substitution. Then the market volume of the industry based on the older key technology product will decline to zero or to a market niche.

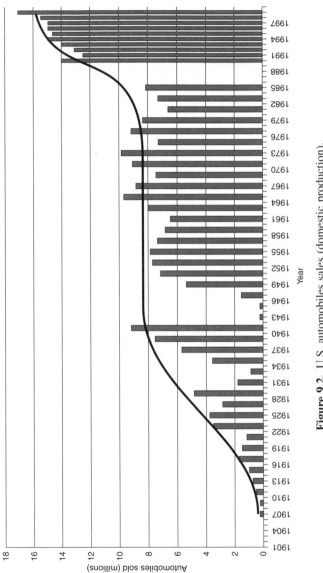

Figure 9.2. U.S. automobiles sales (domestic production).

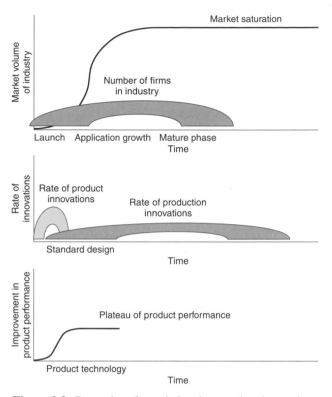

Figure 9.3. Dynamics of new industries cased on innovation.

INDUSTRIAL STANDARDS IN A NEW INDUSTRY

The case of Ford's model T also provided a product design standard for the new U.S. automobile industry.

> **In a new industry a *product design standard* defines the standard features, configuration, and performance of a product line in a new industry.**

Industrial product design standards are critical for the growth of a large-volume market. Sometimes these occur by a design standard set by an innovative product of a competitor, such as Ford.

> **A product design standard is necessary for the high-volume growth of a new market around that product.**

Sometimes, design standards occur not from an innovative product but from a deliberate act of cooperation between industry and government to set safety standards. Then the process itself for establishing industrial product standards can be complex. Reaching consensus agreement between industrial firms and government about industrial standards is never either simple or easy.

For example, Bailetti and Callahan (1995) examined how public standards in communications were managed by industrial firms. They pointed out that managing industrial standards was a difficult problem for several reasons, such as the need for collaboration among competitors in setting standards, the problems of complexity and uncertainty in standards when the technology is changing rapidly, the need for international cooperation in setting standards, and so on. They suggested that a systematic procedure facilitated standard setting, which could be build on three kinds of groups in the standards process: information management, commercial exploitation, and standard development. These are standard-making groups that focus on information formulation, product standards setting, and maintenance and evolution of standards.

FIRST MOVERS IN A NEW INDUSTRY

The illustration of Ford (innovating both the model T and assembly line production) is an example of an innovative firm in a new industry that moved first to establish competencies to dominate the market. Chandler (1990) used the term *first movers* in an industry to indicate the kinds of firms in a new industry that move first to dominate the industry: "Those who made first made . . . large investments—the companies I call 'first movers'—quickly dominated their industries and continued to do so for decades. Those who failed to make these investments rarely became competitive at home or in international markets; nor did the industries in which they operated" (p. 132).

The necessary investments for a firm to become a first mover in a new industry are:

- Investments in advancing technology
- Investments in large-scale production capacity
- Investments in national (and international) distribution capability
- Investments in developing management talent to grow the new industry

Investments in large-scale production capacity and a national distribution system are necessary for an emerging firm to gain a dominant market share in the new national market. Also now in a global economy, a first-mover firm must also move to establish presence in international markets. Developing management talent to run a large, growing firm is also a necessary investment. For example, a frequent kind of failure of many firms to have long-term survival after an initial success has been the failure of the founder of the firm, as an entrepreneur, to build a management team that can succeed the founder. Thus in moving first toward dominance in a new industry, it is necessary:

1. To devise a business strategy that exploits a technology plan to gain economies of scale and scope
2. To make the appropriate investments in the business system to gain these as advantages before competitors

Competitive advantages arise from a balanced combination of skilled management and improving and producing a high-quality product at low cost which captures a dominant market share. A first mover benefits from lower costs while competitors are struggling to build plants of comparable size. The first movers have already begun to debug the new scaled-up production processes and are moving down their learning curve, as competitors are still designing and constructing their plants. First movers also gain brand recognition and market share and establish distribution systems before their competitors. Managers of first movers gain experience and refine their procedures as competitors are just beginning to hire staff and build their organizations.

Economies of scale can lower production cost when the technologies of production provide an improvement of efficiency with size. Not all technologies provide increased efficiency as a function of size. Generally, there will be an optimal scale of production for a particular class of products and production processes—depending on several factors, such as the nature of the product, the nature of the transformation processes, the availability of resources, the possibilities and benefits of automation, and the costs of labor.

Economies of scope can lower costs when production capacity can be used to expand access to more markets by producing a variety of products. When a variety of products expands the markets available to the production facility, the fixed costs of production and the investments in gaining market access can be spread over several products, thereby lowering the fixed and invested cost per unit of a given type of product.

> **In a new-technology industry, a first-mover firm is the first to complete its national business system to gain economic advantages of scale and scope.**

CASE STUDY:

Numbers of U.S. Auto Firms over Time

After the introduction of the model T, the improvements in automobile product technology were incremental—for a long time. For example, in 1921, Hudson marketed a car with an enclosed steel panel body that had steel panels over a wooden body frame. This technical change made automobiles more durable to weather. Later, Studebaker shaped the steel panel for stiffness and eliminated the wooden body frame. Chrysler streamlined the steel body shape for greater airflow efficiency. Just before World War II, General Motors innovated the automatic transmission. As innovation slowed in the automobile product, competition focused on style, price, and quality of product.

After World War II, the U.S. automobile industry entered a period of stable technology. Management focus from 1950 turned to refining existing technology, automating production techniques, styling, and reducing numbers of parts. Only in the 1980s did technology change in the automobile system begin to change again incrementally with the introduction of electronics into control systems.

In 1909, in the new U.S. auto industry, there were 69 auto firms, but only half of these survived the seven years to 1916 (Abernathy, 1978). In 1918, Ford's new model T began putting many of these out of business as the new design standard for automobiles captured the majority of the auto market. Competitors had to re-design their product offerings quickly to meet the quality of the model T and its price. By 1923 in the United States, only eight firms succeeded in doing this and remained, with about 26 firms failing in the four years from 1918 to 1923. The eight remaining firms were General Motors, Ford, Chrysler, American Motors, Studebaker, Hudson, Packard, and Nash.

The depression of the 1930s and World War II interfered with the normal growth of the auto industry in the United States, but after that war, the market growth of the U.S. auto industry resumed (Figure 9.2). The average annual sales of cars in the United States peaked around 1955 at about 55 million units sold per year. By then, General Motors had attained close to a 50% market share. By 1960, four domestic auto firms remained: General Motors, Ford, Chrysler, and American Motors.

After GM overcame Ford's innovative lead in the late 1920s, GM continued to dominate the U.S. automobile market up to late in the twentieth century. In the 1950s and 1960s, GM grew to have about 54% of the automobile market but then, with poor design and engineering, began to yield sales to the world's rebuilding auto industry, particularly in Japan and Europe. By 1995, GM's share had de-clined to 34% and continued declining to 29% by the year 2000.

The 1970s were also the beginning of significant U.S. market share taken by foreign auto producers. During that decade, gasoline prices jumped due to the for-mation of a global oil cartel. American producers did not meet the demand for fuel-efficient cars. In 1980, U.S. auto producers faced a desperate time, with ob-solete models, high production costs, and low production quality. During the 1980s, the foreign share of the U.S. market climbed to one-third, and there were three remaining U.S.-based auto firms: General Motors, Ford, and Chrysler. Chrysler was bought by the German firm of Daimler-Benz in 1998.

The rapid decline of GM's market share was due to the expansion of Japanese auto sales in the United States from the late 1970s. By the year 2000, foreign-based automobile firms (i.e., Japanese and European) had captured over one-third of the U.S. market, primarily because of GM's lower quality. This was the major cause of GM's loss of dominance: poor quality in auto manufacturing and design.

COMPETITION IN AN INDUSTRIAL LIFE CYCLE

Figure 9.3 shows how the number of producing firms in an industry rises and then declines as the technology in an industry matures in an industrial life cycle. The number of firms peaks around the time of design standardization in the industry and declines over time to, finally, just a few firms. This is exactly the pattern we see in the example of the U.S. auto industry.

The early phase in a new manufacturing industry based on a new basic technol-ogy is an exciting time as product improvements continue and market growth is rapid. Then many new companies spring up like weeds. They are started by people entering the industry and by employees leaving and starting up their own firms. At this time, all the new firms are small and change is swift.

The product-design-standard phase is the time when the rapid winnowing out of the many new firms occurs. This happens even as the market grows dramatically, with only a few firms capturing this growth. Survival depends on gaining market share.

As the industry enters a technology mature phase, the number of domestic firms declines to a very small number. Also in the technology mature phase, international competition becomes very important and firms struggle globally for international markets. Utterback and Suarez (1993) charted over time the numbers of competitors in several industries: autos, television sets, typewriters, transistors, supercomputors, calculators, IC chips. Each of these industries showed the same pattern of a large number of competitors entering as the new technology grew and then the numbers peaking and declining dramatically due to intense competition as the technology progressed and matured. For a technology mature industry, market levels may remain relatively constant, since without new technology, market level is determined only by replacement rates and demographics.

CASE STUDY:

Commodity Products in the Chemical Industry

This case study illustrates what happens to an industry when its core technologies become mature. The historical setting for this case was at a major technological turning point for the world's chemical industry as the twentieth century ended. During that century, many of the chemical industry's product lines had resulted from high-tech processes and were therefore high-value-added (i.e., large profit margin) products. However, technological innovation in many areas of chemistry, producing ideas for new chemical product lines, had slowed by the 1970s. Moreover, the world chemical industry was facing the possibility of new competition from chemical plants based in the Middle East. European and U.S. chemical firms were seeing a future chemical industry with low profitability in commodity-type products and excess world production capacity, where earlier they had always been used to innovative and high-profit margin products.

The chemical industry originated with gunpowder manufacturers. For example, the largest chemical firm in the United States, E.I. DuPont deNemours, began with an emigrant French family manufacturing gunpowder for the rebels in the American Revolution in the 1770s. In the 1850s, artificial dyes were invented from the new science of chemistry. This was followed by innovations in artificial fertilizers. Then after 1900, innovations in artificial materials were developed: Bakelite, cellophane, neoprene, nylon, and so on. Innovations in the areas of dyes, fertilizers, and materials drove the technological innovation and economic expansion of the chemical industry for over 100 years, from 1850 through 1970.

However, by the 1980s the industry was lacking growth through a dearth of innovation. In the later 1970s, the lesser industrially developed countries with substantial oil reserves began to build petrochemical production facilities to compete in petrochemicals in world markets. They were able to do this since production techniques for bulk chemical processes had been invented in the period from 1880s through the 1950s. In the 1970s, these technologies were well known and widely disseminated through education and technical literature.

As another example, in the year 2002, DuPont sold its nylon business. But earlier in the late 1930s, it was the DuPont innovation of nylon that fueled the growth of DuPont for the second half of the twentieth century.

HIGH-TECH AND COMMODITY INDUSTRIES

High-tech products are products that are rapidly improving in technical performance due to continuing technical progress in the core technologies of an industry. But as the core technologies in an industry mature, products become relatively undifferentiated in technical performance. Then price and quality of product become the primary competitive factors. In contrast to high-tech products, performance-undifferentiated products are called *commodity-type products,* since they are all technically alike.

Industries are called *high-tech industries* when the core technologies in their products and production are changing rapidly. Industries are called *mature,* or *commodity industries,* when the technologies in their products and production are no longer changing rapidly or much at all.

> **Businesses in high-tech industries can compete on superior technical performance of product, whereas businesses in commodity industries can only compete on price and quality.**

We can see the direct relationship between technology finiteness and industrial maturity in Figure 9.3. There we show charts for industrial life cycle, rates of product and process innovations, and technology S-curves for core technologies of the product system. Looking along the time dimension, we see that the major growth in a market in the applications growth phase begins when the design is standardized for the product. Eventually thereafter, the industry begins a mature technology phase for its product and the product becomes a commodity-type product as opposed to a high-tech product. The technology S-curve for the core technology of the product levels off as product innovations stop.

Yet during this time, technology innovation continues in the production process, and a commodity-type product still uses high-technology technological change for its production. During this time, market volume can still grow as production innovations reduce the cost of producing the product. As the process innovations decline, the technology S-curve for the production processes levels off. Only after both product and production innovations have declined does one have a totally technology-mature industry. Then until a substituting technology occurs for the core technology of the product or for the core technologies of production, the technology-mature industry can continue at a market level determined by market replacement rates and/or market demographics.

> **Because over time, core technologies mature, all high-tech industries eventually become commodity industries.**

The profit margin for any firm's product lines is determined by two directions of conditions, internal and external. The *internal conditions* consist of the efficiency of the firm and the strategy of management. The *external conditions* are the balance of supply to demand in the industry. The external factors include the number of competitors, how rapidly technology is changing, and how rapidly the market is growing. When an industry reaches technological maturity and products are commodity type, the profitability of any firm in the industry is bounded by the external conditions of the number of competitors and production capacity of the industry. Until technological maturity, the internal conditions of innovation can create uniquely high-performing products enabling large profit margins.

There are two basic business strategies concerning technological innovation: first a high-tech product strategy and later a commodity product strategy, The high-tech strategy is time-limited. At first all businesses in a new industry are high-tech businesses, but over time all businesses in that industry become commodity businesses. As an industry matures from high tech toward a commodity industry, the survivors of the industrial shake-out are those companies that are low-cost high-quality producers and have established national and national distribution capability.

However, when a new core technology is invented and substitutes in a commodity industry, the previous producers in the industry become technologically obsolete. Any industrial sector that then continues producing a technically obsolete product line will become industrially obsolete. The societal function of an industry never becomes obsolete, but the core technologies of an industry can become obsolete for that function.

> **An industry becomes obsolete and dies when a substituting technology occurs to replace the core technologies and the industry does not change to the new substituting technologies.**

For example, in the nineteenth century, the ship industry had to change as core ship technologies changed from wooden hulls to steel hulls and sail changed to mechanical propulsion. Ships survived as an economic functional system, but businesses in the shipbuilding industry had to change or die.

SUMMARY AND PRACTICAL IDEAS

In case studies on

- U.S. auto industry
- Numbers of U.S. auto firms over time
- Commodity products in the chemical industry

we have examined the following key ideas:

- Core technologies
- Industrial design standards
- Industrial life cycle
- Industrial competition
- High-tech and commodity industries
- Industrial obsolescence

These theoretical ideas can be used practically to understand the dynamics of industrial change or lack thereof due to technological progress. Competitive conditions in a new industry provide high margins and growing markets, due to the improving technical performance of products. Competitive conditions in mature technology industries provide low margins and stagnant markets, due to lookalike commodity products and saturated markets.

FOR REFLECTION

Identify an industry that was made obsolete in its key technology. When and where did it begin? When was the technology of the industry most innovative? How did the industry end? Do remnants of the industry survive?

10
NEW HIGH-TECH BUSINESSES

INTRODUCTION

We have seen how technological progress creates new industries, and historically, most radically new technologies have been innovated by new small businesses. Accordingly, the topic of new businesses is important in understanding technological innovation. We examine the challenge of starting and managing a new high-tech business. The particular challenge to a new high-tech business is to design a new product and develop a new market for it, both at the same time. But the importance of the new technology justifies the business risks. For example, in studies of the success of new business ventures in the United States in the 1970s, new high-tech business ventures showed twice the survival rate than that of new ordinary business ventures (Vesper, 1980). We review the twin topics of new ventures and entrepreneurship, in terms of the basic ideas of (1) entrepreneurs and (2) dynamics and planning of new ventures.

CASE STUDY:

Cymer Inc.

This case study illustrates the start of a new high-tech company in which the entrepreneur's product vision correctly met a new market need. The historical setting was the middle of the 1990s, when the world's IC semiconductor chip industry continued its pace of rapid technological progress in increasing transistor density on a chip by reducing feature size. We recall that technical progress in chips proceeded exponentially through a series of next-generation technologies: "Over the past 30 years . . . chip makers doubled the number of transistors on a microchip every 18 months. For example, Intel's Pentium, vintage 1993, had 3.2 million transistors; its Pentium Pro, released about two years later, uses 5.5 million transistors" (McHugh, 1997).

IC chips were produced by projecting circuit diagrams onto photographic coatings on silicon wafers and then etching the photographic pattern as physical morphological features. As feature sizes of transistors approached the wavelength of

171

the projecting light, the light could no longer image accurate features. (This is the result of the wave nature of light, whose diffraction patterns around objects ultimately limit the resolution of the images at a given wavelength of light.) In the 1960s, chip makers used visible light. By the late 1990s, chip makers used the shorter wavelengths of invisible ultraviolet light, produced by hot mercury gas, getting chip feature size down to 0.35 micron. Next, chip makers hoped to jump from mercury light down to x-rays, which have much smaller wavelengths, but by 1997, x-rays were still too costly and difficult to use for chipmaking. The new light source for chip photolithography was excimer laser light at a 0.25-micron wavelength. In 1997, these lasers were produced by a relatively new firm called Cymer Inc. (McHugh, 1997)

Back in 1985, two recent Ph.D.'s from the University of California at San Diego, Robert Akins and Richard Sandstrom, were considering their future. Akins's thesis research was in optical information processing which uses lasers, and Sandstrom's was in laser physics. They were working for a defense contractor, HLX Inc., on esoteric projects that used lasers (such as laser-induced nuclear fusion and satellite-to-submarine laser communications). One day they were relaxing on a beach at Del Mar, tossing frisbees and drinking beer. They both wanted to make real money, and they speculated about opening a business. They even discussed fast-food outlets. But they decided to use their special talents. They founded Cymer Laser Technologies in 1986 to build excimer lasers.

Excimer lasers could produce laser light by using a mixture of gases, krypton and fluorine, pumped into a 2-foot aluminum tube, which is then zapped by a 12,000-volt charge across two electrodes inside the tube. The voltage creates a charge in the gases for 75 billionths of a second that excites the krypton atoms to couple temporarily with the fluorine atoms. This forms a krypton fluorine molecule as an excited dimer (hence the term *excimer*), but the unstable molecule breaks apart as soon as the voltage drops. In doing so, it releases a burst of deep-ultraviolet laser light of 0.25-micron wavelength.

In 1986, Atkins and Sandstrom began building the prototypes of this eximer laser in the research labs of the university (for which their new company owed the University of California $250,000 for the use of their facilities). Cymer also succeeding in winning research funding from the U.S. government's defense agencies to continue development work on the laser. The tough technical problem was getting the laser to run dependably for months at a time while handling the 1000 jolts per second of 12,000-volt charges. During this time of product development, Akins and Sandstrom took out second mortgages on their homes to keep Cymer running.

In 1988, Cymer received its first outside investment from a venture capitalist, Richard Abraham. Abraham had been a researcher and factory manger at Fairchild Semiconductor, Motorola, and Texas Instruments. He understood the importance of the deep-ultraviolet laser to the future production needs of the IC chip industry. Abraham's condition for investing was that "Cymer had to focus entirely on semiconductor applications for its lasers" (McHugh, 1997, p. 156). Later in that year, further investments came from Canon and Nikon after teams of their scientists and executives made visits to Cymer. Canon and Nikon manufactured the photo equipment, called *steppers,* which semiconductor manufacturers use to step the photo of a circuit over a silicon wafer and would be customers for Cymer's laser. They bought 6% of Cymer.

In 1995, semiconductor manufacturers began the switch to eximer laser light for the production of chips: "Demand from chip makers forced the stepper companies (such as Cannon and Nikon) right into high-volume buying soon after Cymer launched its $450,000 eximer laser model" (McHugh, 1997, p. 156). Cymer's sales exploded: $18 million in 1995 and $65 million in 1996. In September 1996, Cymer sold stock to the public at $9.50 a share and in December at $42 a share. It raised $80 million in the two offerings. In February 1997, Cymer stock traded at $50. Akins's stake was 2% and worth $2 million. After all, he hadn't needed to sell hamburgers.

TYPES OF NEW HIGH-TECH VENTURES

New high-tech ventures can market high-tech products or high-tech services. High-tech products can be new materials or new components or new final products or new process equipment. These can have large mass markets or small specialized markets. The potential size of a new high-tech firm depends on the size of its potential market. Moreover, the larger the market, the greater the need for capital to grow to dominate it, and the greater the potential of large companies entering it with large capital resources. If the markets are large, the new venture has an initial possibility of growing into a large firm. If the markets are small and specialized, the new venture has only the possibility of becoming a niche producer or being acquired by a larger firm.

Software products of new high-tech software firms can provide either mass-market software or specialized and customized software. Offering a mass-market software product, a new firm has the potential of growing large, but offering specialized customizing software, a new firm generally will also require regional location.

High-tech services can be professional services: consulting, training, services requiring high-tech devices or procedures, or repair and maintenance. New firms providing high-tech services generally require a regional presence and therefore usually have a limited growth potential unless some form of national delivery of services can be constructed. For service firms to grow large usually requires some type of partnership form (e.g., large law firms or consulting firms) or franchising form.

Another very important factor influencing the initial success of a new high-tech venture is intellectual property. A basic patent on a new technology can be an important source of an idea for a new businesses and can also provide early protection from competition.

INGREDIENTS OF A NEW HIGH-TECH VENTURE

What does a new high-tech venture require? Five key ingredients are:

1. A new product or process or service idea based on new technology—*a high-tech product or process or service*

2. A focus on a potential market for the new high-tech product/process/service
3. The *organization* of a new firm to produce and market the new product/process/service to its focused market
4. An initial *venture team* to launch the new firm, which includes technical, business, and marketing experience
5. The *capital* necessary to launch the new firm

We recall that technology is not itself a product or service. Launching a new high-tech venture requires understanding both the new technology and a customer who could use the technical functionality in a product, process, or service. This is the creative concept of a new venture—the synthesis of the ideas of technology and product and customer.

But it is not enough to have a vague idea of a product and customer. A precise market focus is necessary for a successful new venture. What is the class of customers? In what kind of application will the customer use the new product/process/service? What are the performance requirements and features of such a new product/process/service to be useful in such an application? What must be the price of the product/process/service to be purchased by such a customer? How can that customer class be reached by advertizing and distribution? What will be the nature of the sales effort required? What then must be the maximum cost of producing the product/process/service? How fast can the market be grown?

With the product and market in focus, the next idea to make clear is the nature and size of the organization required to produce and sell the product to the market. What must be the beginning form and size of the organization to enter the market with the new product? When will competitors enter with a competing product/process/service and at what price? How fast must the organization expand to grow the firm to competitive advantage over competitors?

With the product, market, and organization in focus, the next idea to fix is who are the initial management team for the new venture. The team must include persons experienced and skilled in (1) technology and product/process/service design, (2) production management, (3) marketing management, and (4) capital and financial management. Thus the ideal new venture team should consist of four persons. One partner should handle product development, design, and testing. A second partner should handle production development, organization, and management. A third partner should handle marketing and sales organization and development and management. A fourth partner should handle capital development and management and development of financial control and reporting procedures. Of course, new teams are often begun with less than four persons, but each of the four areas of product, production, sales, and finances needs to be covered even if initially it must be by one or more persons doing double duty.

The fifth key ingredient in a new business venture is the requirement for capital. Capital provides the ability to gather and organize resources necessary to begin or expand a productive operation. A successful productive operation will eventually create profits to return the capital, but the time delay between starting or expanding a productive operation and the later accumulation of profits to return the capital is the basic reason that capital is fundamental to business. Capital represents the "time" aspect of value-adding economic activities. Accordingly, all capital is evaluated not only on amount but on timing. Accordingly, in estimating the requirement for capital, the new venture team must carefully forecast its requirement amounts and when required and the time line for returning their capital to investors.

Leadership can span many aspects of management activity; but there are three central and critical attributes about organizations that only the top leadership can provide: vision, resources, and delegation. Vision establishes the direction of organizational activities, resources the base for activities, and delegation the team to carry out the activities. In all organizations, these three attributes are the primary responsibility of the leader. If the leader does not fulfill these responsibilities, the organization will be directionless, undercapitalized, or lacking an effective team.

A successful new high-tech venture is often a team effort including:

1. One member with prior market experience
2. One member with research/technical experience
3. One member with production experience
4. One member with financial control experience
5. One member with leadership as to vision and resources

It is true that in many new ventures a single entrepreneur may play several or all of these roles. However, technical, production, marketing, and finance all require quite different skills and talents. Moreover, if a new venture grows rapidly, several persons will be required to control operations. One should try to assemble the new venture team as early as possible in order to increase the chances of a successful venture.

CASE STUDY:

Al Shugart and Disk Drives

This case study illustrates the opportunities and hazards that new ventures encounter. The historical setting was the decades of 1960 to 1990, when the hard-disk drive was innovated for computers. Al Shugart joined IBM the day after he finished college in 1951. In the early 1950s IBM was just beginning to make electronic computers and a key component was permanent memory. Hard disk drive memory storage was innovated in 1961, and Shugart led IBM's development of the technology. But in 1969, Shugart resigned from IBM: "Weary of Big Blue's bureaucracy, he [Shugart] left for Memorex Corp., taking along some 200 engineers" (Burrows, 1996, p. 18).

But only a few years later, in 1972, Memorex had financial troubles, and Shugart left (again with loyal followers) and founded Shugart Associates. Shugart had the innovative idea of pioneering the floppy disk. But in 1974, the new company lagged in its product development for the floppy disk, and Shugart's venture backers ejected Shugart: "Nearly broke, he [Shugart] moved to Santa Cruz, opened a bar with some friends, and bought a salmon-fishing boat. 'I had a tough time meeting my Porsche payments,' he deadpans" (Burrows, 1996, p. 18).

Four years later, in 1978, an old colleague, Finis Conner, went to visit Shugart in Santa Cruz. He suggested making small hard drives for the new personal computers just beginning as a market. It was a good idea. They started Seagate, and the company rode the growth of personal computers, growing to $344 million in sales by 1984. But then price wars began as many producers marketed small hard disk drives. Shugart and Conner argued over production strategy. Shugart insisted on making drives from scratch; Conner argued that buying parts was less risky because it tied up less capital. Dissatisfied, Conner left and launched Conner Peripherals, with capital backing from Compaq Computer. In the first year, Conner made $133 million in sales, since purchasing components is a fast way to build production. In the short term, Conner was right. But the long term was not yet.

Because of the intense price competition in the personal computer hard disk drive market, Shugart switched to the mainframe disk drive market (the one in which a decade earlier Memorex had stumbled against IBM). Shugart bought Control Data Corp.'s Imprimis Technology, which produced mainframe disk drives. Shugart survived by having two product lines. As the PC market demand grew rapidly, key components grew scarce, raising component prices. Now it was Conner's company that was in trouble. Along with other PC hard disk drive manufacturers who only assembled purchased components, Conner had no control over manufacturing costs. He and other assemblers lost money. In 1993, only Seagate made a profit in PC hard disk drives. By 1995, Conner was in deep enough trouble that he needed to sell his company. His old colleague Shugart offered to buy it and to give Conner a job at Seagate. Shugart's make versus buy production strategy proved to be right for the long run. In 1996, Seagate projected revenues of $9.2 billion, up 104% from the preceding year, and profits could surge 130%, to $600 million.

Shugart's entrepreneurial style included correct technical vision and correct long-term cost strategy. He also had the ability to generate staff loyalty: " 'He's the most up-front guy I've ever worked with,' says vice president Stephen J. Luczo" (Burrows, 1996, p. 73).

ENTREPRENEURIAL STYLE

Beginning a new high-tech business venture requires a style of management that is entrepreneurial. Accordingly, we need to review the concept of entrepreneurship; and in the many studies about it, three themes have emerged:

1. There is an *adventure* theme. The entrepreneur is a kind of business hero, creating new firms and causing economic change. The entrepreneur has the qualities to be admired as the hero or heroine of an adventure: vision, initiative, daring, courage, and commitment.

2. There is an *organizational transition* theme. The entrepreneur makes a good manager for starting a firm but is not necessarily good at building organization and growing a firm. Because the entrepreneur in the early stages of a firm plays many roles, the transition to building an effective large organization is often difficult for the entrepreneur.

3. There is an *organizational dilemma* theme. To what extent and how can entrepreneurship be encouraged within a large, stable firm to manage change? Are bureaucratic management and entrepreneurship styles eternally at odds with one another?

High-tech entrepreneurs usually come from larger organizations in which they first gain experience, training, or research capability. In fact, most successful high-tech entrepreneurs start firms with innovative product or service ideas for the markets of the large firms in which they earlier served. An important experience that entrepreneurs can bring to a new high-tech venture is a detailed and experienced understanding of a market. A second important experience is an innovative idea for a new product or service for such a market. A third important experience that entrepreneurs can bring to a new venture is prior knowledge of production. A fourth important experience is in finance, budgeting, and control. These kinds of experiences come from entrepreneurs having prior experience in the marketing or manufacturing or engineering or research or financial control functions of a large firm. Sometimes entrepreneurs have tried to get their superiors interested in the new idea and failing that have left and begun their own companies. Sometimes the innovative ideas have originated in a university or government research lab and researchers then started a new firm.

The customary use of the term *entrepreneur* is compared to the term *bureaucratic manager*. These distinguish two different types of leadership in (1) starting economic activities as opposed to (2) running ongoing economic activities. Entrepreneurs start new firms; managers run ongoing firms. There are important differences in the optimal leadership styles for starting versus running. Often, good entrepreneurs make poor managers of ongoing operations, and good managers often make poor entrepreneurs when presented the challenge of dealing with substantial change.

For example, Stevenson and Gumpert (1985) compared managers along two dimensions: desire for future change and perceived ability to create change. They argued that the entrepreneur is the kind of manager desiring future change and perceiving the ability to create such change, and that the entrepreneur asks the following kinds of questions:

- Where is the opportunity?
- How do I exploit it?
- What resources do I need?
- How do I gain capital control over the resources?
- What structure is best to exploit the opportunity?

In contrast, Stevenson and Gumpert believe that managers of ongoing operations are more concerned with stability and efficiency than with change, and accordingly ask a different set of questions:

- How can I improve the efficiencies of my operations?
- What opportunity is thus appropriate?
- What resources do I control?
- How can I minimize the impact of others on my ability to perform?
- What structure determines our organization's relationship to its market?

The term *bureaucrat* need not connote ineptness but can connote *rationality*. There can be good or bad managers, whether entrepreneurial or bureaucratic. We will use the concept of the manager as a bureaucrat, the bureaucratic manager, to emphasize a managerial concern with rational order for an organization—the "good" bureaucrat. The tools of rational order are formalized procedures that routinize organizational efficiency. When organizational change is needed, an entrepreneurial management style is useful, but when organizational stasis for rational efficiency is needed, a good bureaucratic management style is useful.

PSYCHOLOGY AND SOCIOLOGY OF THE ENTREPRENEUR

Many have studied the psychology of entrepreneurs hoping to learn why some people are more likely than others to become successful entrepreneurs. Researchers have listed several attitudes and values they found typical of the entrepreneur. A typical list (Vesper, 1980, p. 9) includes such attitudes as:

- A desire to dominate and surpass
- A need for achievement
- A desire to take personal responsibility for decisions
- A preference for decisions with risk
- An interest in correct results from decisions
- A tendency to plan ahead
- A desire to be his or her own boss

You will note that all these attributes are essentially positive and reflective of the "heroic" myth of the entrepreneur. But there may also be negative attributes. For example, the entrepreneur may also be self-centered, temperamental, authoritarian, and unable to stand grooming a competent successor. One can see both positive and negative attitudes in the biographies of entrepreneurial people.

Another approach to understanding the entrepreneur has not been psychological but sociological. For example, Quinn (1985) viewed the entrepreneur as occupying a kind of role encouraged by an *individual entrepreneurial system,* a capitalistic system that encourages and supports individual initiative. Quinn identified several characteristics of an entrepreneurial system that encourages technological innovation:

- Fanaticism and commitment
- Chaos acceptance
- Low early costs
- No detailed controls
- Incentives and risks
- Long time horizons
- Flexible financial support
- Multiple competing approaches
- Needs orientation

Quinn viewed the single-minded dedication of the entrepreneur as a kind of fanaticism, and an economic or organizational system should tolerate the type of ruthless, dedicated purpose required of an entrepreneur. The contexts of such single-mindeness will appear chaotic and disorganized because the entrepreneur is fixed on the goal and will use whatever means or expediency that proceeds toward that goal. The economic or organizational system should tolerate this apparent chaos, which includes little detailed control in the early phase of a new venture. New ventures originate in an opportunistic, cost-cutting, shortcutting way to a single-minded, clear-cut goal. The system should provide appropriate rewards for the risk-taking in entrepreneurship, and rewards must be structured for long-term horizons since it takes time for anything really new to become a success. Because of the experimentation and learning that goes into new ventures, it is important for the system to provide flexibility in financing from many sources and for multiple and competing approaches. In the early days of any radical innovation, new ways are being tried out, and only down the line will an optimal configuration emerge for the standard design of a new technology. Need orientation should always be the goal of entrepreneurship, since satisfying a customer's need is the sure way to economic success. Systems that encourage the fulfillment of needs of a marketplace stimulate innovation that endures and is economically important.

One can see the logic in this list. Individual entrepreneurs need to be encouraged to show commitment to start new activities and push them to success. An economic system should be loose enough to tolerate chaotic activity, allow for low costs of startup, and so on. A very important institutional factor for new high-tech ventures are sources of venture capital.

Not all systems encourage entrepreneurship, and there can be many organizational barriers, such as:

- Perceived irrelevance of an innovation
- Punishment for taking a risk and failing
- Rewards only for short-term achievements

- A culture of not-invented-here
- Stodgy and not changeable conventions

For innovation to be encouraged in a large firm, the company culture toward innovation is very important. Particularly important are the criteria about the perceived irrelevance of a new business venture and about the attitude of the firm toward risk-taking. One of the striking facts about innovation in large firms is that in some firms very little innovation occurs from very large research efforts.

CASE STUDY:

Rise and Fall of Osborne Computer

Timing is particularly critical to new businesses, as they are usually thinly capitalized. This case study illustrates the importance of timing upon commercial success or failure during innovation. The historical context of this case study was in the early days of the new personal computer industry, which began toward the end of the 1970s and the beginning of the 1980s. It was just after Apple's initial success and just before IBM entered the new market. In July 1981, Osborne Computer was started as a new company in the then brand new personal computer industry. During its first two months of production, it booked its first $1 million in sales. (Not bad for a new company, for in 1981, $1 million was still a considerable amount of money.) In the second year of operation, its net revenues jumped to $100 million (even today, $100 million is a considerable amount of money). Yet only six months later, Osborne Computer was bankrupt (Osborne and Dvorak, 1984).

In 1980, Osborne had sold his computer-book publishing company to McGraw-Hill and had joined that company. But Osborne then decided to create his own brand of personal computer, so he left McGraw-Hill and started his own firm. He had seen a market opportunity in the new personal computer industry. Then they were all being sold as components: computer, disk drive, monitor, printer, software. Osborne decided to package them as a portable computer and to sell the package cheaper than competitors' equivalent component sets. He hired Lee Felsentein (an engineer) to design the electronics for the computer. Osborne presented his ideas for the new venture to Jack Melchor (a venture capitalist). Melchor invested $40,000. The new computer was an instant success, and by the end of the first year, sales soared to one-third of those of the then leader, Apple. Osborne had created a new market niche in the new personal computer industry.

Quickly, competitors entered the new market, focusing on the Osborne's visibly weak features—a small screen with a narrow column width. A new competitor, Kaypro, pounced on these weaknesses, introducing a similar model but with a 9-inch screen and full 80-character width. Kaypro's sales soared and Osborne's sales collapsed.

Osborne had planned a public offering to generate additional capital. He intended it for the summer of 1982 but put it off until early in 1983. But at that time the brokerage firm decided not to make the offer because of the sales slump. Customers were either buying Kaypro or waiting for Osborne to offer a larger screen. In that summer, the sharp sales drop created a financial crisis. October, November, and December passed without significant sales. January, February, March, and April also passed without significant sales. The Osborne Company was hemorrhaging cash. It could not get additional bank loans, and its planned public offering could not be made with its lack of sales. Osborne made several attempts to raise new capital privately, but without success. There was not enough cash to carry the company, and in September 1983, the Osborne Company declared insolvency under Chapter 11 of the bankruptcy law. All the millions of equity on paper, only a year before, had vanished! Yet though the 1980s, the market for portable computers continued to grow, and they became the most rapidly growing segment of the PC market.

DYNAMICS OF NEW HIGH-TECH VENTURES

In launching a new venture, timing is critical. Integrating technology change and commercial application must occur at a time of market opportunity. The financial risks are several: (1) having the capital required to exploit the market opportunity, (2) timing the exploitation to recover the invested capital, and (3) generating sufficient working capital to meet competitive challenges.

The competitive advantages of a strategic technology occur within a window of opportunity followed by competitive challenges to this opportunity. When technology leaders introduce a new product that creates new markets, competitors will enter by focusing on the obviously weak features of the innovative product. Thus, competitively, technological innovation is never a one-time activity. Innovation requires a sequence of activities that follow up in creating and sustaining a competitive advantage. Proper timing of the financial backing for a technological innovation include:

1. Having the capital required to perform the R&D necessary to design the new product
2. Having the capital required to produce the new product
3. Having the capital required to market the new product to the point of break-even on the investment
4. Having the capital required to meet follow-on competitive challenges as the product and market evolve

Timing is critical when a new product is introduced, when production is expanded, when the product is improved, when competition enters, when production cost is reduced, when working capital is created. The dynamics of the growth or death of new firms center around timing.

> **Product ideas, production expansion, marketing, controlling costs, creating working capital, meeting competition with improved products—the timing of these determines the dynamics of new ventures.**

Forrester (1961) introduced analytical techniques for examining problems that arise from timing of activities in a firm, which he called *systems dynamics* of organizations. Forrester then applied the technique to the startup and growth of new high-tech companies. Figure 10.1 shows the four most common patterns that Forrester found in new companies. The first curve, A, is an ideal pattern that all companies hope for. Initial success is rapid and exponential in sales growth (as happened at first at Osborne), and then growth slows but continues to expand as the firm develops from a small firm to a medium-sized firm to a large firm. But curve A is relatively rare. The harsh facts are that few new high-tech firms ever become large firms.

Another successful pattern is curve B, in which a problem occurs soon after rapid expansion but the problems are solved and growth resumes. Curve B is even rarer than curve A because severe problems early in the growth of a new firm usually kill it, because the working capital of new firms is always thin and fragile (as happened at Osborne). For new companies encountering troubles, the most common pattern is curve D, when (like Osborne) the new firm's capital cannot sustain a period of losses. But for high-tech new ventures, pattern C is frequent. Here growth levels off as competitors enter the market, but the company successfully establishes a market niche for itself and continues as a small to medium-sized company, or is purchased by a larger firm.

What organizational factors determine these dynamic patterns? Figure 10.2 shows a simplified version of Forrester's organizational systems dynamics modeling approach. In Forrester's approach, it is important to distinguish organizational activities (boxes) from information states (circles). Organizational activities transform

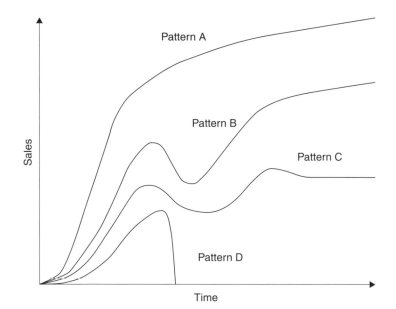

Figure 10.1. Sales patterns for new ventures.

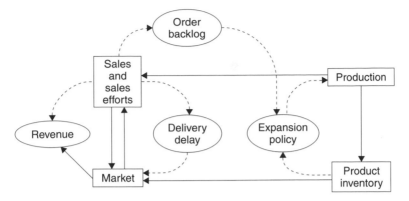

Figure 10.2. Systems dynamics model of a new venture.

organizational inputs to outputs, while information states report on the results of transforming activities. In any productive organization, there will be squares identifying general transforming activities of parts inventory, production, production inventory, sales, product delivery, market, and revenue. In addition, important circles of information states include profits, costs, sales strategy, product-order backlog, and production strategy.

The dynamics of organizations occurs from differences in timing of the activities in the various squares. For example, in an ideal case, sales strategy (c) will have accurately predicted the rates and growth of sales, and based on these projections, production strategy (e) will have planned and implemented adequate production capacity and scheduled production to satisfy sales projections and to meet projected product delivery schedules. Accurate timing will minimize parts inventory and product inventory and delays in delivering product to market. In turn, the market will pay promptly, providing the revenue to pay costs and show a profit.

The dynamics occurs when all these activities are not timed correctly. For example, if sales strategy overestimates actual sales, production will produce too many products, increasing costs from excess parts inventory, product inventory, and production capacity. If, on the other hand, sales strategy underestimates actual sales, the production capacity will be too small to meet demand, resulting in delivery delays which in turn result in lost sales to competitors, thus reducing revenue and lowering profits.

If the scheme shown in Figure 10.2 is expressed in a simulation model (organizational systems dynamics model), what–if games can be run to understand the sensitivity of product strategy on production strategy and then on sales strategy regarding revenues and profits. Then the patterns of organizational growth as depicted in Figure 10.1 can be plotted for a given organization. Pattern D is, of course, a very sensitive pattern and always results from a cash flow crisis. When the cost rate overwhelms the revenue rate long enough to exhaust working capital and the ability to generate immediate new working capital, such a firm will go bankrupt.

In the short term, *cash flow* is always the critical variable for survival.

The critical delays in the activities that impact cash flow dramatically are:

- Sales efforts that lag behind sales projections
- Production schedules that significantly exceed product demand
- Sufficient delay in delivery of products to lose sales
- Significant delays and/or failure of market revenue to pay for sales

CRITICAL EVENTS OF HIGH-TECH VENTURES

Given these kinds of organizational dynamics of new ventures, there are nine critical activities in the startup of new high-tech firms—critical milestones in the path to success:

1. Acquisition of startup capital
2. Development of new product or service
3. Establishment of production capabilities
4. Sales growth
5. Production expansion
6. Meeting competitive challenges
7. Product improvement diversification
8. Organizational and management development
9. Capital liquidity

Acquisition of Startup Capital

The first milestone to be accomplished is to obtain the startup capital. Capital is the resource necessary to begin and operate a productive organization with potential profitability until revenues can sustain the operation and provide profits. Startup capital is required for establishing a new organization and hiring initial staff, developing and designing the product/service, funding production capability and early production inventory, and funding initial sales efforts and early operations. Startup capital is seldom sufficient for rapid growth, and further capital requirements are therefore usually necessary for commercial success.

Product/Service Development

The second milestone is to complete new product (service) development and design. A new firm is high-tech when its initial competitive advantage is in offering the technology advantage of new product or service—with new functionality, improved performance, or new features over existing products and services. Sometimes new high-tech firms can be started with alternative high-tech production processes for existing types of products or services.

Developing and designing the new product or service requires capital and will be a major drain on the startup venture capital. Ordinarily, development and design should be far along before startup capital can be attracted. However, development problems or design bugs that delay the introduction of a new high-tech product or service make serious problems for the survival of a new firm. Such delay eats into the initial capital. If the delay is so long that competitors enter the market with a similar new high-tech product or service, the advantage of first entry into the market is lost.

Establishment of Production Capabilities

The third milestone is to establish the capability to produce the new product or service. In the case of a physical product, parts or materials may be purchased or produced and the product assembled. The decision to purchase parts/materials or to produce them depends on whether others can produce them and whether or not there is a competitive advantage to in-house production. Establishing in-house production capability of parts or materials will require more initial capital than purchase but is necessary when the part or material is the innovative technology in the product.

However, the establishment of any new production capability will also create production problems, problems of quality and scheduling, and problems in achieving on-time delivery. Capital will also be required to debug any new production process.

Sales Growth

Initial sales and growth constitute the next critical milestone event. The larger the initial sales and the faster sales growth, the less room there is for competitors to enter. An important factor influencing initial market size and growth is application of the new product or service and its pricing. Another marketing problem is to establish a distribution system to reach customers. Distributions systems vary by type, accessibility, and cost to enter. Planning the appropriate distribution system for a new product or service, the investments to utilize such, and its cost influence on product/service pricing is important for the success of new ventures. Generally, it costs less to reach industrial customers than to reach general businesses or consumers. This is one reason that a large fraction of successful new high-tech ventures are those in which industrial customers provide the initial market. These are usually industrial equipment suppliers or original equipment manufacturers selling to large manufacturing firms. This allows a new firm to get off to a fast start but eventually limits the size of the firm and makes the firm vulnerable to backward integration by a customer. Moreover, a small firm with only a few industrial customers is very sensitive to cancellation of orders from any of them.

In the general business and consumer markets, distribution system infrastructure will usually consist of wholesale and retail networks. In these, access to the customer will depend on wholesaler and retailer willingness to handle the brand offered by a new firm. Establishing brand identity and customer recognition of the brand is then an important problem and a major barrier for a small new firm to overcome. Moreover, in some retailing systems, under-the-counter practices (such as buying shelf space or generous holiday gifts to purchasing agents) may also be barriers to overcome. As a market grows, the long-term success of a new high-tech venture becomes increasingly important in gaining access to and maintaining access to national and international distribution systems.

Production Expansion

As the new market grows and early sales are made successfully, production expansion must be planned and implemented in a timely manner or future sales may be lost to competitors because of delivery delays in providing products. Production expansion will usually require a second round of capital raising, for the initial capital seldom provides enough for expansion.

Meeting Competitive Challenges

In a very few areas and rare cases, a patent on a new product or process is basic and inclusive enough to lock out competitors for the duration of the patent. This is true in the drug industry and occasionally elsewhere. However, most new high-tech ventures are launched with only partial protection from competition by patents, and competitors soon enter with a me-too products. The me-too products or services are likely to be introduced with improved performance or features and/or at a lower price. The entrance of competitors into the new market is the critical time for new ventures. They must at that time meet the competitive challenges or follow pattern C into a niche or pattern D into bankruptcy.

Product Improvement and Diversification

Also after competition appears, a new firm must upgrade its first-generation products with new products to keep ahead of competition in product performance and features. It must also be prepared to lower its prices to meet or undercut new competition. Improving production to lower its cost of production is a critical event. It must also begin to diversify its product into lines to decrease the risk that a single product problem will kill the firm. The round of capital raised for production expansion also needs to provide for product and production improvement.

Organizational and Management Development

As an organization grows in size to handle the growth in sales and production, it is important for the firm to develop organizational structures and culture and to train new management. This is an important transition, as the early entrepreneurial style of organization and openness and novelty of culture needs to mature toward a stable but aggressive large organization. In a small firm, coordination is informal and planning is casual. In a large firm, both coordination and planning need formalization.

Capital Liquidity

The final step for success in a new firm is to know when and how to create liquidity of capital assets and equity. One means is to go public, and another is to sell the firm to a larger company. Liquidity of capital enables the founders of the firm and early employees to transform equity into wealth.

NEW VENTURE BUSINESS PLAN

All these kinds of milestones need to be anticipated in an initial business plan for the new venture. To deal successfully with critical events, a good business plan anticipates their challenge and prepares for them. For example, one of the hard lessons in the dot.com stock bubble of the late 1990s was that a lot of venture capital money was thrown away on entrepreneurs who had a new idea but not a good business plan.

A business plan for a new business venture should depict the strategy for the business and lay out a plan to implement the strategy. Strategy requires two visionary components: a vision of the transformation of the new enterprise, and a vision of how that transformation adds value to a customer of the new enterprise. Management's vision of the enterprise should provide answers to the following strategic questions:

- What businesses should we be in?
- What are the basic transformations of these businesses?
- What are the core technologies of these basic transformations?
- Who are its customers, and how do we reach them?
- How do these transformations add value for the customer?
- What products or services provide that value?
- In adding value, what competitive edge should we have over competitors?
- What core technologies provide the transformations in product/service and in production?
- What technology strategies can provide competitive advantages?
- How do we staff, organize, and operate productively?
- What capital requirements do we have, and how do we obtain them?
- What are our goals for return on investment, and how are they to be met?

In the value-adding transformation of an enterprise, subsets of strategy must also be addressed, such as:

- *Practice of quality:* How good is good enough?
- *Practice of customer service:* How much and what kind of attention should be paid to customer requirements?
- *Practice of profitability:* What are minimally acceptable returns on investment?

- *Practice of gross margins:* How much overhead can we afford to operate?
- *Practice of rewards:* How much of the profits should be divided among management, labor, and shareholders?
- *Practice of control:* What tools should we use to control corporate performance?
- *Practice of innovation:* Do we compete as a technology leader or follower?

A strategy policy list should be constructed by the new venture team that addresses and answers the foregoing questions. Then the team can write a business plan that will implement the new enterprise strategy and be used to raise capital for the new venture. Accordingly, the purposes of the business plan are (1) to chart the course and identify the resources needed for the new venture, and (2) to attract venture capital.

A standard format for business plan should include the following topics:

1. Executive summary
2. Enterprise strategy and technology strategy
3. Product/service strategy
4. Competition and benchmarking
5. Manufacturing and marketing strategy
6. Management team and organization
7. Financial plan

Since the new business venture does not yet exist and therefore the plan cannot be tested against past performance, *clearly identifying assumptions in the plan is very important.*

> **The basic value of a good new-business plan lies in clear and comprehensive identification and discussion of the critical assumptions about all the business opportunities and risks.**

An interesting example in the 1990s in training to encourage entrepreneurship and the writing of new business plans was an annual student competition in business plans held by the Massachusetts Institute of Technology. In 1996–1997, the MIT Enterprise Competition offered a prize of $50,000 to the student submitting the best business plan for a new company. Then information about the competition was posted on the World Wide Web (*http://web.mit.edu/50k/www*). The competition was limited to MIT students, but nonstudents could team with an MIT student to submit a business plan. The competition announcement nicely summarized the purposes of a business plan and its audiences:

1. Conceptualizes the totality of a significant business opportunity for a new venture
2. Presents the organizational building process to pursue and realize this opportunity
3. Identifies the resources needed

4. Exposes the risks and rewards expected

5. Proposes specific action for the parties addressed

The audience for a business plan includes the founding team, potential investors and employees, customers, and suppliers or regulatory bodies (MIT 50K, 1996).

VENTURE CAPITAL

Venture capital, a form of capital invested in new businesses, is a financial essential in funding new business starts. Several stages of funding are needed for new businesses: seed capital for beginning, production capital for building production and distribution capability, and expansion capital for expanding production and distribution capability. Venture capital is a higher risk investment than later forms of business investment because few business assets exist at the time of venture funding; correspondingly, returns on successful venture capital investments can be very high, because prices of equity shares are much less before a firm proves itself.

Sources of venture capital include investments by the entrepreneurs themselves, by technically oriented persons who have become wealthy (sometimes called an "investment angel," (e.g., Richard Abraham in the Cymer case), by venture capital firms, by venture capital funds, and by venture capital investments from large firms looking for new technology (e.g., Canon and Nikon in the Cymer case). [Guides to sources of venture capital in the United States are published periodically (e.g., Prat and Morris, 1984).] In launching a new high-tech business, it is usually difficult to attract venture capital investment until the product prototype is well under development. In the Cymer example, the entrepreneurs had to use university facilities, government contracts, and second mortgages on their homes to get started. When they had a demonstration prototype to show, they then were able to attract startup capital, first from a person and then from user firms, Canon and Nikon. The company went public just after sales of their product exploded, with the second offering generating more capital than the first offering.

The general criteria on which firms evaluate business proposals are product differentiation, market attractiveness, managerial capability, environmental safety, product liability, and cash-out potential. Venture capital firms also differ as to what they will invest in. For example, Roberts (1991) pointed out these firms differ by the technologies in which they prefer to invest and also by preference for the stage of venture investment. Few venture capital firms will provide seed money for start-ups, most firms preferring funding later in the cycle, to reduce their risks.

In the second half of the twentieth century, the venture capital industry in the United States played a major rule in fostering new industry. For example, in 1992, venture capital investment was $4 billion, and it exceeded $7 billion in 1995. Seventy percent of the 1995 investments were in information technology. Earlier venture investments had created large U.S. firms, including Sun Microsytems, Intel, and Microsoft.

Venture capital as a distinct industry slowly spread globally. In 1995, Europe had about 500 firms that invested in 20,000 companies, for a total of $9 billion.

However, more than half of these investments were not to start new companies but to finance changes of ownership, mostly as management buyouts. In Asia, venture capital was still mostly corporate investment by giant conglomerates and family-run businesses (*Economist,* 1997).

SUMMARY AND PRACTICAL IDEAS

In case studies on

- Cymer Inc.
- Al Shugart and disk drives
- Rise and fall of Osborne Computer

we have examined the following key ideas:

- Ingredients of a new venture
- Entrepreneurial style
- Psychology and sociology of entrepreneurs
- Dynamics of new ventures
- Critical milestones in new ventures
- New venture business plan
- Venture capital

These theoretical ideas can be used practically to think through how to start a new business venture and what needs to be done to ensure the commercial success of the venture.

FOR REFLECTION

Identify a few new firms that went public in the last five years. Find their initial offering prospectuses and trace their stock prices since then. Have any encountered problems? What were the problems, and why did they occur?

11
TECHNOLOGY AND ETHICS

INTRODUCTION

We look next at the relationship of the technological imperative to ethics. Ethics is a complex topic and beyond the scope of this book, but we can look specifically at the relationship of technology to ethics by viewing technology as a kind of means to the ends of action. In general, ethics is about choices of end of action, and technology is a means of action. Many of the ethical concerns regarding technology are about the ends to which technological systems are means. As means, technologies are judged not as good or evil but as efficient or inefficient toward ends. It is the ends, the applications of technology, that are good or evil.

Accordingly, we can view technology ethically as a means that is acceptable or unacceptable. As an acceptable means, technology is judged merely as being efficient or inefficient. As an unacceptable means, technology is judged as never usable, intrinsically evil. We will look at two cases of technology to compare these: technology and environment, and technology and human cloning—technology as an acceptably efficient or inefficient means, or technology as an intrinsically evil means.

CASE STUDY:

Monsanto's Strategy for Sustainable Economy

In 1997, Robert B. Shapiro was chairman and CEO of Monsanto Company based in St. Louis, Missouri. Shapiro had taken the initiative to reorient Monsanto toward a future business in environmentally sustainable economies (Magretta, 1997). It was a major strategic reorientation for Monsanto, since much of its traditional chemical business was in agricultural chemicals. As an example of economy and environment, Shapiro used current potato production practices in the United States. For farmers to grow an annual crop of potatoes from seed potatoes, they had needed to use pesticides against insects and viruses that might otherwise damage the crop. To produce the quantity of pesticides used annually required. Starting with 4,000,000 pounds of raw materials and 1500 barrels of oil to produce

3,800,000 of inert ingredients and 1,200,000 pounds of insecticide (plus 2,500,000 pounds of waste), which provided 5,000,000 pounds of formulated insecticide product (in 180,000 containers of packages), along with 150,000 gallons of fuel to distribute and apply the insecticide. Of this 5 million pounds of insecticide, only 5% actually reached the target pests of the potato crop. The other 95% (4.75 million pounds) of insecticide had a nonproductive impact on the environment. One of the major U.S. potato-growing areas where these insecticides were being applied was in Colorado, used to control the Colorado potato beetle.

How could this kind of useless environmental impact be changed? This was Shapiro's charge to his colleagues at Monsanto—to find alternative technologies that could be environmentally friendly. Monsanto's early investment in the new biotechnology in the 1980s provided a new technological approach. Monsanto's scientists used biotechnology to develop what Monsanto called its NewLeaf potato. This potato had been bioengineered to defend itself against the destructive Colorado potato beetle, and in 1995 it was used on potato farms in Colorado. Adding to the NewLeaf potato's resistance, Monsanto engineered it to include resistance to leaf virus.

Thus, Shapiro's positive attitude toward environmental challenges was practical. Based on new business opportunities that a new technology, biotechnology, was making possible, Shapiro reported: " 'We can genetically code a plant, for example, to repel or destroy harmful insects. That means we don't have to spray the plant with pesticides. . . . If we put the right information in the plant, we waste less stuff and increase productivity. With biotechnology, we can accomplish that' " (Magretta, 1997, p. 82).

Monsanto's strategy was to move from being primarily a supplier of pesticides to being a supplier of biotechnology-engineered seeds and seed plants. The capability to see new business opportunities in a future of environmentally sustainable economy was due to a succession of visionary business leaders who thought strategically about the future. As Shapiro commented: " 'My predecessor, Dick Mahoney, understood that the way we [Monsanto] were doing things had to change. . . . Dick got way out ahead of the traditional culture in Monsanto and in the rest of the chemical industry. He set incredibly aggressive quantitative targets and deadlines. . . . In six years, we reduced our toxic air emissions by 90%. . . . Dick put us on the right path' " (Magretta, 1997, p. 84).

Back in September 1983, when Mahoney became president, Monsanto was a $6 billion company. Mahoney formulated new strategy to reduce costs in Monsanto's traditional chemical-based businesses and to position it in new biotechnology-based products (Labich, 1984). Then Mahoney had foreseen major changes in the chemical industry. In the early 1980s, the world's chemical industry was facing a slowdown in technological innovation in its traditional businesses (i.e., explosives, dyes, fertilizers, plastics). Also, a new set of competitors on the horizon were in Middle Eastern countries making investments for future basic-chemicals businesses. Oil-producing companies were expected to build new chemical plants with a competitive cost advantage because of their enormous oil and gas resources. Mahoney began formulating new technology strategy for

Monsanto through biotechnology as a source of new high-tech products. In 1983, a technical journal, *Chemical & Engineering News* (1983, p. 10) published an interview with Mahoney:

> *C&EN:* Where do you believe Monsanto is going in the next decade?
>
> *Mahoney:* We've set out on a pretty clear course that will see us evolving over time into a company that will be, in some ways, like we are today, and, in other ways, quite different.

The traditional businesses of Monsanto had been in basic chemicals, whose technologies of production had been relatively stable for years, with no new radical inventions. Consequently, basic chemicals had become commodity-type products. Monsanto sold some of these traditional businesses that were not profitable in the United States and abroad (including some fiber operations, a petrochemical plant in Texas, nylon operations in Europe, and a subsidiary chemical and plastics firm in Spain). Monsanto then turned its technology strategy toward biotechnology:

> *Mahoney:* Probably the one [area] that gets most press at Monsanto is the biological component. Here I'm talking about our existing business in agriculture and a new business in nutrition chemicals, which we started a few years ago to move into animal health and nutrition, and what we hope will be a substantial effort in the human health care area.

Monsanto began a large R&D investment in biotechnology:

> *Mahoney:* We have perhaps $100 million of the $300 million per year in discovery—basic research. And $200 million in one form or another or applied research. Of the biotechnology area, virtually all of it is a component of the $100 million (basic research)—say $30 million or thereabouts. So about one third of our basic science is going into biotechnology.

It was the vision of this earlier CEO, Mahoney, that provided the knowledge base for the next CEO, Shapiro, to build upon for Monsanto's strategic future: " 'Ultimately, we'd love to figure out how to replace chemical-processing plants with fields of growing plants—literally, green plants capable of producing chemicals. We have some leads: we can already produce polymers in soybeans, for example. But I think a big commercial breakthrough is a long way off. . . . I am not one of those techno-utopians who just assume that technology is going to take care of everyone. But I don't see an alternative to giving it out best shot. . . . The market is going to want sustainable systems, and if Monsanto provides them, we will do quite well for ourselves and our shareholders' " (Magretta, 1997, p. 84).

This vision by Monsanto's CEO illustrated the important contribution that technology-savvy leadership can make toward a firm's long term future: "Sustainable development is going to be one of the organizing principles around which

Monsanto and a lot of other institutions will probably define themselves in the years to come" (Magretta, 1997, p. 84). Yet Monsanto's vision had not taken into account the possibility that some portions of society would view agricultural biotechnology as a intrinsical evil, because the scale and extent of the possible impacts on the environment were not easily foreseen. Later, when Monsanto released a genetically altered corn, some worried that genes in the corn would affect butterflies and humans. The U.S. government banned the corn for use in human food. Monsanto also had a strategy for placing a sterility gene in all Monsanto bioengineered seeds to ensure sterility of any plant growing from the seed. This strategy was to ensure that farmers would have to buy all future seeds from Monsanto. This strategy was viewed as an intrinsically evil technology to make life deliberately sterile. After public opposition to this strategy, Monsanto renounced it. Monsanto's introduction of biotechnology-altered agricultural products hit a public relations roadblock in the 1990s. Technology slammed into ethical worries.

GREEN STRATEGY

Historically, early practices of industrialization and agricultural mechanization did not address their impact on the environment. Since the physical basis of all economy comes from nature, how we have used nature in industrialization and agriculture has dramatically altered the environments of the world. As matter is conserved in the universe (except when it is converted to energy), materials taken from environments for industry and agriculture are returned eventually to environments in some form. Also, industrial processes of extracting energy from nature alter physical environments. Now the question is: How sustainable in nature are the present forms of industrialization?

We know that the extensive clearing of forests for agricultural land has affected soil conservation and the oxygen–carbon dioxide cycles of the Earth. We know that industrialized economies' emissions of CO_2 to the atmosphere make a significant contribution to global warming. We know that the vast expansion of human population has produced massive species extinctions in modern times. We know that certain industrial and agricultural chemicals have devastating effects on living creatures and even on the ozone layers in the upper atmosphere of the Earth. We know that extensive irrigation results in increasing salinity of the soil. We have seen major regional contaminations due to toxic wastes, nuclear waste, and nuclear accidents. A major challenge to the world's continuing industrialization and agriculturalization is to create naturally sustainable industrial economies and biologically diverse environments. These will require new and improved industrial and agricultural technologies and policies. If the industrialization of the world is to continue as a positive societal force, an ethical good, the creation of technologies for industry and agriculture that *preserve* the environment should be one of the highest priorities of future technological innovation.

In the 1990s the adjective *green* became popular to indicate business strategies that included environmental concerns, such as the phrase *green manufacturing*.

Green was seen as a good color because plants were green and the use of technology in industrialization had been uprooting nature's plants—the degreening of the environment. We use this adjective to characterize business strategy that includes a focus on environmental concerns: a *green business strategy*. We call a technology strategy that includes environmental concerns a *green technology strategy*.

To create sustainable economies on Earth in the long term, it will take a lot more green thinking in future industrialization strategies than in those of the past. We should wish our posterity to occupy an industrialized Earth also green with forest and meadows and farms and wild preserves and fertile seas. After the 1992 United Nation Earth Summit in Rio de Janeiro, the idea of green long-term economic development began to gain currency among business leaders. In the 1990s, there were about 5.8 billion people in the world, with population growth projected to double over the next 40 years. Over half of the people live in abject poverty, a bitter kind of subsistence living worse than in preindustrial tribal conditions.

For example, Stuart L. Hart emphasized three types of economies in the 1990s: a market economy, a survival economy, and a nature economy (Hart, 1997). The *market economy* is the world of commerce, comprised of the economies of both developed and developing nations. *Survival economies* exist in developing nations in rural areas and on the outskirts of cities, with people struggling to grow food and find water and fuel; they participate in the local market economy in marginal ways. A *nature economy* is the natural environment in which primitive peoples have survived. In a natural economy, the environment's natural systems and resources are utilized by primitive tribes without substantial alteration of the environment. The market and survival economies both have a severe impact on the naural environment, often changing it very negatively. The Earth Summit emphasized that a future world economy should include an understanding of the various economies coexisting in the world market and survival economies and their impact on the natural environment (the natural economy).

One way to measure the impact of a market economy on the environment has been called an *ecological footprint* (Meadows, 1996). For example, in the United States it took 12 acres of natural economy to supply an average U.S. citizen's needs compared to 8 acres for a citizen of the Netherlands and 1 acre for a citizen of India. The ecological footprint was one way of expressing the fact that different nations with their different standards of living and different uses of technologies were using different levels of the earth's resources.

Both market and survival economies were being very hard on the natural systems of the Earth's environment. For example, some 15% of topsoil in the United States was lost to erosion over the decades of the 1970s and 1980s. Intensive irrigation of land, such as in the western United States, was increasing the salinity of soil. In survival economies, cutting down trees and shrubs for fuel denuded watershed, and overfishing of offshore areas depopulated fish. In market economies, cutting down of forests for timber and agriculture was eliminating vast areas of forest. Excess commercial fishing in the second half of the twentieth century depopulated the world's fish resources. Some have estimated that 40% of the planet's net primary productivity was already being utilized by then current technologies.

Moreover, the extinction of life's many diverse species has proceeded at a terrible rate in the twentieth century—as large or larger than previous mass extinctions as seen in the fossil record.

Historically, the driving factors behind the impact of economies on the environment were both sociological and technological. The primary sociological factor is the explosive growth of the Earth's population due to (1) the reduction in infectious diseases and improved public health, (2) lengthening of the average life span due to improved medical practices, and (3) high birth rates. Population dramatically increased as people lived longer and high reproductive rates continued. The primary technological factor was the development of industrial technology without regard to environmental impact.

NATURAL ECONOMY AND SOCIETAL NEEDS

The concept of natural economy recognizes that all societies require nature to provide their basic physical and biological needs. Basic societal needs include the following:

- Air
- Water
- Food
- Shelter
- Energy
- Materials
- Transportation

- Communication
- Health
- Education
- Recreation
- Defense
- Justice

This list shows the variety of needs for which different technologies are used. Accordingly, one can understand that there are no obsolete societal needs, only obsolete technologies and industries that provide these needs. Green technologies, industries, and economies provide the basic societal needs in ways that alter the environment less than do polluting technologies, industries, and economies. When a societal need can be better satisfied by substituting a superior technology, the industry for which the inferior technology is a core technology will become obsolete. Thus the idea of natural economy is one toward which the market economy must progress if sustainable economy is to become possible in the future.

The ideas of wealth and natural economy need not be in conflict. What matters is how technology affects the kinds of wealth in the natural economy, which includes humanity. Technology can have an impact on human wealth in many ways. One kind of wealth for humanity that technology makes possible is *functional capability*. For example, in ancient times, travel was expensive, time consuming, and dangerous. Today, with modern commercial jets, ordinary citizens and businesspeople may travel between Europe, Asia, Africa, and the Americas relatively inexpensively, quickly, and safely. Thus, technology can increase societal wealth through new functionality.

A second kind of wealth for humanity that technology makes possible is the *availability of resources*. For example, in the ancient civilizations of the Greek city-states and the Roman empire, only the kings were allowed to wear purple clothing. This was because purple dye came from marine animals and was economically rare. Similarly, in ancient China, the color yellow was restricted to use by the Emperor or Empress. After the invention of artificial dyes in Germany in the nineteenth century, all colors became available cheaply. Today, anyone can wear purple. Thus, technology can increase societal wealth by lowering the cost of resources.

A third kind of wealth for humanity that technology makes possible is the *time required to accomplish tasks*. The economic's term for this is *productivity,* the amount of labor required to produce a unit of economic good or service. For example, one impact of the early industrial revolution was the application of mechanical power to the spinning and weaving required to produce cloth. The quantity of cloth output that an individual worker could produce was increased extraordinarily by the introduction of powered textile machinery. Over time, the cost of clothing decreased with respect to other goods and services because of less labor was required to produce cloth. Thus, technology can increase societal wealth by improving productivity.

A fourth kind of wealth for humanity that technology makes possible is an *increase in knowledge*. Technology improves the range and accuracy of perception through the creation of research instrumentation. Technology improves the theory of knowledge through the computational capability to model or simulate the dynamics of systems. Thus, technology can increase societal wealth by providing more knowledge and culture.

A fifth kind of wealth for humanity that technology can make possible is to *improve the environment* in sustaining the scope of life. For example, soil and forest can be preserved through proper technologies of cultivation and harvest and proper management for long-term sustainability. Regions of the world abundant in natural plant and animal life can be set aside as natural parks, economically sustainable through properly controlled tourism.

The problem of the natural and market economies is therefore not one of inherent conflict but rather, a multidimensional value problem in strategy. Sustainable economy strategy should be formulated as a many-dimensional problem in seeking technologies that simultaneously improve societal capabilities, resources, productivity, knowledge, and environmental quality. This multidimensional strategy problem for green technological innovation is not impossible—just a greater challenge for management, government, and research leadership.

INDUSTRIAL PRODUCTIVITY AND ENVIRONMENTAL REGULATION

Technological innovation can improve productivity and/or product quality while reducing the impacts of production and products on the environment. One principle for green technological innovation is that any pollution from industry equals industrial inefficiency. For example, in the Monsanto case, to produce 5 million pounds of insecticide product, 2,500,000 pounds of industrial waste was generated. Any reduction in such waste (potential pollution) would increase productivity.

A second principle of green technological innovation is that products can be redesigned to improved quality when the environment is part of the concept of product quality. For example, in the Monsanto case, redesigning the potato genetically as the NewLeaf potato eliminated the need for insecticide against the Colorado potato beetle. This new potato was a higher-quality product than a potato lacking resistance to the potato beetle.

Scientific techniques from the environmental sciences and biology provide many tools for analyzing the impact of industrial and economic and population activities on the environment. For example, cycles of such materials as carbon, oxygen, and nitrogen can be used to measure and predict the impact of economic activities. Combinations of material and energy cycles in techniques such as biomass systems can also be used to estimate the impact of energy use and extraction on biological resources.

Environmental analysis should be a routine component in formulating industrial technology strategy. The benefits can be of several types, including:

- Lowering production costs by reducing waste
- Lowering environmental cleanup costs by reducing pollution
- Increasing the quality of brand recognition by being a good corporate citizen
- Increasing competitive advantage through higher-quality products
- Investing in conservation of long-term resources

The form of governmental regulation can have important effects on how industry contributes to environmental improvement. For example, Porter and van der Linde (1995) argued that properly formulated environmental regulation can foster the kind of technological innovation that improves both economy and the environment: "The need for regulation to protect the environment gets widespread but grudging acceptance: widespread because everyone wants a livable planet, grudging because of the lingering belief that environmental regulations erode competitiveness. . . . This static view of environmental regulation . . . is incorrect. . . . Properly designed environmental standards can trigger innovations. . . . Ultimately, enhanced resource productivity makes companies more competitive, not less" (p. 120).

Porter and van der Linde suggested several criteria for a proper form of environmental regulation that promotes positive and effective innovative responses by industry for green technologies:

1. Create pressure that motivate companies to innovate.
2. Allow time for phasing in innovations.
3. Educate companies about resource inefficiencies and potential technological improvements.
4. Support research leading to technologies that promote environmentally friendly products and production.
5. Level the economic playing field for commercial transitions to environmentally friendly technologies

Poorly administered environmental regulation may not substantially improve the environment. For example, in 1986, the U.S. government passed the Superfund law for pollution cleanup of the environment. However, in a 1992 study, the Rand Institute for Civil Justice learned that 88% of the money paid by insurers on Superfund claims between 1986 and 1989 went for legal and administrative costs; only 12% was used for actual site cleanups (Acton and Dixon, 1992).

CASE STUDY:

Human Genome Project

Next we turn to the technology–ethical issue of human cloning. Use of this technology was made possible as the human genome became decoded and understood. We recall that the mechanism of reproduction of life was understood during the course of the twentieth century through the discovery of the structure and processes (physical morphology) of the DNA molecule. We also recall that the sequences of bases in the DNA's double helix structure could be interpreted as having a functional value to life as a "language" of instruction for DNA replication and protein synthesis. Biology needs representations of both mechanism and function to describe living phenomena. The next big project undertaken at the end of the century was to write down (decode) the set of instructions for replication of an entire human being, the human genome.

The human genome project was pursued by two independent and competing scientific groups in the 1990s. One group was an international public consortium consisting of the National Center for Biotechnology in the United States, the European Bioinformatics Institute, and the DNA Data Bank in Japan. This group was funded for 15 years at a total of about $3 billion. The competing group, operating on private industrial funding, was the company Celera Genomics in Rockville, Maryland. In June 2000, both groups announced a first draft of the chemical sequences in all the DNA of the human genome. A refined version was expected to be completed by 2003.

We recall that the mechanism of transmitting heredity in animate objects (living organisms) is the structure and operation of the DNA molecule as a double-stranded chain in the form of a paired helix. The two strands are each sequences of four nucleotide units, or bases: adenine, thymine, and guanine, cytosine (A, T, G, C). These bases bind to each other in pairs, with adenine (A) across the helix from thymine (T) and cytosine (C) across from guanine (G). This base pairing in chemical attraction is the physical mechanism for structuring the DNA molecule. It is also the mechanism for reproducing the cells of an organism and for creating the proteins and operations of an organism. The mechanism of the structure of the DNA and the mechanistic process of the chemical attraction between A and T and between C and G provide the physical process for living reproduction. Use of this process by the living cell is the functional value of the process to the cell: the function of cellular reproduction. The DNA strands are ordered as functional sequences of genes that control a biological function. The complete genome of a human consists of 23 contiguous sequences of genes as chromosomes. For example, when a cell divides, each daughter cell receives one set of chromosomes as single strands of DNA and constructs on each a complementary strand, so that each new cell ends up with a complete set of chromosomes consisting of continuous genes on DNA helixes. Moreover, the base pairing of a DNA strand chemically creates a complementary RNA molecule, which then constructs a protein. These proteins provide the material for the cell and its operations. In sexual reproduction, the mother provides an egg with one set of chromosomes as a set of single-stranded DNA chromosomes that mate with the father's one set of chromosomes. Then the child has a new genome with half contributed by the mother and half by the father.

The success of the human genome project depended on scientific technology, automating the sequencing of DNA. The scientific technique of DNA sequencing, the polymerase chain reaction (PCR), was invented in 1977 by Frederick Sanger at Cambridge University in the United Kingdom. DNA sequencing begins with the preparation and replication of a segment of DNA to be sequenced. Next, many partial but shorter fragments of the segment are also replicated, each part with one less base than the prior part. Then the ending base (A, T, G, or C) is tagged with a fluorescent chemical. These fragments are then all separated according to length and each recorded as to its length and ending base. Finally, the sequence of the bases in the original DNA segment can be reconstructed by ordering the different lengths of the DNA fragments with each identified base. The DNA segment sequence is the sequence of ending bases according to position in the DNA sequence by length of the replicated fragment.

In 1989, the sequencing of DNA was not automated and required much technician labor, with a cost of about $2 to $5 for identifying a base in a DNA sequence. Automation was essential, as the planned budget of $15 billion for the U.S. Human Genome project had been formulated at an estimate of $0.50 per base identification. Then Daniel Cohen, director of a scientific laboratory, Genethon, in Paris, Francis, had begun to made significant progress in automation. Cohen implemented 18 machines, each of which could output identification of 8000 bases per day. He had designed custom robots to place 450,000 samples of DNA per day on a membrane for reactions. Genethon published its physical map of the human genome in 1993 as an ordered set of large blocks of DNA (yet unsequenced within a block). This was used to break the human genome down into a set of blocks.

Progress in DNA sequencing machinery for automating the sequencing technique advanced rapidly, In the year 2000, a DNA sequencing machine could automatically produce 330,000 bases per day. Once a fragment of a DNA sequence was automatically analyzed as to the order of its bases, a computer program could assemble these bases into the DNA sequence. The assembly software was developed by Phil Green and his colleagues at the University of Washington Genome Center in Seattle, Washington. One program, Pred ("Phil's read editor"), looked at the fluorescent signal output from the DNA sequencers, identified the base according to the fluorescent signal, and analyzed the quality of the signal to estimate the accuracy of the identification of the base. A second computer program, Phap ("Phil's revised assembly program"), assembled the DNA segment sequence by overlaying the fragment pieces to get an end-to-end sequence of the segment. A third program, Consed ("consensus visualization and editing program"), allowed a human finisher to see and edit the sequence of the DNA for storing in a DNA library of the genome.

In sequencing DNA automatically, the new technology made possible the historic race between the human genome project and Celera. There were about 30,000 human genes. Sequencing DNA was a first step to understanding how DNA controls the construction and operation of a human being. Next, the genes within each chromosome needed to be identified. Then gene functionality must be established. This is still a very complex work in progress. In 2003, information on the genome project could be found at the Oak Ridge National Laboratory website http://www.ornl.gov/ngmis/project/hgp.htm

HUMANITY DESIGNING HUMANITY

In addition to the sensible use of technologies as useful means, some technologies provide means so horrible that any use can be seen as evil—a technology as an intrinsically evil mean. Examples of this in military applications include the use in warfare of poison gas, biological diseases, and nuclear weapons. For example, in the early 1960s, the physicist Edward Teller, famous for his key role in the U.S. invention of the nuclear fusion bomb, was sometimes called the "father of the hydrogen bomb." The H-bomb was invented independently and nearly simultaneously in the 1950s in the United States and in the USSR, with Edward Teller and Stanislas Ulam as the U.S. inventors and Andre Sakharov in the USSR. After Teller retired from directing the U.S. hydrogen bomb research and design center in Livermore, California, he became a professor at the University of California at Davis. At the time he proposed a program for peaceful uses of the atomic bomb, such as excavating large canals. But the public climate of the time was that an exchange of nuclear bombs between the United States and the USSR would exterminate life in the northern hemisphere. The idea of an atomic explosion was so repulsive that the U.S. public did not want to use atomic bombs for anything. Atomic explosions were viewed as an intrinsically evil means.

In our review of the ethical basis of technology, we can look at technology as both ethically neutral and intrinsically evil means. As an example of an issue regarding technology as an ethically neutral means, we looked at sustainable economy. As an example of an issue of technology as a possibly intrinsically evil means, we looked at human cloning.

One of the enduring truisms about life is that power corrupts and absolute power corrupts absolutely. The ethical dilemma of our modern times is that human power over nature has been approaching the absolute. Technologies used in industrialization and agriculture have been transforming the environments of Earth, even affecting global weather patterns. Now biotechnology is providing human beings with the technical power to design life itself, even human life. As the human genome is understood, human DNA can be deliberately altered, then inserted into a human cell and cloned into a human being.

We recall Figure 1.2, here reproduced as Figure 11.1, which assumed that the relationship of humanity to nature through technology could be viewed as separable. Humanity was a separable part of nature. Now we can see that the idea of human separateness from nature is really too simple. How can humanity be separable from nature if humanity can design nature itself? This is the ethical dilemma in the idea of the biotechnology of human design. Might cultural evolution in the form of biotechnology alter permanently the biological evolution of the human species? Will this be a means toward an improved nature and an improved humanity? Or is it just a nightmare of an intrinsically evil technology?

The intellectual problem arises from the simplicity of one of the great ideas of modern science—the essential simplicity of the biological idea of evolution. The idea of evolution presupposes that the environment is separate from the species. In the theory of biological evolution, species adapt to environments. But what happens to the scientific principle of evolution when a particular species can alter the environment instead of adapting to it genetically? We have called this *cultural evolution*. But are the two ideas of biological evolution and cultural evolution really separable? What happens to the scientific principle of evolution (biological or cultural) when one a particular species can alter its own evolutionary genetic history?

As the twentieth-first century began, laws were being formulated in the United States to ban human cloning. The cloning of embryos for stem cell research was restricted by the U.S. government. Techniques for altering genes in an attempt to cure some diseases were being formulated as *gene therapy*.

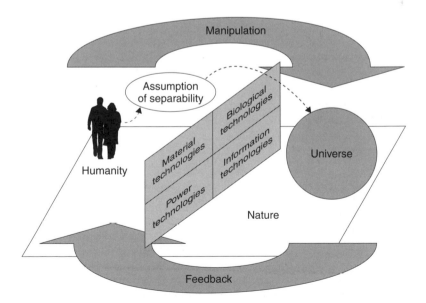

Figure 11.1. Humanity interacts with nature in the universe through technology.

The ideas on technological progress that we have been using in this book are the basic ideas of biological and cultural evolution. But now they turn out to be not so simple. Biotechnology has coupled what historically was separable—biological evolution from cultural evolution. The future of technology and society will be a time when evolution is mixed up—biological and cultural evolution interacting and raising ethical questions.

SUMMARY AND PRACTICAL IDEAS

In case studies on

- Monsanto's strategy for sustainable economy
- Human Genome Project

we have examined the following key ideas:

- Technology as ethically-neutral or intrinsically-evil
- Green industrial strategy
- Natural economy and societal needs
- Environment and wealth
- Industrial productivity and environmental regulation
- Humanity designing humanity

The practical use of these ideas provides approaches to considering the ethical implications and public policy issues of technological innovation.

FOR REFLECTION

Select an industry and list the core technologies in production. From a government or industrial association source, find the types and volumes of industrial waste produced. What kinds of technological innovations would be desirable to reduce waste, and how much savings could be created by reduced waste?

12

RADICAL INNOVATION

INTRODUCTION

We recall that there are three scales of technological innovation: (1) basic innovation, (2) incremental innovation, and (3) next-generation-technology innovation. Both basic and next-generation technological innovation are radical innovations and have dramatic disruptive impacts on industrial structure. For this reason these have been called technological *discontinuities* in technological progress. We also recall that discontinuous technological change creates new industrial structures (from basic innovations) and alters existing industrial structures (from next-generation technology innovations). This is in contrast to the kind of smaller change provided by incremental technological innovations that tend not to alter industrial structure. Incremental technological change reinforces an existing industrial structure. We have also emphasized how the modern era was built on a new research capability of scientific technology. Now we look at the process of innovation for a technology discontinuity—the radical innovation process. Radical innovation brings new knowledge from science to market in four key stages: scientific feasibility, technical feasibility, functional prototype, and product life cycle.

Nuclear Fission Energy

We now look at a dramatic case of radical innovation—nuclear energy. It is a famous case but a terrible one. Nuclear energy put humanity on the edge of extinction—and there, precariously, we still live. The events in this case occurred in the middle of the twentieth century and involved new scientific studies that were quickly focused on technology applications in the military. The nuclear research story began in 1933 with the scientists Curie and Joliot in France. It moved quickly to research by the Italian scientists Fermi and Segré. Further research was performed in 1938 by the German scientists Meitner and Hahn. In 1939 the story goes to the United States, where research done for the Manhattan Project resulted in the first atomic bomb.

In 1933, in France, Irène Curie and Frédéric Joliot, two university scientists, discovered that radioactive decay could be *stimulated* in a chemical element after allowing the element to be bombarded by alpha particles emitted from the natural radioactive decay of another element. (An alpha particle is the nucleus of a hydrogen atom, a proton and neutron bound together by a strong nuclear force.) That radioactivity could be induced in a chemical element by the absorption of particles was an exciting discovery. Immediately, other university scientists, including Enrico Fermi in Italy, began investigating the new phenomenon: "At the beginning of 1934, Enrico Fermi in Rome had started using neutrons from radon-beryllium sources, in lieu of alpha particles, to activate many common elements" (Segré, 1989, p. 38).

A chemical element is determined by the positive charge of the nucleus, the number of protons in the nucleus. An isotope of an element is determined by its nuclear weight, the total number of protons and neutrons. Members of Fermi's scientific team included Edwardo Amaldi, Oscar D'Agostino, Franco Rasitte, and Emilio Segré. The experiments of Fermi's team established that the nucleus of an element could be transformed into a different element (or a different isotope of the element) by the absorption of a neutron and the emission of protons (or neutrons). Fermi's team next bombarded the elements of thorium and uranium with neutrons from a decaying source. They were particularly interested in uranium, because at the time, it was the heaviest of the naturally occurring elements. They hoped to create new elements heavier than uranium when a uranium nucleus absorbed neutrons.

After a neutron was absorbed, radioactive reactions were observed by seeing their decay products. But this induced radioactivity was not much greater than the natural radioactivity of uranium. Thus, it was very confusing to understand just what was happening, as Segré (1989) later wrote: "In Rome, we immediately found that irradiated uranium showed a complex radioactivity with a mixture of several decay periods. We expected to find in uranium only the previously observed (radioactivity) backgrounds, and so we started looking for an isotope of element 93 [a next element above uranium's proton count of 92] produced by an neutron-absorption and gamma decay reaction on the 238 isotope of uranium, followed by a beta decay (p. 39).

They hoped that the 238 weight (238 neutrons and protons) of the uranium element would absorb a neutron to transmute to a 239 weight, and they hoped a neutron would decay to a proton, by emitting an electron. Thus the element of uranium (with 92 protons and weight 238) would be transmuted to a new element (of 93 protons and weight 239). Thereby they hoped to be the first to create and discover a new element created by transmutation of another element. This was the old dream of medieval alchemists—the transmutation of matter. But scientists are human and can make mistakes: "Here we made a mistake that may seem strange today. We anticipated that element 93 would resemble rhenium (element 75). . . . Products of the bombardment were indeed similar to rhenium" (Segré, 1989, p. 39).

They were looking for a new element of atomic number 93 in the products from the experiment that would have a chemistry similar to the known element of rhenium. This expectation was a mistake. Due to it, they did not recognize that what was really happening was that these were products of nuclear *fission*. They just missed discovering the scientific phenomenon of the breaking of an atomic nucleus, nuclear fission. As they were later to learn, some of the fission products were isotopes of technetium (element 43), which also a chemistry similar to that of rhenium. So nature was fooling them. (Of course, scientists never anthropomorphize nature; but if they did, they would note that sometimes in research nature appears to have cooperative moods for scientists and sometimes uncooperative moods.) In this case, the confusion meant that Fermi and his group didn't discover nuclear fission.

Yet the scientific puzzle was soon to be solved by other scientists. In 1935, after Fermi's group had reported their results to the world, Otto Hahn and Lise Meitner at the Kaiser Wilhelm Institute began work to confirm the Fermi group's results. At first, they also were not able to discern the fission phenomenon: "Hahn and Meitner, working with a neutron source about as strong as the one used in Rome and Paris, started by confirming our Rome [the Fermi group's] results. This is some what surprising because they applied quite different chemistry. Their early papers are a mixture of error and truth, as complicated as the mixture of fission products resulting from the bombardments" (Segré, 1989, p. 40).

Meanwhile, in Paris, Irène Curie was working with Hans von Halban and Preiswerk on bombarding neurons on thorium. They found a decay product that chemically lanthanum resembled. But they did not realize that it was lanthanum of atomic weight 141, which was a fission product. They thought that it might be an isotope of actinium, and they just missed discovering fission. By May 1938, the puzzle remained. None of the Italian, German, or French teams had discovered nuclear fission.

Hahn (German) and Joliot (French) met in Rome at the 10th International Chemistry Conference. They discussed Curie's results, and Hahn concluded that something was wrong in the suggestion that the decay product was actinium. Back in Germany, Hahn and Meitner decided to repeat some of Curie's experiments.

In March 1938, Hitler's German army marched into Austria. Since Lise Meitner was an Austrian of Jewish ancestry, she lost the protection of her Austrian citizenship. In danger of arrest by Hitler's Gestapo, Meitner escaped to Holland. Hahn remained in Austria and continued work (with the help of Strassmann). They concentrated on identifying the radioactive component Curie had thought was actinium (but which was really lanthanum). Finally, by December 1938, Hahn and Strassmann had concluded that the radioactive decay products that Fermi and Curie had thought were radium, actinium, and thorium were really barium, lanthanum, and cesium. This was important, for in these fragments were the clues to nature's behavior—the splitting of the atom. Hahn wrote to Meitner about the results: "He and Strassmann . . . were coming 'again and again to the frightful conclusion that one of the products behaves not like radium but rather like barium' " (Stuewer, 1985, p. 50).

After receiving Hahn's letters in December, Meitner continued to think about the puzzle of the fragments. She traveled to Sweden to spend the holidays with friends and with her nephew, Otto Frisch, who was also a physicist. (Frisch also had been forced to leave Hitler's Germany in 1933, going to Denmark to work in Niels Bohr's physics institute; Bohr was the scientist who created the first quantum-mechanical model of the atom.) Frisch arrived in Sweden to holiday with his aunt. On December 28, Meitner received a third and key letter from Hahn; and she showed her nephew all the letters. In the last letter Hahn had asked: " 'Would it be possible that uranium [of atomic weight 239] bursts into barium [with weight 138] and masurium [with weight 101 and now called technetium]?' " (Stuewer, 1985, p. 50).

The key feature that Hahn pointed out was that the atomic weights added correctly: 138 nucleons from barium added with 101 nucleons from masurium totaled the 239 nucleons in uranium. Yet the puzzle continued! The atomic number did not add up correctly. Uranium had 92 protons, barium 56 protons, and masurium 43 protons. The sum of 56 protons and 43 protons was 99 protons instead of the 92 of uranium. Hahn had also written: " 'So some of the neutrons would have to be transformed into protons. . . . Is that energetically possible?' " (Stuewer, 1985, p. 50).

Excited about these speculations, early on the following day the aunt and nephew breakfasted together and went outdoors for a hike. Frisch pushed along on skis, and Meitner walked. They talked about physics. Frisch suggested that a liquid-drop model of the nucleus might be used to explain Hahn and Strassmann's results: " 'I worked out the way the electric charge of the nucleus would diminish the surface tension and found that it would be down to zero around $Z = 100$ (atomic number of 100 nucleons) and probably small for uranium. Lise Meitner worked out the energies that would be available from the mass defect in such a breakup (the splitting of a liquid-drop like mass of uranium with 239 nucleons into fragments with liquid-drop like sizes of 138 and 101 nucleons). . . . It turned out that the electric repulsion of the fragments would give them about 200 MeV (million electron volts) of energy and that the mass defect would indeed deliver that energy' " (Stuewer, 1985, p. 50).

This was their idea. Like a drop of water might split into two smaller water drops, the uranium atom could split into two halves, pushing away from each other by the repulsive positive charges of the protons in each half. Meitner and Frisch's idea was the concept of nuclear fission. Of course, in actual fact, as Frisch and Meitner knew, their idea of a liquid-drop model of a nucleus (nucleons packed together as water molecules pack into a drop of water) was only an analogy. Sometimes in new territories scientists have to explore using analogies. An atomic nucleus contain nucleons that have both positive electrical charges in protons (and therefore a repulsive force between protons) and an attractive strong nuclear force between protons and neutrons that bind them all together. Still overall, when the number of nucleons is small, it takes a particle of large impacting energy to tear the nucleus apart. But as the nucleus gets very large, it may be unstable and undergo spontaneous fission; then even particles of only small energy can distort the nuclear sphere into fissioning halves. And the uranium nucleus is up there in size where large nuclei become unstable. This is why elements with much larger nuclei than those of uranium do not occur naturally. On January 1, 1939, Meitner wrote to Hahn: " 'We have read and considered your paper very carefully, [and] perhaps it is indeed energetically possible that such a heavy nucleus bursts' " (Stuewer, 1985, p. 50).

Very excited, Frisch returned immediately to Copenhagen to tell Niels Bohr about the bursting of the uranium nucleus. Later Frisch wrote to his aunt that upon hearing the news, Bohr had burst out: " 'Oh what idiots we all have been! Oh, but this is wonderful! This is just as it must be!' " (Segré, 1989, p. 42).

RADICAL INNOVATION PROCESS

How does a firm innovate a new basic technology product or a next-generation technology product? This is through a radical innovation process, whose logical steps go all the way from science to economy. Radical innovation begins with the latest in science—new scientific knowledge. In radical innovation there is a direct connection between how scientists experimented with a new scientific phenomenon and how inventors created new technology using the phenomenon. The instruments of science have often become the devices of new technology (the computer, the maser and laser, and recombinant DNA techniques are some examples). Thus, historically, in radical innovation there has often occurred a kind of linear transfer of knowledge and technique from science to technology. We can represent this linear logic schematically as one kind of innovation process for radical innovation, as represented in Figure 12.1. The steps that can link the passing of information from science to economy are listed down the middle of the figure as:

- Discovery and understanding
- Scientific feasibility
- Invention
- Technical feasibility prototype
- Functional application prototype
- Engineering prototype
- Manufacturing prototype
- Pilot production process
- Volume product production
- Product improvement

The first step in moving from nature to product is the *discovery* in nature of an interesting phenomenal object, process, property, or attribute. The next step is a sufficient *understanding* of the phenomena to know how variables in the phenomena influence other variable to intervene systematically and consistently and alter the course or state of the phenomena. With an awareness of the existence of nature and understanding of intervention in such existence, one can *invent* a way to

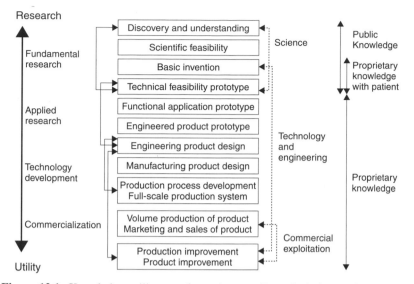

Figure 12.1. Knowledge–utility transformation as a linear-logic innovation process.

manipulate the nature that may be useful to humanity. For example, in the invention of the laser, knowledge of the discrete quantum states of atoms and the transitional photon yielded between states provided a knowledge base for Townes and others to invent ways of stimulating the generation of coherent light—light amplification from stimulated electronic radiation (laser).

Invention is an idea and must next be reduced to practice; that is, it must be shown that the technical idea is feasible and can be demonstrated. This is the purpose of a *technical feasibility prototype*. This is also sometimes called a *proof-of-principle technical demonstration*. However, the mere technical feasibility of something that works is still not enough to proceed to a product. One must next establish that the technically feasible invention can work well enough for an application. Working well enough means that the performance of the invention is sufficient to be useful for a specific purpose, and this is the point of next creating a *functional prototype*.

After a functional prototype shows sufficient performance, a product design can begin. The first goal of a novel product design is to establish completeness in product requirements. In addition to sufficient performance, a product idea must embody necessary features, appropriate size and shape, be safe and convenient to operate, and be maintainable and repairable. The goal of the *engineering prototype* is to design and demonstrate that the product can be engineered as a complete product. The next step on the way to product production is the creation of a *manufacturing prototype*. In this prototype the engineering design is modified for manufacturability, considering tolerances, assembly, quality, and cost. (If a new production process must be innovated to produce the product, that process should be tested as a *pilot production process*.) After the manufacturing prototype is tested and pilot production processes are developed and tested, *volume production* of the product begins.

There are several points to note about this spectrum of research, which connects nature to product:

1. There is an overlap of research activities between science, engineering, and technology.
2. There is an overlap of research motivation between basic, applied, and developmental research.
3. There is an overlap of intellectual property interests between proprietary and nonproprietary research.

These overlaps make science and engineering and technology dynamically interacting processes. Science is a general approach to understanding nature. Technology is a generic way of providing functional capability of doing things. Engineering grounds technology in scientific understanding and general principles. Product development focuses on a specific embodiment of technology and engineering understanding and principles for a business purposes of providing value-added sales to customers. Accordingly, for the purposes of forecasting and planning technological change, management must understand the differences between science, engineering, technology, and product and how to gain cooperation among these activities. And from a management perspective, the problems for technology strategy begin with the fact that product development is in the domain of the strategic business units, and the research basis for a next-generation technology is in the domain of the corporate research laboratory, and these are positioned differently in the logic of innovation.

> **Unless there are good and close working relationships between corporate research and product development in the strategic business units, product development may not properly be led by visions of next-generation technology.**

Also, the logic of the process of radical innovation indicates that one cannot make a simple distinction between research for engineering and technology and for product development without also taking into account the intellectual property rights of the research. Information in product development about product design and production processes are very important proprietary information to the firm designing, producing, and selling the product. It is in this product-specific knowledge wherein lie the competitive differences between firms. Yet next-generation technologies occur at an industrial level. All firms in an industry share the generic knowledge about technology and engineering that underlay product and production design and control. The generic level of engineering/technology knowledge is generally nonproprietary. The overlap between proprietary and nonproprietary research usually occurs in the functional prototyping stage. It is here where the critical decisions about the next generation of technologies, the directions and pace, occur.

> **Management of strategic technologies over the long term requires firms to participate in the national development of technology progress.**

Nuclear Fission Energy

The discovery and understanding of nature was indeed the first step toward nuclear power. Finding the phenomenon of the splitting of atoms and release of energy in the splitting was new nature, never before seen directly by humanity. The discovery was unexpected and tricky. The atom weights of the split nuclear fragments added up, but the atom number did not. Neutrons were converted to protons. Another new discovery! This is what makes science exciting—surprise! Not a simple surprise, but a complex puzzle in which the sophistication of nature is discovered. The discovery of the fission of an atom from bombardment with nuclear particles was that it also released energy. In chemical terms, this was an exothermic reaction—giving off energy. What kinds of technologies can use natural processes that gave off energy? A bomb, for one thing! The differences between the discovery of nuclear fission and its transformation into bomb technology can illustrate the differences between science research and engineering research. We next follow the research toward the development of the atomic bomb.

A few days after Frisch informed Niels Bohr about the discovery of fission, Bohr left for the United States to attend a physics conference, carrying with him Frisch's momentous news. Bohr sailed with his wife on the ship *Drottningholm,* and accompanying them was a colleague, Leon Rosenfeld, who later recalled the voyage: "As we were boarding the ship, Bohr told me he had just been handed a note by Frisch, containing his and Lise Meitner's conclusions; we should 'try to understand it.' We had bad weather through the whole crossing, and Bohr was rather miserable, all the time on the verge of seasickness. Nevertheless, we worked very steadfastly and before the American coast was in sight Bohr had got a full grasp of the new process and its main implications" (Stuewer, 1985, p. 52).

In North America at the Fifth Conference on Theoretical Physics in Washington, DC, on January 26, 1939, Bohr announced the news of fission. The conference began in the afternoon in Room 105 of Building C of George Washington University, where "today a plaque outside the lecture room commemorates this historic meeting. It was there . . . that the bombshell burst" (Stuewer, 1985, p. 54). The report that came from this conference read: "Certainly the most exciting and important discussion was that concerning the disintegration of uranium of mass 239 into two particles each of whose mass is approximately half of the mother atom, with the release of 200,00,000 electron-volts of energy per disintegration" (Stuewer, 1985, p. 54).

This was an enormous release of energy per atom! Leo Szilard, a young physicist attending that conference, began thinking about the military implications of nuclear fission. Having left Germany after the Nazi takeover in 1933, Szilard emigrated to Britain and then to the United States. Szilard worried that the Germans would be the first to turn the discovery into an atom bomb. Szilard thought American action urgent. He drafted a letter to the President, Franklin Delano Roosevelt. But first he took the letter to Albert Einstein. Albert Einstein was the most famous scientist in America, and Szilard hoped the President might read a letter from Einstein about the potential development of nuclear weapons. Roosevelt did and responded by creating the Manhattan Project, which was to build the world's first atomic bomb.

In historical perspective, one can see that Szilard's sense of urgency was warranted. When nuclear fission was announced, anyone who understood the physics with its immense release of energy could immediately imagine its application as a technology of power. The implications of new science for new technology frequently has been obvious to scientists at the time of the discovery. Scientists in Germany also saw the explosive potential of a nuclear fission bomb, and the Nazi government in Germany began an atomic bomb project at the same time as the Manhattan Project in the United States.

The Manhattan Project was organized under the U.S. Army, and General Grove was assigned to manage it. With advice from the scientific community, he chose Robert Oppenheimer of the University of California to head the research activities. Oppenheimer was one of the few American-born physicists capable in nuclear studies. Many of the other scientists in the Manhattan Project (such as Szilard, Fermi, Bethe, Teller, etc.) were foreign born and therefore considered by Grove as ineligible to run the secret war project. A new and secret U.S. government research laboratory was built in Los Alamos, New Mexico.

For nuclear fission power, the first key was how to experiment with the fission phenomenon, that is, how to set up a self-sustaining nuclear chain reaction. This assignment was given to Enrico Fermi, who was then on the faculty of the University of Chicago, having left fascist Italy in 1938.

As often happens in the history of technology and science, an inventive idea occurs to more than one inventor. The invention of the nuclear reactor was conceived independently by several people. But it turned out that the form of the reactor would make the difference between the American success in developing the first atomic bomb and the German failure. Comparing these two efforts illustrates the difference between success and failure in research. Part of the explanation of this difference is that with the exception of Werner Heisenberg (who was to lead the German atomic bomb effort), Otto Hahn (who was to help in the German atomic bomb effort), and Niels Bohr (who remained neutral in Holland), all the other really great atomic scientists of the time had left Germany, escaping Hitler's anti-Semitic policies. It was the émigré scientists who helped Oppenheimer build the U.S. atomic bomb.

In Germany after Hahn had learned of Meisner and Friche's conclusion about the fissioning of the uranium atom, Hahn told a colleague, Wilhelm Hangle. In April 1939, Hangle delivered a lecture in Germany on "the possible applications of nuclear energy making use of a uranium–graphite pile" (Walker, 1990, p. 53).

After hearing Hangle's lecture, Georg Joos (one of Hangle's colleagues at the University of Gottingen) transmitted a report of the lecture to the German Ministry of Education, which passed it on to Hitler's Reich Research Council. Also, independently, Nikolaus Riehl (who was an industrial physicist at the Auer Company in Berlin and a former student of Hahn and Meitner) brought nuclear fission to the attention of the German Army Ordinance Office. Also independently,

Paul Harteck and Wilhelm Groth, physical chemists in Hamburg, wrote to the German Army about the possibility of making nuclear explosives. So the idea of an atomic bomb was in the air once nuclear fission was discovered.

> **Often, a possible application of a scientific phenomenon has been envisioned immediately (even by different people) after the discovery, and this is the link between the innovation process steps of (1) discovery and understanding, (2) scientific feasibility, and (3) invention.**

The German Army responded promptly, holding two workshops on uranium in April and September 1939, followed by two more conferences in September and October. These conferences resulted in the German Army planning research for an atomic bomb. Like the Manhattan Project, the German Army project planned first to create a sustained nuclear chain reaction. The Army Ordnance Office distributed work on nuclear fission among several institutes: "Throughout the war, Gerlach, Bothe, Heisenberg, Harteck, Hahn and Klaus Clusius—all scientists of first rank—were involved in the scientific work and the administration of the German fission project" (Walker, 1990, p. 54).

The problem was the demonstration of the technical feasibility of a nuclear reactor machine. The idea of such a machine was envisioned as a mixture of uranium interspersed with a material moderator to control the diffusion of neutrons between uranium atoms. However, there were technical issues. What kind of uranium and what kind of moderator? In what geometry should the uranium and moderator be arranged? These kinds of research issues are not scientific but technological—*engineering research issues* about the composition and configuration of the new technology. They formed the basis of early research planning for both the U.S. and German atomic bomb projects.

In the German research institutes, several experiments were begun, using different moderators, including graphite and heavy water. Walther Bothe tried using graphite as a moderator. Paul Harteck in Hamburg tried using dry ice (carbon dioxide) as a moderator, with uranium powder arranged in layers within a sphere. Later, others tried using paraffin as a moderator. Bothe's early measurements and calculations led him to conclude that graphite would absorb too many neutrons to allow the reactor to work. Subsequent experiments by Wilhelm Hanle showed that was wrong, because there had been impurities of boron and cadmium in the graphite that were really the strong absorbers of thermal neutrons. Then in August 1940, Robert Dopel in Leipzig demonstrated that heavy water made an excellent moderator for thermal neutrons. Heisenberg calculated that a uranium machine relying on graphite would require much more uranium and much more moderator than a machine relying on heavy water.

German Army Ordnance, based on Heisenberg's recommendation, decided to construct a nuclear reactor using heavy water as the moderator. This was a critical and incorrect *engineering* decision. It was the reason that the Germans did not succeed in developing an atomic bomb. (The American project under Fermi using carbon as a moderator would be successful.) The Germans made their choice mostly because of economics and supply. Several German scientists were already expert in dealing with heavy water and had been instrumental in persuading a Norwegian company to produce heavy water in Norway. After Germany occupied Norway, the Norwegian company, Norsk-Hydro, was ordered by Nazi officials to increase its annual production of heavy water from 20 liters to 1 metric ton: "Thus the Germans came to focus on developing a reactor fueled by natural uranium and moderated by heavy water" (Walker, 1990, p. 54).

Light water is ordinary water, with two hydrogen atoms and one oxygen atom making up the water molecule. Heavy water is rarer, with its hydrogen nuclei containing one neutron as well as one proton. Heavy water occurs naturally mixed with normal water in the very dilute concentration of one molecule of heavy water for about every million molecules of normal water.

Meanwhile, in the United States, Enrico Fermi was building the first nuclear fission reactor for the Manhattan Project at the University of Chicago. He had chosen graphite as a moderator. The U.S. reactor went critical in 1942, demonstrating that a self-sustaining nuclear fission reaction was possible. Two equally great scientists, Fermi and Heisenberg, had led the reactor research, but Fermi had made the right engineering choice, whereas Heisenberg had not.

The U.S. government research effort achieved the atomic bomb by first creating the experimental research means of studying controlled nuclear fission. This was essential knowledge needed to figure out how to make an atomic bomb. From studies on the probability of capture of a neutron by a uranium atom, they could calculate the size of the uranium needed for a bomb—the critical mass of the bomb.

Fission reactors based on Fermi's invention produce heat by splitting nuclei of the fissile isotopes of uranium or plutonium. These are shaped into fuel elements such as rods or pellets. Fission occurs when an atom of uranium absorbs a neutron and fissions with the release of two neutrons (on average). If at least one of these neutrons is absorbed by a fissionable atom, producing fission again, a nuclear chain reaction is continued. But it was uranium-239 that fissioned well with slow neutrons. The technical problem was how to slow the neutrons to increase their probability of capture by another uranium-239 nucleus for fission.

Fermi used graphite to slow the neutrons. He constructed a graphite-moderated reactor burning uranium in a matrix form, with alternating rods of uranium and graphite. The graphite slowed the neutrons without capturing them. To turn the chain reaction on and off, other alternating rods of borium could be inserted between the uranium rods to absorb the neutrons.

Heavy water was another way to slow neutrons down for fission reactions, but it did not slow down enough of the neutrons to achieve a chain reaction. The difference between the U.S. graphite moderating reactor and the German heavy-water moderating reactor was the *higher performance* of the U.S. design in slowing down sufficient numbers of neutrons to create a chain reaction.

Once U.S. scientists measured the cross section for absorption of moderated-speed neutrons by uranium-239, design of the atomic bomb could begin. In Los Alamos, scientists and engineers designed the first bomb as two half-spheres of uranium-239 located in a gun barrel to be blasted together with dynamite—the "Long Boy" atomic bomb. It was dropped on Hiroshima. The second bomb was in a sphere of uranium-239 surrounded by dynamite—the "Fat Boy" atomic bomb, dropped on Nagasaki. (We recall that von Neumann did the calculations for the dynamite sphere of the Fat Boy and understood the need for electronic calculations when he later heard of the Machly–Eckert computer project.)

ENGINEERING RESEARCH

The difference between research in science and research in engineering lies in the stages from science to technology in radical innovation. Recall that the early steps of the radical innovation process are:

* Scientific feasibility (meaning that nature may make it work)
* Invention (a way to make it work)
* Technical feasibility prototype (a way to make it work efficiently)

We recall that we defined *technology* as knowledge of the manipulation of nature for human purposes, that we defined *science* as the discovery and explanation of nature, and that we defined *engineering* as understanding and design of the manipulation of nature for human purposes. In these definitions, one should note that science is not focused by human purposes, as are engineering and technology. Science is relevant to human purposes but is not focused by it. Science is focused by the ubiquity and generality of existence of nature. (In Chapter 13 we will see that this is the reason that science research, *basic research* as it is usually called in the United States, is difficult to manage strategically in a corporate environment. Corporations need research focused by purpose, whereas scientists are oriented by their profession to disciplinary classes of ubiquitously existent phenomena.)

Although research activities in science focus primarily on discovery and understanding, inventions sometimes occur in scientific research (notably, instrument or materials inventions). When inventions do occur in basic science, they are usually the basis for radically new technologies (such as the laser or transistor).

Next, one should note that whereas technology is the knowledge of how to manipulate nature for human purposes, it does not necessarily imply a deep understanding of nature which allows the manipulation.

> **Engineering differs from technology by focusing on combining the technical knowledge of how to manipulate nature with scientific explanations of nature for design principles of manipulation.**

Technologically focused research concentrates on inventions and their development toward products. Research in the engineering disciplines also focuses on advancing the understanding of nature directly relevant to technology.

> **The better engineers understand nature, the more and clearer design ideas they may have for refining technical manipulation.**

For this reason, engineering research is organized not only on specific technologies but also on areas of phenomena underlying technological sectors, such as mechanical engineering, electrical engineering, chemical engineering, and civil engineering. The importance of understanding the relationships between science, technology, and engineering is that research for radical innovation requires *advancing the underlying science base* of the nature manipulated by the technology, formulated as engineering design principles.

University-based science has often provided the basic invention for a new technology, but industrial engineering research provides the innovation of the new technology. The research spectrum connecting science to technology has been one of the bases for university and industrial research cooperation.

> **The companies that participated in the early stages of a radically new technology have often been the firms that built strong and large organizations to exploit the new products and new markets of the new technology.**

RESEARCH AND COMPLEXITY OF NATURE

In the case of the discovery of fission, we saw that one of the reasons for the difficulties in interpreting the natural phenomenon for fission (and later the great technological barrier) was the small portion of fissionable isotope found in nature. It was only 0.7% of the uranium found in nature. That was nature's doing! Sometimes, nature has been easy to manipulate for technology, and in other cases difficult. The complexities of nature are the reason that research is necessary, so that science can discover and understand nature sufficiently for technological manipulation. The subtle complexities of nature are the sources of all the engineering problems of physical technologies.

> **The more complex the nature used in a technology, the more scientific research is required to understand nature further.**

Even though a scientific phenomenon has been discovered on which to base a new technology and a concept for manipulating the phenomenum as a technology has been invented, there is still a long way to go before a technology can become effective and efficient engineering research. The steps from scientific feasibility toward an effective technology lie in engineering research. Engineering research establishes the basis for an engineering design for using the technology. Only until technical feasibility is established can technology forecasts based only on scientific phenomena become certain. Engineering research is the connection between science and technology. Engineering research deepens and fills in the knowledge base of science necessary to improve an invention that manipulates natural phenomena. Inventions may be made by scientists or engineers of other personnel. But engineering research moves the invention to practical fruition.

For technology, the scientific discovery of nature is insufficient. Technology, as well as science, also requires instrumentation and techniques for experimenting with nature. Experimental instruments and techniques for experimenting with nature have often become the original basis for a new technology (or the basis for production process control).

Until one can experiment with natural phenomena, one cannot manipulate phenomena.

Historically, the invention of scientific instrumentation has led to its later use as technological tools.

Both science and engineering research require developing the instrumental means of observing the phenomena underlying a technology.

CASE STUDY:

Nuclear Reactor for the Navy

After the atomic bomb, the U.S. Navy command turned to the idea of using nuclear fission as a power source for electric submarines. The idea was to use the heat of the controlled fission in the nuclear reactor to warm water to steam to drive a steam turbine, which turned an electrical generator to produce electricity for the electric motors of a submarine. In addition to the neutrons in the nuclear fission reaction, the fission of the uranium nucleus also releases energy in the form of gamma rays (light particles, photons of very short wavelength). This light could be partially absorbed by a coolant that uses the heat to boil water into steam. Thus about a third to a half of the heat from the reaction would be transformed into electricity. The rate of the fission reaction and the rate of heat produced are controlled by a matrix of control materials that (when inserted) absorb the neutrons, preventing their use for fission. This is essentially Fermi's reactor design, surrounded by a coolant to absorb the heat of fission radiation.

After World War II, four engineering alternatives were conceived as possible functional prototypes: (1) the heavy-water reactor, (2) the light-water reactor, (3) the gas-cooled reactor, and (4) the liquid-metal-cooled reactor. Each type is distinguished by different types of coolant, which is essential both to moderation of neutrons and to heat transfer for electrical generation. Because of this, there are also differences in the type of fissionable fuel the reactors could burn and the performance of the reactors:

- The heavy-water reactor could burn naturally occurring uranium ore with its very small percentage of fissionable uranium isotope (uranium-235).
- The light-water reactor could not burn naturally occurring uranium ore but required processed ore that had a substantially increased ratio of fissionable uranium isotope (uranium-235).

- The gas-cooled reactor also required enhanced uranium but operated at higher temperatures, thereby increasing the efficiency of the conversion of heat to electricity.
- The liquid-metal-cooled reactor could not only burn naturally occurring uranium ore but could produce more fissionable material—it could breed new fuel.

Which reactor type was best? It depended on the application. In nuclear fission reactor design, there were two principal applications: nuclear-powered submarines and nuclear-powered electric generating plants. The first application was nuclear-powered subs.

At the end of the war, a new U.S. law, the Atomic Energy Act of 1946, established a new civilian agency, the Atomic Energy Commission (AEC), to control and promote atomic energy. In the meetings of this commission, research for the new technology was planed. In the summer of 1947, the General Advisory Committee of the AEC held a meeting at the Oak Ridge National Laboratory: "The Committee included Conant, DuBridge, Oppenheimer, Fermi, Teller, Rabi and Seaborg, world renowned scientists and participants emeritus in the Manhattan District, who had been appointed by the President (Truman) to advise the new Atomic Energy Commission on the technical goals of its programs. Temporarily assigned to Oak Ridge at the time was a group of naval engineering officers headed by the then virtually unknown Captain Hyman G. Rickover" (Hazzelrigg, 1982, p. 156).

Although all the persons in the room were brilliant, talented, productive people, the committee was composed mostly of scientists. Rickover was not a scientist but an engineer. Rickover had wheedled the privilege of listening to the General Advisory Committee's deliberations by getting permission for himself and one of his lieutenants to sit on one side of the room merely as observers. "When eventually the Committee's deliberations turned to the question of developing useful power from the atom, the consensus after considerable discussion was that useful power generation might perhaps be demonstrated in 20 years, and its application for specific purposes might come some time later. Rickover stood up and, interrupting the discussion, proceeded to berate the distinguished scientists for using a scientific approach to an engineering problem. This broke up the meeting" (Hazzelrigg, 1982, p. 156).

Just after World War II, in the area of the politics of science and technology in the United States, the scientists were then "top dogs." They held the political stage over the engineers. They were not ready to be lectured to about anything—let alone by an engineer and a junior officer. Oppenheimer was the President's Science Advisor (popularly referred to as the "father of the atomic bomb" for his technical leadership in the Manhattan Project). Teller was also a scientist (later to become popularly called the "father of the hydrogen bomb"). Rabi and Seaborg were also scientists and eventually would become Nobel prize winners. Only Conant and DuBridge were engineers, but academically based engineers with doctoral degrees, closer to scientists—engineering scientists.

In contrast, Rickover was a practicing engineer, holding a master's degree in engineering, and a Navy captain. But he had an engineer's technical dedication, with an urgent technical vision to build an atomic-powered submarine. His vision was different from a scientific vision—different as to the application, urgency, and timing of the innovation. Six months before that Atomic Commission had been formed, the Navy had already decided to build nuclear-powered ships. The Navy's General Board had recommended to the Secretary of the Navy, James V. Forrestal, that a study begin for the development of atomic power for propulsion of naval ships. The Bureau of Ships sent a group of officers to Oak Ridge, Tennessee to observe a project that was intended to develop a power rector for civilian use. The senior officer of this group was Captain Rickover (Hewlett and Duncan, 1974; Polmar and Allen, 1982).

By 1948, Rickover was in charge of the Navy program and proceeded to develop the Naval nuclear submarine program. Rickover chose the light-water reactor in a pressurized form for the Navy. Rickover's choice of the pressurized light-water reactor focused on the submarine application, for which it was believed to be the best technical choice. The Navy had evaluated the liquid-metal (sodium)-cooled reactor as being "tons heavier" than a pressurized light-water reactor for the same production of energy (JCAE, 1964). They also saw a major safety problem in this alternative that if seawater flooded a liquid-metal-cooled reactor, the liquid sodium would react with water, producing a terrible explosion. In addition, the control equipment for controlling such a reactor would be very complex and difficult. Finally, the induced radioactivity in liquid metal coolants was long lived, making repair difficult. So the Navy decided against using a liquid-metal reactor for a submarine. The Navy also saw problems with the other two types: The heavy-water reactor was envisioned as just too big and heavy for a submarine, and the gas-cooled reactor was also very big, and its safety and control requirements were more risky and complex than those of a pressurized light-water reactor. Thus, at the time, for a ship with modest power requirements, the weight, size, simplicity, and safety of the pressurized light-water reactor provided the best engineering choice for the Navy.

The nuclear-powered submarine revolutionized naval warfare. At the beginning of World War II, the relatively slow diesel-powered German submarine (the U-boat) had almost strangled the English war effort. After nuclear power, by the 1960s, the submarine became the dominant war boat. Nuclear power enabled submarines to remain indefinitely at sea and underwater at depths and speeds, undreamed of before, the limits of a cruise being only the amount of food that a crew required. For example, in a popular magazine of the 1980s, a civilian reported about the experience of cruising on a modern nuclear submarine: "A profound sense of wonder swept over me as Cmdr. Alfred E. Ponessa, USN, passed the tuna salad. We were seated in the wardroom aboard the nuclear attack submarine *Norfolk,* enjoying the expertly prepared food and the convivial discourse, when I glanced up at the instrument repeater on the bulkhead. This little party was taking place somewhere off the coast of Florida along the 100-fathom curve. Speed? Well in excess of 20 knots. Depth? Much deeper than 400 ft. . . .

Norfolk's nuclear reactor quietly simmered just a few steps aft—converting water into steam, steam into turbine energy. . . . Requiring refueling every 13 years, the power plants' continuous, controlled fission generates on-board electricity drives an electrolysis process that separates hydrogen from water to produce oxygen, operates the lithium–hydroxide scrubbers that cleanse the atmosphere of carbon dioxide, and desalinates and de-mineralizes ocean water" (Cole, 1988, p. 28).

Since most developmental research went first into a military application, the later civilian choice of nuclear reactor built upon it. Thus it turned out that the pressurized light-water reactor became the dominant type of reactor in use, both military and civilian. For the military this turned out to be a good engineering choice. But for the civilian electric power application, this choice turned out to be an economic disaster.

FUNCTIONAL PROTOTYPES

We recall in the logic of radical innovation that the stage from technical feasibility to a functional prototype is in making the technology perform efficiently for a specific application. What we see in this case is that although the scientific feasibility established that nature may be manipulated, the actual technical manipulation can still involve alternatives for an application. There will be alternatives in materials, in processes, and in form.

> **The early stages of engineering research should be aimed at systematically exploring all technical alternatives to find a *technically optimal* solution.**

The criteria for technical optimality include (1) performance, (2) efficiency, (3) cost, (4) safety, and (5) life cycle. Different combinations of these factors will be optimal for different applications. An application prototype should adopt the best engineering solution for a given application context. Again, we recall from the radical innovation process that in establishing engineering feasibility, there are two stages of prototyping of technology systems: technical feasibility prototypes and application functional prototypes.

> **Research for engineering feasibility requires an explicit application and the technical requirements for the application.**

Previously, we discussed the importance of using experts in forecasting technology, but it is important to use a mixture of experts, scientists and engineers, and applications users. Each will view technical issues differently. The view of a technology as a scientific problem and as an engineering problem are different views. It is the responsibility of the engineering perspective in research to establish as early as possible the requirements of an applications user of an innovative technologies— even before users exist.

A technical feasibility prototype demonstrates that the engineering principles do work in at least one configuration. A functional prototype is a selection from the technically feasible alternatives that optimizes the performance criteria for a given application.

> **Different configurations of a technology system perform differently for different applications.**

Early Death of the U.S. Nuclear Power Industry

In the second half of the twentieth century, the U.S. nuclear power industry, utilizing pressurized water reactors, ran into commercial difficulties due to safety and environmental problems. By 1989, the U.S. nuclear industry was at a dead standstill with no new nuclear plants on order.

Several nuclear accidents, including one at Three Mile Island in Pennsylvania, frightened the public about nuclear safety. The pressurized water reactor design was inherently unsafe under loss of cooling water. This unsafe design led to a strong movement protesting the placement of new reactors. Accidents occurred at Three Mile Island in Pennsylvania and at Chernobyl in Russia. The nuclear industry had not replaced a first-generation design with a second-generation design that was inherently much safer.

The second problem was the environment. Nuclear disposal of long-lived radioactive waste had not been addressed effectively or solved by the U.S. nuclear regulatory agency. Accordingly, the public also lost confidence that the nuclear industry could handle the life-cycle problem of what to do with waste. This environmental problem still was not solved as late as 2002, when the U.S. Department of Energy decided to store nuclear waste in Nevada.

As a result of safety and environment problems, the new technology industry ground to a halt in the United States just after it had begun growing. For example, in 1966, 16 plants had been ordered, of which 12 were built and 14 canceled. In 1971, 21 plants had been ordered; 8 were built and 13 canceled. In 1973, 41 plants had been ordered; 9 were built and 32 canceled. In 1975, only 4 plants had been ordered and all 4 were canceled. In 1978, only 2 plants were ordered and both were canceled. From 1979 to 1989, there were no new orders for nuclear power plants in the United States. Both safety and environmental factors are essential to the long-term commercial success of a technology.

PRODUCT LIFE CYCLE

In a new technology, research should be focused not only by the technology system but also by the application system. One important difference between the concepts of the technology and applications systems is in the notion of a *product life cycle*.

> **The product life cycle concept in engineering design requires consideration of all the technical requirements for producing, maintaining, supplying, and disposing of a product.**

Technologies relevant to a product include:

- The technologies embedded in the *design*
- The technologies relevant to *maintaining and repairing* the product in use
- The technologies required to provide the product with *resources* (such as fuel or other supplies) while in use
- The technologies required to *dispose of or recycle* the product after use

Research for the innovation of a new technology should include all technologies for the product life cycle.

> **Next generations of a new technology should not only improve performance and lower costs but also improve the product-life-cycle issues of safety and environment.**

SUMMARY AND PRACTICAL IDEAS

In case studies on

- Nuclear fission energy
- Nuclear reactor for the Navy
- Early death of the U.S. nuclear power industry

we have examined the following key ideas:

- Scientific feasibility
- Technical feasibility
- Functional prototype
- Product life cycle

These theoretical ideas can be used practically in understanding research, particularly the relationship of science research to engineering research in R&D. All national S&T infrastructures operate to advance both science and technology. It is engineering research that provides the critical link between science and technology.

FOR REFLECTION

Identify an area of science and list the kinds of new technology that came from it. Who did the engineering research that linked the science to technology?

13

RESEARCH FUNCTION

INTRODUCTION

We have reviewed why technological innovation creates an imperative in societal change. Now we begin examining what capabilities a business must have to be innovative, and the first is a research capability. We examined how research can contribute to radical innovation, and now we look at how research can be organized in a business as a corporate function. In the case of nuclear energy, we saw how research could make a life-and-death difference in the military conflicts of nations. Similarly, corporate research can make the difference in corporate survival. The development of corporate research capability was one of the major institutional advances in the twentieth century. We will examine the research function in terms of its *personnel, objectives, organization, funding,* and *culture.*

Traditionally, the different areas of operations necessary in a business are called *business functions.* Any business must have the functions of (1) producing a product, (2) selling the product, (3) organizing to run the business, and (4) controlling the operations of the business. The four traditional business functions are *production, marketing, administration,* and *finance.* But these four functions are not sufficient for a modern innovative business, which must also have the capability to develop new technologies and new products. An innovative business must have a *research function* and an *engineering function.* In the second half of the twentieth century, information technology became pervasive in all businesses, and this required businesses to have an *information function.* Therefore, a modern business must have seven functional capabilities:

1. *Production function*
   —For producing products/services
2. *Marketing function*
   —For selling products/services
3. *Administration function*
   —For organizing and staffing operations

4. *Finance function*
   —For controlling business operations
5. *Engineering function*
   —For designing products and production
6. *Research function*
   —For inventing new technologies
7. *Information function*
   —For facilitating communication and decisions in operations

CASE STUDY:

General Electric's Corporate Research Center

Let us return to the early case of the technical progress in incandescent lamps to see how GE's response to the challenge of a new kind of lamp technology was to create a research function in GE. We recall that after traveling to Germany in 1900, Charles Steinmetz (GE's chief engineer) recognized the intrinsically superior technical efficiency of Hewer's lamp (the first fluorescent lamp). From its physics, Steinmetz understood that fluorescent lamps would always be more energy efficient than incandescent lamps. We also recall that the incandescent lamp had been innovated by Thomas Edison and that GE later bought Edison's lamp business. To meet this potential competitive challenge to GE's lamp business, Steinmetz urged GE to do fundamental research and to build a research function in GE.

It was on September 21, 1900, that Steinmetz wrote a letter to Edwin Rice, vice-president of GE's engineering and manufacturing functions. Steinmetz proposed a new research and development center for GE. Rice and other key technical leaders of GE, Albert Davis and Elihu Thomson, agreed with Steinmetz's proposal. Rice obtained approval for the new lab from GE's president, Charles Coffin. Davis later wrote: " 'We all agreed it was to be a real scientific laboratory' " (Wise, 1980, p. 410).

Rice hired Willis Whitney, a young professor at MIT, as the director of the new lab. This industrial idea of using science to explore the basis for new technologies was important, as the historian Wise (1980) noted: "He [Whitney] was not the first professional scientist to be employed in American industry, or even in General Electric. Nor was his laboratory the first established by that company or its predecessors. . . . Whitney's effort marks a pioneering attempt by American industry to employ scientists in a new role—as 'industrial researchers' . . ." (p. 410). To defend GE's lamp business, Whitney formulated a research strategy to explore all suitable elements in the periodic table systematically to find an improved substitute for the carbon filament in the incandescent lamp. Whitney began hiring scientists to carry out his strategy, He assigned one possible element to each newly hired scientist, and one of these was William Coolidge, a fresh Ph.D.

Coolidge had been born in 1874, the son of a farmer and a shoe factory worker. He attended a one-room elementary school, then common in rural North America. Graduating from a small high school as an outstanding student, he went off to college in 1891. He enrolled in a new electrical engineering program at the Massachusetts Institute of Technology (Wolff, 1984). Coolidge graduated with a bachelor's degree and decided to become a scientist. But in the 1980s the new universities in the United States were not yet training scientists. To get a real scientific education, students had to go to Europe. Coolidge went to Germany and returned in 1905 with a doctorate from the University of Leipzig. He had a $4000 debt for his graduate education. He took a position at MIT as an assistant to Arthur Noyes (a notable U.S. scientist). Coolidge's pay as an assistant was $1500 a year.

Earlier as a professor at MIT, Whitney had taught the young Coolidge in a chemistry class. Now at GE and learning that Coolidge had returned with a German doctorate, Whitney offered Coolidge a job in GE's new laboratory. Whitney offered $3000, which was double Coolidge's MIT salary. Since this would allow Coolidge to pay off his education debt quickly, Coolidge was attracted by the offer. Yet he worried whether he could conduct scientific research in industry. Whitney assured him he could, as this was the mission of the new GE laboratory. The only constraint was that the scientific research needed to be related to GE commercial interests.

Coolidge accepted the offer and moved from Boston to Schenectady to work at GE's facility. There he was pleased at the atmosphere for science that Whitney was creating: " 'Dr. Whitney had already successfully transplanted a lot of academic atmosphere from MIT to the new laboratory. . . . The misgivings I had about the transfer all proved to be unfounded' " (Wolff, 1984, p. 81).

Whitney had a research strategy to find a replacement for the carbon filament, and he assigned to each of his researchers the responsibility for exploring each metal in the periodic table with a melting point higher than that of carbon, and Coolidge received tungsten as his assignment. Whitney had explained the commercial importance of the research, and Coolidge wrote to his parents: " 'I am fortunate now in being on the most important problem the lab has ever had. . . . If we can get the metal tungsten in such shape that it can be drawn into wire, it means millions of dollars to the company' " (Wolff, 1984, p. 84).

The technical promise of tungsten was that it had a higher melting point and therefore might be usable to incandesce at a higher temperature, thus physically providing a higher lumen/watt efficiency. We recall that most industrial research is developmental; some is applied research and very little is basic research. Coolidge's problem with tungsten was applied research. He knew the attractive feature of tungsten for the lamp application and the need for tungsten to be drawn into a wire for this application. But how to do that? Tungsten was too brittle to be drawn into a wire.

The applied problem of using tungsten as a filament was its natural ductility. It had a high melting point but was too brittle to draw into the thin wire shape needed to perform as a lamp filament. In June 1906, Coolidge had his first research break, observing that mercury was absorbed into hot tungsten and, on cooling, formed an amalgam. Next, Coolidge experimented with making other amalgams of tungsten. He observed that he could make one with cadmium and bismuth absorbed into tungsten. Moreover, this amalgam could be squeezed through a die to make a wire! Coolidge was on the right path. On March 1907, Coolidge discovered that by heating this amalgam to 400° Fahrenheit, it could be bent without cracking.

Coolidge had a potential candidate material for a new filament; the next step was to figure out how to manufacture it. He decided to visit wire and needle producing factories and toured several in the New England region. There he saw the swaging techniques used for wire making. Swaging was a process of gradually reducing the thickness of a metal rod by repeated use of dies to draw smaller and smaller diameters. May 1909, back in Schenectady, Coolidge purchased a commercial swaging machine and altered it to swage his tungsten amalgam into wire. It did make tungsten wire, but when the wire was heated in a lamp as a filament, it broke. Coolidge researched the failure and learned that it was due to crystallization of the tungsten wire upon heating.

This problem was solvable. Coolidge knew that in a similar problem of making ice cream, glycerine was added to prevent the forming of ice crystals as the milk freezes. Using this analogy, Coolidge tried adding another substance to the tungsten amalgam, thorium oxide, to prevent crystallization. It worked. On September 12, 1910, Coolidge had achieved a technical success. He had developed a new tungsten wire for a lightening application and a manufacturing process to produce it.

GE's management was pleased and immediately innovated the new material and process. By 1911, GE had thrown out all the earlier lamp-making equipment and was producing and selling lamp bulbs with the new tungsten filaments. The research effort had cost GE five years and $100,000. By 1920, two-thirds of GE's $22 million profit came from the new lamps.

A century later, General Electric Research and Develop Center still existed, located on a site overlooking the Mohawk River at the outskirts of Schenectady, New York. In 1982, the GE center employed about 2100 people, and its budget was about $150 million. GE's total research and development budget was then $1.5 billion. In 2002, GE's new CEO emphasized once again the importance of GE's research capability to GE's long-term future.

INSTITUTIONALIZATION OF SCIENTIFIC TECHNOLOGY

The inclusion of science into the functions of business as a corporate research laboratory was a key event in institutionalizing scientific technology. The first industrial research laboratories were created in the new technologies of the electrical and chemical industries. In addition to the GE example, many other now famous corporate laboratories in the United States were begun in the early twentieth century, such as the DuPont Laboratories, the Bell Laboratories of AT&T, the Dow laboratory, GM's technical center, and many others. Industrial laboratories were also established in Europe, such as at I.G. Farben and Siemans. As the twentieth century proceeded, research laboratories became an important feature of high-technology firms in all the world. For example, in 1989, the NEC Corporation in Japan spent 10% of sales on research and development, performing R&D in the corporate research laboratory and in divisional laboratories distributed in the 190 businesses of NEC (Uenohara, 1991).

In 1992, the United States had about 1,000,000 R&D engineers and scientists, with Japan having about 800,000 and Germany about 600,000. The fraction of gross domestic product spent for R&D was around 2.5% for the United States, Japan, and Germany (Geppert, 1994). In the United States, an industrial organization concerned with R&D, the Industrial Research Institute, periodically sponsored surveys of the state of industrial R&D, such as that of Whiteley and co-workers (1994).

The idea to employ scientists in industry as industrial researchers was the beginning of the institutionalization of scientific technology as a way of life in modern industry. It was a second step in institutionalizing a modern S&T infrastructure. The first step of institutionalizing a modern S&T infrastructure had occurred a century earlier at the beginning of the nineteenth century, when Germany began reforming its medieval universities into research universities. The third step of institutionalizing a modern S&T infrastructure did not occur until the middle of the twentieth century when modern national government agencies began systematic funding of basic research at universities. The modern S&T infrastructure, consisting of research universities, industrial research centers, and government research-sponsoring agencies, was not completed until the second half of the twentieth century. The S&T infrastructure is a very recent cultural evolution of modern civilization, and managing this is a challenge.

For example, Herbert Fusfeld identified three critical historical periods in the U.S. evolution of industrial research: 1870–1910, World War II, and the new period of declining technical self-sufficiency starting in the 1970s. From 1870 to 1910, new industrial research organizations in the United States began to give the new science-based chemical and electrical industries timely technical support: "The companies no longer had to depend on the unpredictable advances generated externally. From this start there was a steady growth of industrial research up to World War II" (Fusfeld, 1995, p. 52).

Next, the experiences of military research in the United States during World War II had consequences for subsequent industrial research: "A great reservoir of technical advances became available for further development in commercial areas; public expectations were raised for the potential of science and technology in new products; new techniques of systems development were successful in planning and conducting complex technical programs that required the generation of new knowledge as an integral part of the planned program" (Fusfeld, 1995, p. 52).

After that war, U.S. research and industry dominated the world until the war-devastated economies in Europe and Japan were reconstructed. By the 1970s, industrial competition and progress was again worldwide. U.S. management realized that technical self-sufficiency within a U.S. company was no longer practical: "For the 30 years following World War II, [U.S.] corporations were able to plan growth strategies based on technical resources that existed internally. . . . That situation changed . . ." (Fusfeld, 1995, p. 52).

After the 1970s, there were the new conditions of a global economy, and corporate research needed to (Fusfeld, 1995, p. 54):

- Make strategic use of external resources from outside the company
- Disperse corporate technical activity beyond a central laboratory
- Emphasize the effective integration of research into total corporate technical resources
- Have a major effort on the organized pursuit of technical intelligence worldwide
- Have extensive foreign-based R&D activity to support growth in international markets
- Improve the integration of technical strategy into corporate business strategy

In 1995, Lewis Edelheit, GE's senior vice-president for corporate R&D, commented: "How can a corporate-level R&D lab renew its role as a vital part of a winning business team? My answer is . . . cost, performance, speed, quality. In the past, innovation often began with a performance breakthrough made at the corporate lab. . . . The result was relatively slow development of often costly products that nevertheless might prove winners in the marketplace because they offered capabilities no competitor could match. That does not work today . . . because technology is exploding worldwide. Dozens of labs are at the forefront today whereas yesterday there might have been only one or two" (Edleheit, 1995, p. 14).

CASE STUDY:

William Coolidge as Research Manager

After Coolidge's innovation of tungsten filaments in incandescent lights, he turned his scientific attention to a related but entirely new area: production of x-rays. Coolidge had the idea that tungsten might be a useful material as a target in a x-ray tube. With the help of another GE scientist, Irving Langmuir (Langmuir would later receive a Nobel prize for his discoveries in physics), Coolidge invented the first practical x-ray tube in 1913. Over the course of the next 15 years, Coolidge continued to make technical contributions to x-ray applications. Then in 1928, he was promoted to the post of associate director of the GE laboratory, beginning a research management career.

When the 1930s began with an economic depression in the United States, GE was financially stressed. In April 1932, Whitney decided to retire as director of the laboratory. His health had begun to suffer from worry about keeping the GE lab going and fears that it might be shut down in view of the bad economic conditions. On November 1, 1932, Coolidge succeeded Whitney to become the second director of the GE laboratory. Immediately, Coolidge took action to reduce expenses and save the lab. Coolidge put the laboratory on a four-day workweek and cut the workforce from 555 people to 270 people. He was careful to prune, keeping the best researchers and maintaining the emphasis of the laboratory on fundamental research and the invention of new technologies.

It was tough to maintain morale and to continue the search for new products. By December 1933, he convinced higher management to permit him to add five new chemists to develop chemical-based products. So research continued through the Great Depression in the United States, and many of the inventions of the 1930s provided GE with a base for its war effort in the 1940s and economic expansion thereafter.

Coolidge was seen by his contemporaries as combining a management style of expense control with technical leadership. He interacted with his staff and was accessible. He walked around the lab, looking as some said, "more like a scientist than a laboratory manager" (Wolff, 1984). He was encouraging to his researchers with comments such as: "This is fascinating; anything we can do to help?" One of his colleagues said of Coolidge: "That man oozes optimism of an inspiring brand. You feel in his presence that if all things are not possible, many are. Yet he has plenty of circumspection, with a verifying, sagacious mind that readily isolates what is either impossible or extraneous" (Wolff, 1984, p. 84).

CAREERS OF SCIENTISTS IN INDUSTRY

In the corporate research function, scientists are hired to perform research. But scientists differ from managers in their training and professional orientation. The transformation from a scientist to an industrial scientist is drastic and difficult. Scientists are trained to discover and understand nature, but they are not trained to manipulate nature. Thus for scientists to become industrially oriented, they must learn to value the manipulation of nature as a primary goal, not merely the discovery and understanding of nature. Reorientation of an academically trained scientist to an industrial scientist requires the young scientist to learn two new perspectives: a technologist's perspective on the value of manipulation and a manager's perspective on the economic value of manipulation. For proud people such as scientists, learning perspectives other than science often does not come easily.

Thus, due to the business context of industrial research, the career of an industrial scientist involves a continuing and needed creative tension between science and management. Industrial research personnel must be specially recruited, motivated, retrained, and integrated into the corporation.

> **Science in industry requires the balancing of research with business—focusing fundamental knowledge on invention and upon technical problem solving aimed at improving the productivity of the firm.**

Accordingly, this tension of research to business for scientific personnel in industry requires a dual basis for their performance evaluation. For example, Wilson and co-workers (1994) have emphasized the need to evaluate researchers on two sets of criteria, one technical and the other managerial.

Moreover, to accommodate this tension to individual scientific careers in industry, some corporations instituted what is called a *dual career ladder,* in which there is a technical route for advancement for research engineers and scientists to remain researchers for the duration of their employment or alternatively, a managerial route for research scientists and engineers to broaden to become general managers. Not all researchers should or can become managers. A dual career ladder with both technical and management sides allows some technical people to advance without becoming general managers. However, the effectiveness of the dual ladder approach remains a subject of debate. For example, Allen and Katz (1992) pointed out that it is the Ph.D. group of scientists and engineers who appreciate having the choice of dual career ladders, those who are engaged in fundamental research most likely to choose to remain on the technical career ladder.

OBJECTIVES OF CORPORATE RESEARCH

The general objectives of corporate research are to:

- Have a window on science for long-term threats and opportunities
- Maintain existing businesses
- Create the technology base for new businesses

Corporate research provides a firm with a way to perform relevant science and to transform it into new technologies: a corporate institutional capability of scientific technology. For example, it was from Steinmetz's visit to German research laboratories to spot new science (i.e., the new "glower" lamps) that Steinmetz was able to alert GE's management to future threats to GE's lamp business. Before the GE lab was established, there had been no systematic way for GE to have windows on science, except in individual initiatives such as Steinmetz. Establishing the GE lab provided the critical means for GE to maintain its incandescent lamp business by dramatically improving the filament technology. Cooledge also went on to start an x-ray lamp business for GE on the basis of his new tungsten technology.

> **Properly managed and exploited corporate research prepares a company for its future survival.**

Successive top managers of corporations who (1) fail to manage research properly and (2) fail to exploit successful research will (3) fail to prepare the company for future survival. But this is the challenge of managing corporate research, for it is the sins of failure of past CEOs about corporate research that catch up with present CEOs.

Since one of the principal purposes of corporate research is to create and extend the lifetimes of the company's products, anticipating the need for R&D support for products is an important element of research strategy. When product technologies are changing, product lifetimes will be short. In products with mature technologies, product lifetimes are long. But even in a long-lived product (such as the soap product Ivory Snow), periodic product reformulations do occur to meet changing conditions in market, environment, safety, and so on. Also, to maintain a long-lived product, research and innovation in manufacturing quality must be made continually.

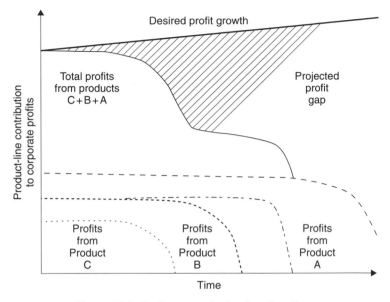

Figure 13.1. Profit-gap analysis of product lines.

A useful way for management to track the attention required for product development is *profit-gap analysis*. Figure 13.1 shows the form of plotting the projections of contributions to a business's profits from each of its product lines. Some product lines, such as product A, can be projected to continue to contribute steadily to profits; others, such as product C, already have dropping sales and will be obsolete due to its aging technology (and the company plans to terminate product line C). In between these extremes, some product lines, such as product B, can be revitalized by redesign. When all the profits from these three types are summed to the total profits of the company, the chart shows the projected profits of the company. If the management plots on this chart a dotted line of where the desired profits should go, the area beneath the dotted line and above the summed profits line shows the gap between desired profits and actual profits. This is the profit gap anticipated unless new products are introduced (Twiss, 1980, p. 40).

> **The profit gap displays the magnitude over time of the business need for new products.**

RESEARCH LABORATORIES

How many and what kinds of laboratories a firm needs to be innovative depend on the business organization of the firm. There are three ways to organize research laboratories in a firm: (1) divisional laboratories reporting to business units, (2) a corporate-level laboratory, and (3) both divisional laboratories and a corporate-level laboratory. Ordinarily, research in the divisional laboratories is focused on next-product-model design and on production improvement, whereas research in corporate laboratories is focused on next-generation product lines and on developing new businesses from new technology.

The type and number of research units depends on the size and diversity of businesses in a firm. A single-business small firm will probably have only an engineering department. A medium-sized firm will probably have an engineering department and a divisional laboratories. A large diversified firm will probably have engineering departments and divisional laboratories in business units and a corporate laboratory for all businesses.

Moreover, research organization varies by industry. For example, the first annual Industrial Research Institute R&D survey in the United States noted: "The art of technology management may be more industry-specific than some observers have realized" (IRI, 1994). Even within an industry, research organization varies. For example, Bosomworth and Sage (1995) surveyed 26 U.S. firms in 1993 as to their management practices. They found that R&D investments ranged from 0.6 to 15% of sales and that research organization varied from only a central research laboratory to only divisional laboratories. Also, research strategies varied from emphasizing defensive to emphasizing offensive technology strategies.

As another example, Chester (1994) studied the research management practices in aerospace and electronics systems companies. He found that most of the 30 largest firms had either one or two corporate laboratories, but there was substantial variance. TRW and General Dynamics (GD) had no central research laboratory, whereas Hitachi had nine central research laboratories. Firms without central laboratories, such as TRW and GD, depended more on outside technology than do firms with central labs. He also found that the corporate research organization reflected the core technologies and science bases of the relevant businesses. For example, Siemens had five technology groups with 24 core technologies.

Is there a "best" organization for research? Alfred Rubenstein (1989) addressed this question after having observed the organization of research in many firms over a long period of years, and his conclusion was that no organizational form alone could provide a best solution. However, the influence of research organizations on research activity is clear. Research organizations consisting only of decentralized divisional labs does encourage a short-term focus mainly on the current businesses of the corporation. Therefore, in this form, management must provide special attention to focus on long-term issues. In contrast, research organization consisting only of a corporate research lab does encourage long-term focus, but at the cost of short-term relevance. Accordingly, in this form, management must pay special attention to making its research relevant to the current businesses of the firm.

Thus, research organizations consisting of both divisional and corporate labs have the potential strengths of focusing on both current and future businesses. The problem, however, is that the research subcultures develop differently in divisional and corporate labs. This difference can foster competition rather than collaboration: "As the operating divisions begin to flex their decentralized muscles and start acting as though they were indeed independent enterprises . . . they begin to become impatient with the level or quality or relevance of the work being done in the central R&D activity. For those division managers who see a real need for strong, direct technical inputs to their division's operation, the central lab seems unwieldy, distant, and not very responsive to their immediate and near-term future needs" (Rubenstein, 1989, p. 41).

For example, research at DuPont from 1950 to 1990 was organized as the combination of a central research lab (central research and development), engineering R&D (part of the engineering department of DuPont), and business-specific R&D conducted in engineering departments in the strategic business units. Then the central research lab became increasingly disconnected from business units: "Where once breakthrough technology was the hallmark of DuPont's research, the focus over the four decades increasingly shifted to incremental improvement of product lines and manufacturing processes. . . . Central R&D became more isolated from business management and their work more speculative and academic" (Titus, 1994, p. 351). In 1992, research at DuPont was reorganized: "In effect, R&D is almost as centralized now as it was forty years ago, but for different reasons. . . . The old paradigms of sequential development are no longer operative in today's competitive global environment" (Titus, 1994, p. 351).

The conclusion about research organization is that a mixed form may not be "best" for everybody's purposes, but it may be necessary for all purposes. Corporate-level research provides the opportunity for cutting-edge science, synergy across different business technologies, and opportunities to begin new high-tech business ventures. But corporate research in isolation is seldom implemented into divisional businesses without each business having a business-level research laboratory. At the business level, research is necessary to fine-tune the engineering opportunities of technological progress. Effective collaboration between research units and business units is very important, as are concurrent practices in R&D.

EVALUATING CORPORATE R&D

Since R&D is an investment in a corporation's future, it should ultimately be evaluated as to return on investment. However, in practice, this is difficult to do because of the time spans involved. There is usually a long time from (1) funding research to (2) successful research to (3) implementing research as technological innovation to (4) accumulating financial returns from technological innovation. The more basic the research, the longer the time to pay off, and the more developmental, the shorter the time. For example, times from basic research to technological innovation have historically varied from 70 years to a minimum of 10 years. For applied and developmental research, the time from technological innovation to break-even has been from two to five years.

In addition to the varying time spans, the different purposes of research also complicate the problem. We recall that research is aimed at maintaining existing businesses or beginning new businesses or maintaining "windows on science." Accordingly, evaluating contributions of R&D to existing businesses requires accounting systems that are activity based and can project expectations of benefits in the future and compare current to projected performance. The evaluation of research needs to be accounted to these purposes of research:

1. *R&D projects in support of current business*
   a. The current products can be projected as to lifetimes.
   b. This product mix is then projected as a sum of profits.
   c. The current and proposed R&D projects in support of current business are evaluated in terms of their contribution to extending the lifetimes or improving the sales or lowering costs of the projects.
2. *R&D projects for new ventures can be charted over expected lifetimes*
   a. Projects that result in new ventures are charted over expected return on investment of the new ventures.
3. *R&D projects for exploratory research*
   a. These projects cannot be evaluated financially and should be treated as an overhead cost. They should be evaluated technically only on their potential for impact as new technologies.

Many scholars have tried to simplify this logic by trying to create analytical shortcuts to the evaluation problem. For example, Foster and co-workers (1985) suggested trying to trace the following logic as an expression:

$$\text{return} = \frac{\text{profits}}{\text{R\&D investment}}$$

This might be expanded as

$$\text{return} = \frac{\text{profits/technical progress}}{\text{R\&D investment}}$$

Although at first this expression may be seem appealing, its problem lies in measurement. Profits and R&D investment can be measured in dollars, but there is no general measurement for technical progress across all technologies. Accordingly, the expression as a whole is not accurately measurable.

There is no really useful shortcut to evaluating R&D. R&D must be tracked in accounting, as it really contributes to business. This requires establishing an appropriate accounting system for the research function. For example, Kuwahara and Takeda (1990) described an accounting method for measuring the historical contributions of research to corporate profitability. In the accounting system, profit from each product had to be proportioned according to productive factors, one of which was a profit contribution factor by the research.

Accordingly, budgeting corporate R&D is difficult because it is a risky investment of different kinds and over varying periods of time to return of investment. For this reason, most R&D is usually deducted as a current operating expense and treated as a part of the administrative overhead. In practice, most R&D budgeting is done by incremental budgeting, increasing research a little when business times are good, cutting research when profits drop.

The level of R&D expenditures often tends to be an historically evolved number depending on many variables, such as the rate of change of technologies on which the corporate businesses depend, the size of the corporation, levels of effort in R&D by competitors, and so on. In areas of rapidly changing technology (high technology), firms tend to spend more on R&D as a percentage of sales than do firms in mature core technologies. For example, high-tech firms have spend in the range of 6 to 15% of sales, whereas mature technology firms may spend 1% of sales or less.

The share of R&D divided between corporate research laboratories and divisional laboratories also differs among industry, but generally with divisional laboratories having the largest share because the direct and short-term nature of their projects contributes to profitability. For example, in high-tech firms, the corporate research laboratory might reach as high as 10% of R&D, but seldom more.

In the corporate research lab, R&D funding is usually of three types: (1) allocation from corporate headquarters, (2) internal contracts from budgets of business units, and (3) external contracts from government agencies. The corporate allocation provides internal flexibility for the corporate lab to explore long-term opportunities. The internal contracts provide direct service to business units. External contracts from government agencies either provide a direct business service to government or additional flexibility for the research laboratory to explore long-term future technologies. Normally, internal contracts will provide the majority of corporate laboratory (unless the corporate laboratory is in the government-contract R&D business).

Many studies have tried without much success to find correlations between R&D investments and profitability. The reason is that not all R&D is good research, and not all good research will even be useful to a firm. The quality of managing the R&D function is more important than the quantity of resources spent on R&D. Thus, some firms that have managed R&D well have received great benefits, whereas other firms, even with large expenditures have not benefitted from it.

CASE STUDY:

Xerox Pioneers the Paperless Office

Not all corporate research is successful, but when it is, it should be exploited commercially. But the problem of successful commercial exploitation of research is a major problem itself. A classic case of the failure of an innovative corporation to exploit its own research successfully was Xerox. Xerox not only innovated xerography but was also the inventor of the networked personal computer—the inventor but not the innovator. In the 1970s and early 1980s, Xerox management pioneered next-generation technology in personal computers but failed commerically. Finally, by 2002, Xerox was on the edge of bankruptcy. It had missed exploiting not only the personal computer revolution but the Internet revolution. That is a sad but true story.

We recall that Xerox was born from Joseph Wilson's (president of Haloid) investment in Carlson's invention of xerography. Up to the 1970s, Xerox's only business was copiers. Then Xerox decided to diversify and bought a computer company. But it could not compete against IBM's dominance of mainframe computer market in the 1960s, and Xerox sold its computer company. Still the need to diversify continued, and Xerox decided to reenter the computer business through innovation, not by acquisition. Xerox established a new research laboratory in Palo Alto, California, the Palo Alto Research Center (PARC), to pioneer new computational ideas for Xerox's strategic technical vision of the "paperless office." Xerox then had three research laboratories, two in technologies for the xerography business and one for the hoped-for computer business: Webster Research Center, Xerox Research Centre Canada, and PARC.

Xerox hired George Pake, a physicist, to head the new PARC laboratory, and he located it next to Stanford University, which was strong both in electrical engineering and the then new discipline of computer science. Pake hired many bright young researchers, and an important one was Alan Kay. As a student in the late 1960s, Kay had been influenced by the ideas of an MIT professor, J. Licklider. Licklider had envisioned an easy-to-use computer, portable and about the size of a book (Bartimo, 1984). At Palo Alto, Kay and his colleagues further developed these ideas into a vision of the future computer system: personal computers linked in a communications network with laser printers, operated with ikons, desktop metaphors, and mouses. Called the Altos system, this was built in 1979 as an experimental computer testbed and prototype.

What did Altos look like? It looked like what one would eventually see after 1990; it was 10 years ahead of its time. Only by 1990 would one see the commercialization of the Altos vision in offices with local area networks of Macintosh computers hooked together with Ethernet coupling using ikons, desktops, and mouses and object-oriented programming software. By 2000, the difference between Altos and present office computer systems has been only the addition of the Internet. This is how advanced Altos was back in 1979.

But did Xerox make a bundle from its visionary investment in computer technology? No, Xerox did not make a dime from it. Instead of producing the Altos system, Xerox produced a workstation that looked like a Wang word processor of 1981, which turned out to be the technical vision of Xerox's office products division. In 1980, the manager of the office products division suffered from visionary myopia, thinking that the Wang word processor was the latest in technology. He saw the office future as a network of word-processing stations (which the Wang company had innovated). Xerox's office products division put out a Wang look-alike product using only some of PARC's inventions. Called the Star workstation, it was a technical and commercial failure. It cost Xerox its entire investment in personal computers and resulted in its failure to capture a market position in the emerging personal computer market.

The failure of Xerox's manager of the office products division was a failure of technological vision to go with PARC's brilliant vision and research. A business reporter, Bro Uttal, commented in 1983: "On a golden hillside in the sight of

Stanford University nestle's Xerox's Palo Alto Research—and an embarrassment. For the $150 million it has lavished on PARC in 14 years, Xerox has reaped far less than it expected. Yet upstart companies have turned the ideas born there into a crop of promising products. Confides George Pake, Xerox's scholarly research vice president: 'My friends tease me by calling PARC a national resource.' " (Uttal, 1983, p. 97).

Eventually all of PARC's Altos system inventions were innovated but by companies other than Xerox. For example, in the early 1980s, Steve Jobs of Apple visited PARC and saw Altos. Rushing back to Apple, Jobs used PARC's vision to design Macintosh personal computers, and this Mac saved Apple in the 1980s. PARC's research had given Apple its technology lead in personal computers from 1985 to 1995.

In the late 1980s, Xerox turned its entire corporate attention to manufacturing copy machines, missing the evolving personal computer industry and the Internet revolution of the 1990s. By 2002, Xerox was deeply in debt and on the edge of bankruptcy. Xerox even offered PARC for sale, and a Xerox CEO was under investigation by the ERC for fraudulent reporting of sales in the late 1990s.

> **To stay high tech, a high-tech company needs continuing success in corporate research and commercial exploitation of research.**

RESEARCH AND BUSINESS SUBCULTURES

It is in the transfer of innovative ideas from research into innovative products that makes money for a corporation. But innovative product design requires a visionary and collaborative partnership between research and marketing. As we saw in the Xerox PARC case, it is possible to create pioneering research without commercializing it successfully. The reason for the difficulty in exploiting research successfully lies in the cultural gap between research and business personnel. For example, Klimstra and Raphael (1992) emphasized some of the basic organizational differences between research and business subcultures: "There are long-standing and deeply-rooted problems involved in integrating R&D and business strategies. . . . R&D and business organizations have conflicting goals and practices."

Among the conflicting goals and practices, they especially noted the differences between the time horizon, financial perspective, product perspectives, and preferred method of innovation (Table 13.1).

R&D is generally a long-term investment. Even the shorter product developments from applied research to development take two to three years. The longer developments from basic research usually take 10 years. Thus, R&D is long-term strategic in its planning horizon. In contrast, business units are always under the quarterly profit accounting system, focused principally on this year's business. Business units are short-term operational in their planning horizon. Yet if an operational component is not included in R&D strategy, it will not integrate properly into a business unit's planning horizons. Conversely, if business units do not have strategic 10-year planning horizons, they have strategic difficulty in integrating the R&D units plans.

TABLE 13.1 R&D versus Business Considerations

Characteristic	R&D	Business
Time horizon	Long term	Short term
Financial focus	Expense center	Profit center
Product focus	Information	Goods/services
Innovation method	Technology push	Market pull

Another differing characteristic is that organizationally, R&D laboratories are expense centers, since they do not produce income directly, whereas business units produce income directly. This can create arrogance in the business unit toward the R&D organizations, making it more difficult for R&D to elicit support and cooperation from the business unit.

The characteristics of products from the R&D center and business units also create important cultural differences in the organizations. Business units products are goods and/or services sold to customers of the firms, whereas R&D center products are information and understanding and ideas that must be communicated internally to business units and then embodied into goods and services. This can generate cultural differences as to how the different organizations value ideas, since the R&D unit must place a higher value on the idea of a future product than does the business unit.

Finally, the methods of the organizational units differ, with the R&D unit valuing scientific and engineering methodology and principles that result in pushing technology from the opportunities of technical advance. In contrast, the business unit, with its direct contacts with the market, will be primarily interested in market demand, meeting the needs of existing markets.

This problem of making research relevant to business requires formal procedures to foster cooperation, and a culture of trust and relevance must be built up by experience. From the perspective of the researcher, cooperation with a business unit will always be problematic. For example, Frosch (1996) nicely summarized the researcher's perspective on their customers for research: "After 40-odd years of working in application-and-mission-oriented research, I have come to believe that the customer for technology is always wrong. . . . I have seldom, if ever, met a customer for an application who correctly stated the problem that was to be solved. The normal statement of the problem is either too shallow and short-term. . . . What really happens in successful problem-solving . . . is the redefinition of the problem."

To tackle these differences in cultures, it is useful to formalize procedures for strategically integrating R&D into business unit activities. For this, Klimstra and Raphael (1992) argued the usefulness of having two parallel sets of formal decision processes in what they called an *R&D product pipeline*. A procedure for such a pipeline involves two parallel sets of activity to integrate research strategy in research labs with business strategy in business units (Figure 13.2). This requires that as research programs move into development projects in product development,

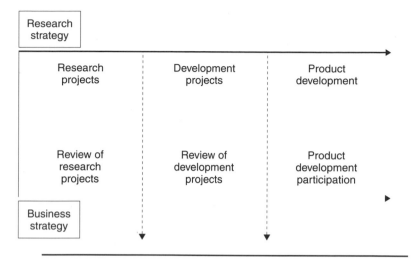

Figure 13.2. R&D product pipeline decision points.

research strategy should be reviewed formally by business units as part of their business strategy.

Business units should participate in the research program reviews. Project selection decisions should be made jointly by research and business units. While development projects are proceeding, business unit participation with the research unit in development project reviews continues to be necessary. Then joint go or no-go decisions can be made by research and business units before product development begins. During product development, business unit personnel should participate actively with research unit personnel in joint product development teams.

SUMMARY AND PRACTICAL IDEAS

In case studies on

- GE's Corporate Research Center
- William Coolidge as research manager
- Xerox pioneers the paperless office

we have examined the following key ideas:

- Institutionalization of scientific technology
- Careers of scientists in industry
- Objectives of corporate research
- Profit-gap analysis

- Research laboratories
- Evaluating corporate R&D
- Funding corporate R&D
- Research and business subcultures

These theoretical ideas can be used practically in understanding the purpose and challenges of performing science and technology in a business, at both the corporate and divisional business levels. It is in the corporate research function that the long-term future of a company can be secured through technological innovation.

FOR REFLECTION

Find a study of a corporate research lab and identify contributions to the corporation's history. Were new product lines begun for research? Were new businesses begun from research? Was production improvement begun from research?

14

ENGINEERING FUNCTION

INTRODUCTION

Next, we look at the business function of engineering. Engineering is the direct link between research and new products. Earlier we defined *engineering* as the understanding and design of the manipulation of nature for human purposes. Engineering does not create new technology but embodies new technology in products, processes, and services. Successful research from the *research function* is innovated as new technology into new products or services by the *engineering function*. We examine (1) the logic of engineering activities, (2) particularly the logic and process of engineering design, and (3) the perceptual differences between engineering and management.

CASE STUDY:

Wang's Word Processor

Designing new products and new production processes are central to the engineering function. An Wang pioneered an early application of the microprocessor to word processing. This case study of early word processors (before personal computers took over this application) illustrates the product design and innovation activities of engineering. It also illustrates how briefly innovative product designs can last when technology continues to change.

In the 1950s, Wang was a professor of electrical engineering at the Massachusetts Institute of Technology (MIT). He invented the first magnetic core type of memory circuit for the then new electronic computer technology. IBM purchased his invention for $1 million. Wang resigned from MIT and started his own electronics firm with the capital. At first Wang designed and built electronic calculators, but larger firms entered the field, driving his small operation out of business. Wang needed a new product and introduced a microprocessor-based word processor in 1971. Then larger companies also introduced microprocessor-based word processors.

On September 28, 1975, An Wang held a strategy meeting with three of his employees (Shackil, 1981). He was worried about their current word processor facing the new competition from IBM, Burroughs, and Xerox. Their word processor had only captured 3% of sales, whereas IBM already had 75%. How were they to survive? Wang said that they needed to make a major technological advance in word processors, a new-generation product model. Early word processors were very difficult to use. Their display showed only one line of text, and to edit, the writer had to calculate position to locate a word in the manuscript. Wang and his engineers imagined a new model word processor that would be easy to use, having a full page display and storing large documents.

With this functionality in mind, the first stages of the project were to define engineering specifications and design the system architecture. The employees began programming the system and the application software. They decided to build the word processor as a distributed system around a minicomputer as the system control device. Each application user (word-processing person) would have their own dedicated terminal for date entry, editing, and recording. Each terminal would be connected to a central minicomputer and printer. Microprocessors would be used in each terminal and in central control and in the printer, to distribute computation for speed and memory handling.

The project was scheduled to show a product at the next International Word Processing Show in New York on June 21, 1976. This gave them nine months in which to design the machine. As one of the project members, Edward Wild, recalled: " 'We worked until 10 pm almost every weekday, many Saturdays and some Sundays. I can remember eating lots of McDonald's hamburgers during those long hours' " (Shackil, 1981, p. 30).

The hardware parts of the system needed to be designed or purchased. The software needed to be written: system and word-processing control. It was the software that was pushing the state of the engineering art in the design and gave the engineers and computer programmers the most problems: "Every new design has its share of 'war' stories, and even though this system's architecture was relatively simple, the designers' unfamiliarity with microprocessor-based designs did generate some new headaches. . . . There were microprocessor documentation problems. . . . There were elusive system problems. . . . Another problem that arose later on concerned the allocation of space of the rigid disk while a read or write was going on" (Shackil, 1981, p. 31).

But the design was finished, the technical problems solved, and the product got to market on time: "The opposition was caught flat-footed. Wang leapfrogged IBM in word-processor sales. From being a nobody, the company shot up to capture more than 50% of the dedicated world-processor market" (Shackil, 1981, p. 29). Introduced in 1977, sales began at $12 million, rose to $21 million in 1978, then to $63 million, $130 million, and $160 million in 1981. But 1981 was the peak and also the beginning of the end. By 1985, the new personal computer, particularly the IBM PC, introduced in 1984, was taking over the word-processor application and subsequently, killed Wang's word processor in the business market. Wang did not switch to making personal computers nor to writing word-processing programs for PCs. The Wang company struggled to redefine its position during the 1980s. In 1990, after An Wang's death, the company went bankrupt. This case illustrates how an innovative product can create a market but also lose that market when a later innovative product substitutes for the earlier innovation. Engineering design of new products is not a "one-shot" activity in a successful business but a recurrent and continuing activity to continually introduce new and improved products.

ENGINEERING LOGIC

The logic of engineering focuses around three types of activities: design, problem solving, and invention. Design is the activity of creating the form and function of products, production processes, or services. Technical problem solving is the activity of making technologies work and work well in products, processes, or services. Invention is the activity of creating a new technology.

Design

In design, the essential logic is to create forms of physical and logical processes to perform function. Form and function provide the basic intellectual dichotomy of the concept of design. The logical steps involved require first determining the performance required for the function for the customer and then creating an integrated logical and physical form for fulfilling the function. Accordingly, in product design the first logical step is to establish *customer needs*. This is a list of:

- The customers' applications of the engineered product
- The functional capability of the product for these applications
- The performance requirement for the applications
- The features desired for the applications
- The size, shape, material, and energy requirements for the applications
- The legal, safety, and environmental requirements for the product
- Supplies for and maintenance and repairability of the product
- The target price for the product

Once the customer-needs list is established, the next logical step is to establish a *product specification set*. These product specs translate customer needs into technical specifications that guide the engineering design of a product. The third logical step is then the design of the product, using ideas from previous product designs and innovative new design ideas to create a product that meets the product specs and customer needs. Although these logical steps sound sequential, in practice, successful design requires concurrent interactions with marketing and finance and redesign loops as both the needs and specs get refined into design details. Thus, in a large organization, design occurs in groups of designers and goes from a conceptual design stage into a detail design stage and back and forth until a final design is realized that is ready for testing.

After a design goes into testing, the design must often be modified to meet disappointments in the product's design. Once a tested design is ready to be produced, the design must again be altered to become manufacturable in volume with high quality and to meet target costs. (As many as the manufacturability criteria as possible should be brought early into the design process, to minimize redesign for manufacturability.)

A 1991 study by the National Research Council Committee on Engineering Design Theory and Methodology emphasized the importance of product design in industrial competitiveness: "Engineering design . . . is the fundamental determinant of both the speed and cost with which new and improved products are brought to market and the quality and performance of those products. . . . It is estimated that 70 percent or more of the life cycle cost of a product is determined during design" (NRC, 1991, p. 1).

Problem Solving

Engineers make the technologies that a firm uses work; and all technologies are touchy. Nature is always more complicated than the use we make of it. Simplifying nature to make it work for us technologically always stimulates technical problems, which engineering must solve.

The logic of problem solving includes:

- Recognition of a problem
- Identification of the problem
- Analysis of the problem
- Solutions to the problem
- Testing of problem solution
- Improvement of solution and/or redefinition of the problem

In large organization, problem recognition may not be simple. This requires that leadership sees that there is a problem and acknowledges the existence of a problem. If leaders will not recognize that a problem exists, personnel cannot work on solving the problem. Problems may not be recognized because the leadership does not have the expertise to recognize the problem or because it is politically inconvenient or embarrassing to leadership to acknowledge that a problem exists.

Once a problem is recognized, the next logical step is to identify the nature of the problem, its location, and the client to whom it causes difficulties. Identification of a problem can then logically be followed by analysis of the problem, its sources, and its causes. Once the source and cause of a problem are known, solutions to the problem can be diagnosed or invented. The proposed best solution can then be tested to see if it solves the problem. If testing shows that the problem is still not solved, further refinement of the solution or alternative solutions may be tried and tested. Sometimes even the problem requires redefinition, when testing shows that the problem was not understood properly initially.

Problem solving is also an essential aspect in the design and invention activities of engineering. There will always be problems in inventions and in new designs that are found only after use in the field. For example, Von Hippel and Tyre (1995) examined problem solving in two cases of designing new process equipment. They found that in about half the problems encountered, information about the potential problem did exist in the users but were not communicated to the designers, as not thought to be relevant. In the other half the problems appeared only after use of the new equipment in the field.

Since technical problem solving is a major activity of engineering, understanding the nature of solutions for problems is a critical skill of engineers. This is why science is a base knowledge for engineering, because science provides the knowledge of nature that underlies technology problems.

Invention

Often, in design or in solving a technical problem, an engineer needs to invent a new way of doing something. Invention is the creation of new technology. We recall that the logic of invention is an idea that maps functional logic to physical morphologies. Invention is always new to the person inventing it. Invention need not be new to everyone as long as the invention solves the immediate technical problem. But if the invention is new to everyone, it may be patented.

CASE STUDY:

Design of the B-52

We now look more deeply at the design activity using a case of airplane design. Figure 14.1 pictures the North American bomber called the B-52. In 1948 the Cold War had begun between the Western democracies and the communist empire of the USSR. General Curtis LeMay, then head of the U.S. Air Force, saw the urgent need of a new long-distance bomber: "The plane [B-52] was designed

Figure 14.1. Boeing B-52.

by Boeing in a fitful weekend in 1948 after Curtis LeMay, the imperious Air Force general, insisted he needed a new heavy bomber that could drop several nuclear bombs on the Soviet Union in one pass" (Zuckerman, 2001, p. A17).

Previous bombers had been the propeller-driven aircraft of World War II, with the longest range of the U.S. bombers being those of the B-17 and the B-50. Both had been designed and produced by the Boeing Company, and Boeing was the company to which LeMay turned to design a new kind of long-range bomber with jet engines. Jet planes had been developed by the Germans in World War II as a fighter jet that could fly at twice the speed of any propeller-driven aircraft. Fortunately for the Allies forces, the German fighter went into service too late to help the Nazi government stave off destruction from Allied armies. The technological future of aviation was clearly jet planes: "The [B-52] plane was one of the first jet bombers and the first large airplane to have swept-back wings, enabling it to fly 650 miles an hour. The design subsequently helped Boeing become the leading maker of civilian passenger jets" (Zuckerman, 2001, p. A17).

The first B-52 bombers were produced in 1951. They had four jet engines and could fly as high as 50,000 feet and as long as 9000 miles, with a bomb load weighing 54,000 pounds: "The specter of B-52s circling the globe 24-hours a day, within minutes of striking the Soviet Union, was part of the deterrent that prevented nuclear war for more than 40 years" (Zuckerman, 2001, p. A17).

B-52s were used to bomb during several U.S. conflicts: the Vietnam conflict, Persian Gulf war, Kosovo, and Afghanistan. Beginning in 1951, a total of 744 planes were built, with 95 still being flown by 2002. Many were destroyed as mandated by the 1991 strategic arms reduction treaty between the United States and Russia.

DESIGN AND SYSTEMS ANALYSIS

All products begin as ideas—ideas of embodying technology into the materials, processes, procedures, or logic of the product.

> **The design of a product is an embodiment of a technological capability to meet a customer's functional need.**

In an engineering text on mechanical design, Suh (1990) described *design* as an interaction between the functional and physical domains of a mechanical product: "The objective of design is always stated in the functional domain, whereas the physical solution is always generated in the physical domain. The design procedure involves interlinking these two domains at every hierarchical level of the design process. These two domains are inherently independent of each other. What relates these two domains is the design" (p. 26).

Hardware design is creating physical form (physical morphology) to perform function. For example, the design of a hammer has the function of driving nails into wood in a customer's construction application. The physical form of the hammer consists of a heavy metal head and a long wooden or fiberglass handle. The schematic logic (function) embodied in the hammer lies in the form of the head of the hammer, with one flat surface for driving nails and a clawed surface for extracting nails, and in the configuration of the hammer, with a long handle to swing the heavy hammer head down onto a nail. The schematic logical combination of the weight of the head and the length of the handle facilitates the force the customer can apply when using the hammer. The operation of the hammer as a product (to use the technology embodied in the hammer's design) is to hold the hammer in one hand and a nail in another and to drive the nail into wood with blows of the hammer: technology–design–form/function–product–application. (In the case of early CPU chips, the transistor density of current present technology always limited the designer in the physical forms of the chips for functional performance.)

All hardware products can be so analyzed as to the form/function in their design—physical morphology mapped to schematic logic in the core technology of the product. This hardware form/function relationship provides the criteria for distinguishing a good design from a bad design. A good design provides an appropriate physical form for a function. A bad design is one whose functional performance by a physical form does not perform adequately or safely and at the right cost. Designs of products that provide poor performance are unsafe and at a high price result in products that do not sell.

Fitting form to function requires an analysis of the function as a system. We recall that all products are systems. The description of a product as a system is constructed through a *systems analysis* of the product design. Embedding a technology system into a product design concept requires a systemic analysis of the schematic logic of the functional transformations of the product.

> **Systems analysis of a product design is a systematic depiction of the sequential and concurrent functional transformations (schematic logic) required for total product performance.**

The design logic of aircraft such as the B-52 uses a "systems" view of the product. Figure 14.2 depicts a systems flow diagram for designing an airplane. The configuration specification expresses operating parameters for the airplane system: load capacity, top speed, fuel efficiency, rate of climb, range, acrobatic stress limits, damage tolerance, maintainability, and so on. These specifications provide targets and constraints for the airframe design and the engine design.

In the airframe design, choices are available to the designer about the shape of the airframe, its structural reinforcement, and the packaging of its exterior surfaces. There are also choices in engine design, power, weight, features, and dependability. Once the airframe configuration and engine design have been made, the next systems design step lies in integrating the two by adding controls. This integration may lead to revisions of either design for final requirements for sizing and weight and performance (payload, speed, agility, stealth, etc.).

A systems analysis for a product/service design lays out (in a system design flow diagram) the sequence of logical transformations as a series of both parallel and

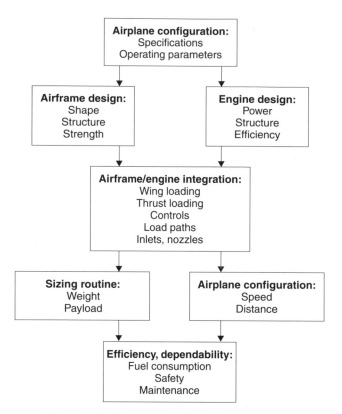

Figure 14.2. Systems analysis of airplane design.

sequential operations of the product system. In that analysis one can identify the locations of technological innovation within the system as:

- Boundary
- Architecture
- Subsystems
- Components
- Connections
- Structural materials
- Power sources
- Control

Accordingly, targets for implementing new technology into the product system should be expressed in a systems analysis of the product system, identifying loci of innovation and estimating the effects on the functional performance parameters of the product system. This analysis establishes the location and results of technical improvement in the product system. The system questions for technical innovation in product design include:

1. What is the loci in the product system for embedding the technical progress?
2. What improvement in functionality and/or performance will the technical advance provide?
3. How will the customer recognize value in applications for this improvement in functionality/performance?
4. How important will the improvement be to the customer?
5. What are the technical risks of innovating a technical advance in the product system?
6. What is the appropriate trade-off between technical advance and cost of an innovation that will be appreciated by the customer?

Another important decision in product design is which parts of a product to make and which parts to buy. This decision applies to products that are hard goods, software, or services. Criteria for the make/buy part decision are competitiveness, timing, cost, quality, and safety. The most important components of a product to make are those that provide a competitive edge in the product for the firm, such as the components that embody the core technologies of the product. Also, the important core technology in a designed/made part must be recognizable by the customer as a product superior to those of competitors.

> **Technical innovations in a product that are invisible to or unused by a customer produce no competitive advantage. Also, if all components are purchased, the business has no technology competitive edge in its product and no way to differentiate its product.**

DESIGN BUGS

Because of the complexity of product and technology systems (whether a hardware or a software product), all innovative product designs will encounter technical problems—*bugs*. In product design it is important to anticipate technical bugs. For example, Clark (1990) described designing Digital's VAX 8800 family of midrange computers announced in 1986: "We simulated the VAX 8800 design extensively before building the first prototype. . . . Engineers simply marked a centrally posted sheet with a tick-mark for each bug they found. . . . We regarded finding and fixing hardware design bugs as our most important activity. . . . Over 96 percent of the total reported design bugs were found before the prototype was built" (pp. 26–27).

Clark emphasized the importance of including bug identification and fixing bugs as a formal process within the design process, arguing that *debugging* (the tracking down and fixing of bugs) is a normal and necessary part of a design process. Software designers have added the word *bug* to our lexicon to indicate any problem with a new product that was not anticipated and thus designed correctly to avoid. In software, bugs are logical errors in the design. In hardware, bugs can be either logical errors or physical problems. Debugging is important in both software and hardware design.

Design bugs arise from the complexities of the technology and product system. The more complex the system, the more difficult it is to fully think out and consider all the aspects of the system and the interactions within a complex system. Therefore, bugs surprise the designer because the full implications of a designed complex system may only be understood in testing out the system. This is particularly true of designs that embody new technology.

> **The more radically new a technology, the more likely that there will be bugs—not from lack of judgment but from lack of experience with the new technology.**

Bugs are inherent in new technology, so when innovating new technology into a product design, the design procedures should include debugging as normal to design, not as a failure of a designer. The important thing in managing technological innovation in product design is to find the bugs early.

> **The later in the product development process that a bug is discovered and fixed, the more costly its solution and the slower its development.**

To identify bugs early, it is useful to design and test products in a modular fashion. Also, the ability to simulate a product design in a computer is very useful to the early identification of design bugs. Design areas most likely to produce bugs include (1) areas of the greatest complexity, (2) interfaces in subsystems when the subsystems have been designed by different people or groups, and (3) parts of the design that are used more rarely in product functioning.

The procedures for product testing are also important to the early identification of bugs. The traditional approach to product testing has been to make a list of tests that test each subfunction of the product and then to perform these tests in sequence. That is the proper way to begin testing, but it may be insufficient, particularly for a complex system. Clark (1990) has suggested complementing testing design with a "bug-centered approach" on the part of product designers. For example, the product design team may set aside some periods (a day, a week, or a month, perhaps) as deliberate bug-finding days. The idea is that finding bugs is a positive contribution to design, not a negative contribution. Prizes might even be given to the team member discovering the most or most challenging bugs. Keeping a chart of the number of bugs discovered per week during the design is also a useful procedure to managing innovation in design. Confidence in the design should be determined not by the number of bugs found but by the *rate* of finding bugs. There is never a guarantee that all design bugs will have been found in a complex product system. However, when the rate of finding bugs decreases toward zero, particularly in a design process that seeks to find bugs deliberately and continuously, the design is approaching robustness and will be ready for final testing. It is also helpful to review the types of bugs being found in the design. For example, is a bug indicative of a more general problem with the design? Why didn't the test methods catch the bug in the first place? Could the method used to find that bug find other bugs? And so on.

In summary, making bug-finding an active, deliberate, and focused part of the design process as a complement to formal testing of the completed design improves the capability of innovating new technology in a new product design.

CASE STUDY:

Iridium

Commercial success on an innovative product depends critically on good design. And a good design properly matchs a *technology system* to a *market application*— successful innovation as technology push matched with market pull. This requires good cooperation between those who understand technology, engineers, and those who understand markets, sales personnel. This turns out not to be easy. *Misunderstanding* between engineering and marketing personnel about just how to embody a technology system into a high-tech product is the *major cause of business failure* in high-tech businesses. We can see this in the sad case of Iridium, a spectacular business failure in the 1990s due to the faulty product implementation of a technology system. Iridium was a new business started by a capable and successful high-tech global firm, Motorola. But the implementation of new technology systems into commercially successful products is a tricky process—so tricky that sometimes even the very competent stumble. This was a big stumble. Iridium lost $5 billion, and Motorola alone lost $2.5 billion of that. But Motorola was big enough to survive.

The historical setting was the year 2000, when satellite technology was continuing to progress. Motorola planned Iridium to be the first global communications system using satellites from ground-to-space-to-ground communication. Motorola planned and managed the new venture, with financial participation by Sprint and BCE of Canada. Motorola built and operated the communication satellite system and designed and produced the handheld phones.

In 1997, Iridium launched its first 88 satellites into low orbits around the Earth. The low-Earth orbits were chosen to minimize communication delay time. If satellites were in high orbit, there would be a noticeable delay when one party was talking to another party. But low-Earth-orbit satellite systems need many satellites to cover the Earth, since low-orbit satellites see less area on Earth than do high-orbit satellites. Yet three years later, after the first satellite launch, Iridium had failed—by 2000 the company was bankrupt. Iridium had spent all its capital of $5 billion, with no hope of recovering its investment. Iridium planned to destroy their satellites: "In the coming months, Iridium, L.L.C., the bankrupt global satellite telephone company, will begin sending 88 giant satellites spiralling in toward Earth, ending one of the colossal corporate failures in recent memory" (Barboza, 2000, p. C1).

Iridium never even came close to getting its planned 1.6 million subscribers by the year 2000: "After spending more than $5 billion on a system that promised to communicate 'with anyone, anytime, virtually anywhere in the world,' Iridium could muster only about 55,000 subscribers, not enough even to pay interest on its start-up costs" (Barboza, 2000, p. C1).

Iridium phones didn't work well because the phones (1) were too large to hold comfortably in the hand and (2) didn't work inside buildings (they had to "see" the satellites directly, without obstruction). As one observer of the time, James Grant, commented: " 'It was a technology that didn't live up to its hype or its billing' " (Barboza, 2000, p. C2).

> **This was a failure in the *engineering vision* of the technology system— too large phones that didn't work inside buildings.**

Moreover, the customer subscription prices had been set too high, at $7 dollars a minute, with an additional cost of $3000 for an Iridium phone. This price was in contrast to that of cellular phones at less than $100 for the phone and pennies per minute for connect time. Cellular phones did have limited range, but were connected to long-distance services. The numbers of corporate customers really needing access from lonely global places and willing to pay that price were few.

> **This was a failure in the *marketing vision* of the technology system—too high per-minute price and too high phone price.**

After Iridium's bankruptcy, Daniel Colussy, a former president of Pan-American World Airways, acquired the assets of Iridium for $25 million.

> **This was a failure in the *financial vision* of the technology system—a $5000 million investment was sold for $25 million.**

In December 2000, the U.S. Defense Department contracted with Colussy's new Iridium Satellite company for $36 million a year to obtain unlimited use of the global phone network: "Iridium Satellite is the only company that can offer encrypted wireless service worldwide, allowing Defense Department officials to discuss classified information anywhere in the world over the company's satellite phones . . ." (Jaffe, 2000, p. A18). This contract provided one-third of the revenue needed for the new Iridium company. It needed another 60,000 customers to pay 80 cents a minute to break even. The billions of investment in Iridium had been lost to Motorola, but Colussy might yet operate and eventually build a more modest system, profiting from that failed investment.

In summary, Iridium was planned and launched with marginal technology and high prices. The reason was a technology systems failure due to the lack of appropriate experience of Motorola personnel. Motorola was a company not in the service business but in the communications equipment business. Accordingly, Motorola's engineers and marketing people did not get the service system right. The technology system was not adequate for a communications service. Since

Motorola had no direct experience in the communications services business, its engineers did not get the system design right, and Motorola's marketing people didn't get the service pricing right.

A technology system has to be just right for a market's performance and pricing requirements. From this case one can see that building a commercially successful system requires some major up-front design decisions about equipment and operating systems. Iridium's strategy failed in engineering, marketing, and finance:

- It was *incomplete in technology strategy* by not being able to design workable small handheld devices with the then-current state of technology.
- It was *incomplete in marketing strategy* by not being able to deliver the service close to competitive wireless phone pricing.
- It was *incomplete in financial strategy* by not having sufficient capital to correct any initial product and pricing mistakes.

INNOVATIVE DESIGN

We recall that the differences between technology and product/service are:

- Technology is *knowledge* of the way to manipulate nature.
- Product is a *designed embodiment* of technologies in a good or process.
- Service is the *application of products* for a functional activities.

Accordingly, innovative product designs can fail commercially in two ways:

1. The product design may fail as its expression of a principal technology system performing functionally less well than a competing product.
2. The product design may fail in its balance of performance and features as perceived by the customer (even if performing technologically as well as a competing product).

Accordingly, there are risks in product design, including making the appropriate trade-offs in product design between product performance and cost, and uncertainty in establishing precisely the set of customer requirements of a product for the various classes of customers. Therefore, key problems in implementing new technology in new products or new services include:

1. Making the appropriate performance/cost trade-off in designing a new-technology product/service
2. Determining the best application focus for a new-technology product/service
3. Determining appropriate requirements and specifications for a new-technology product/service
4. Deciding wherein should lie the proprietary and competitive advantage of a firm's product/service

The makeup of the design team is central to making these design decisions properly.

Ford's Team Taurus in 1980

How can proper cooperation between engineering perspectives and market perspectives be ensured in a design process? Next we look at a case study that focused precisely on the interaction of engineering personnel with other business personnel in product model design—the Ford Taurus project of 1980.

In 1979–1980, the U.S. automobile industry was in a major crisis. After a rapid rise in oil prices in the late 1970s, U.S. cars were too inefficient for customers to buy. Also, the U.S. auto industry was plagued with low productivity, high costs, and long product development times. In 1980 Ford Motor Company began the formation of self-organizing, multifunctional teams for product development for the development of their new Taurus/Sable models. Then they called it a simultaneous approach to product development, or *concurrent engineering*. Lew Veraldi (who had been the leader of the Ford Taurus product development team) described that project: "We're very honored by all the attention about Team Taurus and by what it says about Ford Motor Company. But you must stop and ask yourself, why should a large company like Ford, which has been developing new cars for over eighty years, decide to change the way it does business? . . . The need to change was 'survival.' That gets everyone's attention" (Veraldi, 1988, p. 1).

Veraldi also described the business condition of Ford at that time: "Remember when Taurus began, it was 1979–1980. Ford's image for quality was not very good. In addition, we were in the process of losing over $1 billion for two years in a row" (Veraldi, 1988, p. 1). Ford's product development process at that time was linear in format. It began with (1) concept generation, then (2) product planning, next (3) product engineering, then (4) process engineering, after which (5) full-scale production was begun. This kind of linear product design and development process was then standard in the U.S. auto industry (Clark and Fujimoto, 1988). Veraldi summarized the process then used at Ford: "The old organizational structure [was] where designers do their thing—design then turns that over to the engineers. After the engineers do their job, Manufacturing is told to go mass produce the product and Marketing is then told to go sell it. . . . What you have is each group of specialists operating in isolation of one another" (Veraldi, 1988, p. 1).

The problem was the number of design changes (and redesigns) linear development encouraged and, consequently, the long time it took to complete the development with all changes: "Moreover, what someone designs and styles may be quite another matter to Engineering. And by the time it reaches Manufacturing, there may be some practical problems inherent in the design that make manufacturing a nightmare. . . . Marketing may well discover two or three reasons why the consumer doesn't like the product, and it is too late to make any changes" (Veraldi, 1988, p. 1).

Change in the design process occurred in 1979, under new Ford management: Donald Peterson as president and Phillip Caldwell as chairman of the board.

Earlier in the 1970s, Caldwell had been head of international operations (1973) and was assigned the task of developing the small Ford Fiesta in Germany. Development of the Fiesta cost $840 million and was then the most expensive car development project at Ford. Caldwell had chosen Lew Veraldi, a design engineer, to lead the project. Veraldi had worked 25 years in Ford's design systems. He had experienced the frustration of taking a design to manufacturing only to be told that it wasn't designed correctly to be manufactured. In the Fiesta project, he called in the manufacturing people at the beginning of the design and asked them: "Before we put this design on paper, how do you, the manufacturing and assembly people, want us to proceed to make your job easier?" (Quinn and Paquette, 1988, p. 3).

Instead of using a linear design process during the Fiesta project, Veraldi began trying concurrent cooperation in design between design engineers and manufacturing engineers and liked the benefits of the experience. He understood the importance of early close cooperation between product design and manufacturing, and he saw that he could make that happen. Veraldi was convinced of the real advantages of designing a car simultaneously between product engineering and process engineering and marketing rather than in sequential isolation. It had avoided many late changes or desirable changes that were identified too late to be made, and it saved money in the development of the Fiesta.

When in 1979, Caldwell and Petersen decided to replace Ford's midsized cars with innovative new products, they assigned the Taurus project to Veraldi. Then it was high risk, as Veraldi later commented: "the courage . . . of Mr. Philip Caldwell, then Chairman, and Mr. Don Petersen, then President, to take a risk—in many ways, to bet the Company—to spend $3 billion when we were losing $1 billion per year" (Veraldi, 1988, p. 1).

In the summer of 1980, Veraldi and his group presented concepts for the new car to top management. After Veraldi's previous experience with concurrent engineering in the design of the Fiesta, senior management agreed to create a management team (later to be called Team Taurus) to try out the concurrent engineering ideas on the Taurus development project.

CONCURRENT ENGINEERING

We pause in our case to review the concept of using multifunctional product teams for product design, or as some then called it, *concurrent engineering,* or *simultaneous engineering* as it was called at Ford. A concurrent engineering product development team requires forming multifunctional teams for the management of complete development of a new product. The product development project management team should consist of representatives from the relevant activities of engineering product design, engineering process design, manufacturing, marketing and sales, finance, and research.

The job of this product program management group is to formulate the design requirements and specifications, taking considerations of manufacturing and marketing and finance and research into account as early as possible. The product program management group first conducts a competitive benchmarking of competing products in all price categories and establishes a list of best-of-breed performance and features for the product system. They next establish a product development schedule and early prototyping goals and means. They identify early sources of supplies and draw in suppliers' expertise and suggestions about part and subassembly design. The group also consults and solicits suggestions from dealers and service firms that repair and maintain the product systems, and from insurers, for suggestions on product improvement and feature desirability. The group also samples customers and solicits reactions to current and competing products.

After this, the product program group compiles a want list from all these groups, internal and external, of desirable product performance, features, and configuration, and analyzes these into priorities of (1) must have, (2) very desirable, and (3) desirable if possible within costs. Finally, the group continues to manage the product development process to encourage teamwork and cooperation in developing a product rapidly and of highest attainable quality and lowest cost.

> **Concurrent engineering practices can facilitate the implementation of new technology in products by improving communication and cooperation between research, engineering, manufacturing, marketing, and finance.**

In 1991, Boston University had a conference with manufacturers to summarize lessons that were being learned in speeding products to market (based on a project conducted on the subject by Stephen R. Rosenthal and Artemis March, 1991). The conferees particularly emphasized how important were the early phases of product development on cost and the quality of the product design. The conferees also discussed lessons about collaboration in concurrent engineering:

1. Selection of the team is a critical and complex problem and crucial to eventual project success.
2. Senior managers should concentrate their involvement in the early stages of defining the product concept but should resist calling all the plays, thus eliminating choices for the team.
3. Attention must be paid to overcoming the barriers to communication in the "separate thought-worlds and meanings" that the different cultures of specialists bring with them (such as we discussed in Chapter 13 on the intellectual differences between technical and business functional personnel).
4. Special attention must be paid to the difficulties in creating cross-functional team collaboration in organizational settings that are multisite or organized by functional specialization and that have older, long-established traditions of sequential responsibility in product development.
5. The targets for product success should not be confused with the means, so the product specifications should be established early but the design team should be allowed to design the product.

6. Initially, the importance of using concurrent program management teams is to allow all product targets to be considered simultaneously, for proper trade-offs and balance in setting the product specification.

7. It is important for the technology requirements that the product be explicit and clear very early to representatives of the downstream activities (such as manufacturing and marketing) to enable decision making as to whether the requirements are feasible and achievable within target cost constraints and can be produced by existing facilities.

8. Because of risks inherent in new technology, contingency plans are necessary in case the new technology proves immature, with early product prototyping and early testing to determine whether it is immature.

9. Attention should be paid to thorough testing of new production processes and the training of workers before volume production begins.

10. Cost-reduction targets should be set for a mature product initially (rather than merely an untargeted cost-reduction type of program).

11. It is important to integrate vendors into the concurrent engineering teams.

CASE STUDY CONTINUED:

Ford's Team Taurus in 1980

The Team Taurus group consisted of a car program management group, headed by Lew Veraldi and including such key personnel as John Risk (Car Product Development Planning Director), A. L. Guthrie (Chief Engineer), John Telnack (Chief Designer), and Philip Benton (later, President Ford Automotive Group). The importance of organizing this kind of initially *inclusive* participation was to bring downstream judgments into the design process early. The car program management group set initial management goals for the project (Quinn and Paquette, 1988):

- Developing best-in-class performance for the car
- Building the first prototype to contain 100% prototype parts to test the manufacturing prototype completely
- Improving the focus and timing of key decisions for rapid product development
- Reducing the complexity of the product to improve quality
- Eliminating late and avoidable design changes by bringing manufacturing considerations in early in the design
- Controlling changes to as few as possible late in the design in the program, to avoid extending the product development time
- Creating a small-business-like working situation for the project, to improve commitment

To target the goal of best-in-class, the team began with competitive benchmarking of competing products: "The team sought out the best vehicles in the world and evaluated more than 400 characteristics on each of them to identity those vehicles that were the best in the world for particular items. These items ranged from door closing efforts, to the feel of the heater control, to the underhood appearance. The cars identified included BMW, Mercedes, Toyota Cressida and Audi 5000. Once completed, the task of the Taurus Team was to implement design and/or processes that met or exceeded those 'Best Objectives.' " (Veraldi, 1988, p. 5).

The features on this group of cars provided standards for the features of the Taurus, and Veraldi saw success in meeting many of these standards. Another of the first things the group did was to create a want list from all relevant constituencies, including:

- *Internally:* Ford designers, body and assembly engineers, line workers, marketing managers and dealers, legal and safety experts
- *Externally:* parts suppliers, insurance companies, independent service people, ergonomics experts, consumers

The number of components within which a designer can design a part is often variable. Here information from manufacturing experience made the designers aware of the importance of reducing the number of components. Their want list built up to 1400 items, of which the team concentrated on 500 of the items. Veraldi included Ford's dealers and suppliers in the simultaneous process.

The new product development process enabled the Taurus team to complete the project on time and below budget. Introduction of the Taurus in the Ford Division and its variation the Sable in the Mercury Division cost $250 million less than projected in their design budgets. Both products were successful in the marketplace. In 1988, Lew Veraldi was promoted to a vice-president of Ford, in charge of luxury and large car engineering and planning.

As a historical note, Ford did very well throughout the 1980s and 1990s. But by 1999, Ford was again in trouble. Then Ford's CEO, Jacques A. Nassar, had in three years "spent his way though a pile of cash acquiring Volvo Car and Land Rover and also squandering billions on several e-businesses [as] the company's $15 billion cash hoard dwindled to less than $1 billion" (Kerwin et al., 2002, p. 89). Then in 2000, as Ford ran a $5.5 billion deficit, William Clay Ford, Jr. (a great grandson of Ford) fired Nassar and stepped in to run the company. In addition to good business processes, good leadership is necessary to run a company.

BEST-OF-BREED PRODUCT FEATURES

Product designs are always compromises between performance and cost. Yet from a competitive perspective, it is foolish to design a product system that has any feature distinctly inferior to any feature of a competing product in the same price class. Moreover, there is a competitive advantage in providing, in a lower-price-class product, the quality of features of product systems in higher price classes.

The systematic identification and listing of product system features and their highest quality expression in a product system of any price class is called *competitive benchmarking*.

Competitive benchmarking allows a designer to see the best features in a current product line but does not prevent surprises that a competitor may have in store in their next model. In performing a competitive benchmarking effort, the design team should also use *technology forecasting* to guess what competitors might be able to add in the next model. For competitive purposes, those features whose quality are most noticeable to the customer are those that should take priority in design for providing best-of-breed product features.

The design into a product system (of a given price class) of the highest-quality features of all product system price classes is called a *best-of-breed product design*.

Of course, given cost constraints in a lower-cost product system, one can seldom attain all the best-of-breed features. But here is a good place to use technological innovation in product design.

Technological innovation that allows best-of-breed designs in lower price classes adds product differentiation competitiveness to price competitiveness in a product.

MULTIFUNCTIONAL PROJECT TEAMS

Within a company, engineering is often organized as a separate division which fosters the specialization of expertise in engineering. However, technical projects that involve engineers should be cross-functional by adding other personnel. In organic types of organizations run in a matrix mode, an engineering department may provide only personnel slots, without providing budgets. So what makes it difficult to create and manage multifunctional project teams? It is the cultural gulf between engineers and mangers in business. Engineers, the personnel responsible for the technical activities of the firm, and managers, the personnel responsible for the business activities of the firm, have been trained very differently and experience very different aspects of the business system, technical and commercial.

Engineers and managers perceive and live in two really different worlds: the *worlds of matter and of money*.

Technical personnel see and live predominantly in a world of matter, physical nature. Since the training of technical personnel is in the physical sciences, they come to appreciate that the most important thing about the world is that it is made up of matter and energy and forces—physical phenomena. Other business functional personnel live in a world of social contracts and money. The training of business personnel responsible for marketing sales, accounting, and production often is in economics and business. They appreciate primarily how much the world consists of human cooperation and competition, in which money measures, facilitates, and gauges human interactions.

Which perception is correct—the world as physics or the world as money? Both, of course, as the human world has both physical and social natures. But personnel trained primarily in one view do have difficulty appreciating the alternative view. They have been trained in different logics of thinking. For example, personnel trained in science and engineering are taught to think with concepts such as design, analysis, synthesis, problem solving, technical service, and research and development. In contrast, personnel trained in economics and business are taught to think with concepts such as strategy and planning, capitalization and budgeting, organization, staffing, control, evaluation and profitability, expected utility and discounted futures, and supply–demand curves.

To reduce risks in innovation, it takes special attention and effort to integrate these different worlds of the engineer and the manager. The technical concerns of engineers must be translatable and integratable into the strategic concerns of management. The business concerns of managers must be capable of handling technical risk and exploiting technical progress. Achieving communication and cooperation between technical and business functions and integrating business and technology strategy is not easy. And it is not simply just teaching some management principles to technical personal, or conversely, teaching a little technology to financial personal. Both approaches have been tried in some engineering or business schools, without much success. The problem goes deeper. It has to do with having a refined perception, sensitivity, and commitment about the role of technology in the enterprise system.

> **The problem in managing technology is to *refine and adapt* the logic of management so as to deal effectively with the logic of technology.**

The different backgrounds of engineers and managers require special attention to creating and operating cross-functional product teams. For example, in a case study on how engineers interact with marketing personnel, Workman (1995) listed typical complaints the two groups make of each other, reflecting their different subcultures. For example, engineers frequently complain that customers don't know what they want, so what good is marketing analysis; marketing does not have the needed expertise to specify technically sophisticated products; and marketing's time horizon is too short. In its turn, marketing complains about engineering that engineers lack perspective, do not appreciate prior customer investments, and do not appreciate the diversity of the market segments.

Building cross-functional product teams that include engineers and marketing and other business personnel is difficult but essential. For example, Swamidass and Aldridge (1966), in considering the management of cross-functional teams, emphasized the importance of obtaining consensus in decisions and of keeping a cross-functional team focused on deadlines. They proposed that project leaders take special care to develop a team perspective and concern for meeting time and budget targets. Thus in managing the engineering function, special attention must be paid to getting effective cross-functional cooperation operating in effective teams.

ENGINEERING AS A PROFESSION

Engineers are not only a different kind of folk, but they also regard themselves as different. They think of themselves as professionals, belonging to the *engineering profession*. A *profession* is a trained body of people who practice the application of a body of knowledge. The "professing" part of the concept of a profession is that there is a body of knowledge to profess. Professions organize formal education to master the body of professed knowledge, and they also organize the certification of practitioners in the profession.

Engineers organize themselves professionally into engineering disciplines centered around generic sets of technologies and their science bases. In the engineering schools of the United States, the traditional engineering disciplines include mechanical, industrial, chemical, electrical and computer, and civil engineering. Other disciplines in engineering schools may include specific technologies, such as nuclear engineering.

In addition to a professed body of knowledge, all professions have codes of ethics that have to do with practicing their profession responsibly and safely. For example, medical doctors express their ethical standards in the Hippocratic oath. State medical boards can discipline doctors practicing within their state for unethical or incompetent medical practice. Engineering ethics stress safety in the engineering design. For example, in civil suits involving injury from product use, consulting engineers may testify as expert witnesses about the soundness and safety of a product design.

There thus arises a dual set of loyalties for an engineer: loyalty to professional standards and loyalty to the employing firm. A conflict between these can give rise to a practice called *whistle blowing,* when an engineer perceives that management decisions have resulted in unsafe design of products or operations that endanger public safety. The engineering profession expects the engineer to oppose all unsafe technical practices.

Because of the dual orientation of engineers, other studies of work performance of engineers have emphasized that management should encourage a "creative tension" for work performance. For example, Lee (1992) suggested that such a creative tension, pushing both technical challenges and business goals, is very important in the development of young engineers, affecting both their early and subsequent job performance. Baugh and Roberts (1994) looked at a sample of engineers in terms of their perceptions of conflict or complementarity between their professional and organizational commitments. They emphasized the importance of an employing organization encouraging its engineers to see the complementarity of their professional commitments with the organization, and they argued that an important way to do this is for the organization to encourage its engineers to be active in their profession.

Based on technical backgrounds, engineering careers can follow two paths: a technical path in engineering (or management within the engineering function) or a general management path. In the technical path, an engineer stays within the engineering function of a firm. In the general management path, an engineer moves from the engineering function into other business functions. The career movement of an engineer into general management requires that engineers, both by experience and continuing education, gain as sophisticated an understanding of other business functions as that of engineering.

SUMMARY AND PRACTICAL IDEAS

In case studies on

- Wang's word processor
- Design of the B-52
- Iridium
- Ford's Team Taurus in 1980

we have examined the following key ideas:

- Engineering logic
- Systems analysis in design
- Principles of design
- Design bugs
- Product design process
- Best-of-breed product features
- Concurrent engineering product design
- Multifunctional design teams

The practical use of these ideas provides a way to understand the function and management issues of engineering in innovative firms. All products have finite model lifetimes, and engineering has the responsibility of designing new products and new product models and improving the production of products. Engineering is directly responsible for technological innovation.

FOR REFLECTION

Choose an engineering profession, and describe the change in the profession as science and technology progressed from the time of the origin of the profession.

15

INFORMATION FUNCTION

INTRODUCTION

In addition to the business functions of research and engineering, innovative businesses also need an information function. The corporate information function is responsible for the use of computer and communication technologies in business operations. Information systems in a business are complicated sets of technologies involving both hardware and software. Change in information technology in a business can involve change in any part or all of the information system:

- Change in *computer system*
  —Architecture
  —Hardware
  —Software
  —Peripherals
- Change in *communication system*
  —Architecture
  —Hardware
  —Software
  —Peripherals
- Change in *skills of using*
  —Computers/communications systems

The *overall technical scheme* of the information system is called the *architecture* of the information system. The technical part of information strategy consists of changes in the technical architecture of the information system. But this book is about the management of technology, and accordingly, we focus not on the technical architectures of information but on the *business processes* using information. How should information technology be used to improve business processes? We call this use a *business architecture* of an information system.

> **The purpose of the information function of a business is to construct and operate a *technical architecture* of an information system which facilitates an effective *business architecture*.**

The impact of *information architecture on business architecture* was summarized nicely by Miller (2000), who pointed out that progress in information technology (IT) had:

1. Altered the competitive dynamics of both products and services, leading to the new importance of dominant designs and platforms in product/service strategy
2. Flattened organizational hierarchy (or even dissolved boundaries into networked forms, such as *virtual enterprises*)
3. Affected management style through introducing (at the same time) a transparency of the business model to all levels of employees while making their jobs more complex through increased need for teaming and direct attention to the bottom lines of business goals

We will examine the concept of a *business architecture* based upon the key ideas of: (1) *economic value-addeding,* (2) *enterprise systems,* (3) differences in *manufacturing, service,* and *retail architectures,* and (4) *speed* as a competitive factor.

CASE STUDY:

Information Technology and Productivity in the 1990s

In the second half of the twentieth century, investments in computers and communications became the major category of business investment. For example in 1992, a business journal observed: "Though barely out of its infancy, information technology is already one of the most effective ways ever devised to squander corporate assets. Year after year, the typical large business invests as much as 8% of revenues in telecommunications, computer hardware, software, and related high-tech gear. Information technology soaks up a dramatically growing share of corporate spending, accounting for over 14% of existing U.S. capital investment as of last year [1992] versus 8% in 1980" (*Fortune,* 1993, p. 15).

By 1999, businesses were making 35% of all equipment investments in information technology. During the last decade of the twentieth century, business investment in IT rose annually from $0.6 trillion to over $1.2 trillion and accounted for 12% of the gross domestic product. The U.S. business market of information technology in 1999 was $400 billion (Uchitelle, 1999). It was in the 1990s that improvements in economic productivity from IT began to be documented. The business magazine *Fortune* reported on the information strategies of five different companics: "We [Fortune] identified five—Ford Motor, Dell Computer, Caterpillar, Chris-Craft, and American Express—that have successfully applied information technologies to familiar business problems" (*Fortune,* 1993, p. 15).

In 1993, Ford was using information strategy to improve its product planning capabilities in its global organizational structure. Ford had integrated its computer-aided design tools and communications to connect several design studios located around the world. In new designs of vehicles in a global context, they were achieving "substantial cost savings and smarter allocation of scarce resources, both human and mechanical. . . . These computers transform designer's sketched lines into mathematical models of a vehicle's entire surface. . . . The computers . . . can produce the numerical codes to control the milling machines that carve Styrofoam or clay into full-size mockups of cars. Plugged into the automaker's $35-million-a-year-global telecommunications network, the system allows designers in seven studios all over the globe to work together . . ." (*Fortune,* 1993, p. 16).

In 1993, Dell had applied new information strategy to their marketing strategy. Employees receiving "800" calls from customers about purchases were linked by network to a parallel Tandem computer. This computer-archived information about a customer in a single database was accessible by all employees in their internal information network. Dell personnel had used the database to increase sales and customer satisfaction: "For example, the rate of response to its small-business mailings rose 25% once Dell customer feedback was used to refine its pitch. Experience from the database guides the sales representatives who receive calls. Product developers rely on the database to help them shape new offerings. . . . Routine analysis of sales information helps Dell spot such consumer trends as the shift to larger disk drives" (*Fortune,* 1993, p. 22).

In 1993, Caterpillar had used new IT to improve its production planning capability: "Caterpillar, which returned to profitability this year [1993] after seven quarters of losses, is typical of old-line manufacturers that re-engineered key processes in response to reverses in the marketplace. Gored in the mid-1980s by Japanese competition, rising manufacturing costs, and declining market share, Caterpillar launched a $1.85 billion program to modernize its 17 factories. The goal was to increase product quality and plant flexibility while slashing inventories and production-cycle time. The company recouped its costs last year, and since then the return on its investment has been running at a 20% annual rate" (*Fortune,* 1993, p. 22).

In 1993, Chris-Craft was using IT to improve its product planning capability. Earlier, in 1988, the U.S. pleasure boat market sales had peaked, and Chris-Craft had then been plunged into bankruptcy: "That vortex sucked into bankruptcy one of the world's best-known boatmakers, once renowned for the sleek mahogany speedboats it sold to movie stars. Chris-Craft emerged chastened, with a new corporate parent, Outboard Marine of Waukegan, Illinois, and a new plan for survival . . . with a strategy to introduce new products to gain share in a market that is now essentially flat. To do that, Chris-Craft laid out $300,000 for a computer-aided design system. . . . The primary payoff from CAD is a dramatic cut in product-development cycles and lower manufacturing costs" (*Fortune,* 1993, p. 24).

In 1993, American Express was annually spending over $1 billion on IT. American Express was also allowing its most reliable agents to work from their homes.

It cost American Express a one-time expense of $1300 to connect home-situated employees to its phone and data lines. Then calls bounced seamlessly from its reservation center to home, where agents looked up fares and booked reservations. Supervisors could monitor agent's calls for control: "The productivity gains so far have been astounding. The typical agent—they're all women so far—handles 265 more calls at home than at the office, resulting in a 47% average increase in revenue from travel bookings, or roughly $30,000 annually each" (*Fortune,* 1993, p. 28).

This historical case illustrated how information strategy emerged as important for *all operations of a business.* Information technology was used to improve a firm's operations through:

- Product planning and design (Ford and Chris-Craft)
- Production planning and design (Caterpillar)
- Service design and delivery (American Express)
- Marketing and sales strategy (Dell)
- Control of operations (Caterpillar)

By the late twentieth century, information technology had affected businesses strategically in a variety of ways:

- Some businesses were *in the information technology business,* providing information technology goods and services (such as Hewlett-Packard).
- Some businesses *used information technology as a core technology* in their production of goods and delivery of services (such as Amazon).
- Some businesses *used information technology as a supporting technology* in their design of products/services (such as Ford and Caterpillar).
- Some businesses *used information technology as a marketing tool* to attract customers to its products/services (such as the CNN News Web page).

BUSINESS ARCHITECTURE AS AN ENTERPRISE SYSTEM

How can one construct a business architecture? This is essential to formulation of the proper information architecture of a business. This can be done in a systems perspective which views a business as a system to add economic value. The idea of adding value is a traditional economic term called *economic value adding* (EVA). But there are two different ideas in EVA: two different meanings, as viewed from two perspectives: that of a firm's customers or that of a firm's managers and shareholders. The two meanings of EVA are:

- *Product value:* the value as a value added to resources that a business transforms into a useful product for a customer's application
- *Product profit:* the profit obtained in the marketplace—the product's sales price to the customer minus the cost of producing the product

From the perspective of the customer, EVA is the functionality and performance a product provides for a customer's applications of a product. From the perspective of management, EVA is the gross margin of the product. The management perspective on EVA as a margin is also the view of the shareholder on the EVA of a company. From the shareholder's perspective, the true cost of equity (stock) is what shareholders could be getting in price appreciation and dividends if they invested in a portfolio of companies similar in risk to the firm.

Although EVA is a basic concept, it is not always measured in traditional accounting: "Incredibly, most corporate groups, divisions, and departments have no idea how much capital they tie up or what it costs" (Tully, 1993, p. 38). Measuring the use of capital in business operations is necessary to measure product profit as value. For example, in 1993 at the railroad company CSX, the CEO John Snow began using EVA: "Snow has lots of capital to worry about, a mammoth fleet of locomotives, containers, and rail cars. His stiffest challenge came in the fast-growing but low-margin CSX Intermodal business, where trains speed freight to waiting trucks or cargo ships. Figuring in all its capital costs, Intermodal lost $70 million (EVA) in 1988. Snow issued an ultimatum: Get that EVA up to breakeven by 1993 or be sold. Freight volume has since swelled 25%, yet the number of containers and trailers—representing a lot of capital—has dropped . . ." (Tully, 1993, p. 39).

> **To use the idea of EVA as capital utilization for constructing a business architecture, one needs to model a business as an _enterprise system of economic value adding_.**

All the productive EVA transformations of business processes together constitute the enterprise system of a business. In 1985, this view of modeling a business as EVA was called a business value chain by Porter (1985). As seen in Figure 15.1, the EVA activities of a business form a sequence of transforming operations—a chain of value-adding activities. A business takes resources from the economy, transforms them into products and sells them back into the economy, adding economic value to the original resources as product value to customers and profit value to

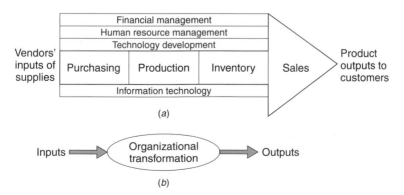

Figure 15.1. (*a*) Value-added and (*b*) open-systems models of a business enterprise.

shareholders. This concept of a value-adding chain is the same concept as in organization theory wherein all productive organizations can be conceived of as open systems (Betz and Mitroff, 1972). Also illustrated in Figure 15.1, any organization can be seen as receiving inputs from its environment, transforming these into outputs, and sending the outputs back into the environment. As a transforming open system, a business (1) acquires material, capital, and personnel resources from the economy; (2) transforms these into goods and/or services; and (3) sells the goods/services into the markets of the economy.

To model a business as an EVA system, one needs to add to Porter's idea of productive transformations the idea of the flow of controling information. Earlier, Forrester (1961) had proposed this as a systems dynamics modelling approach. Combining Porter's and Forrester's ideas, one can construct a general form for EVA modeling of a business architecture (Figure 15.2). In EVA business architecture modeling, one needs to describe three planes of enterprise operations for production transformations, overhead operations, and information control: *a middle transformation plane supported by a support plane and controlled by a control plane.*

> **An EVA operations model of an enterprise requires three planes of description: support activities, transforming production activities, and control activities. An EVA operations model provides the business architecture for designing the information architecture of a business.**

We examine how this EVA operations model of business architectures can be applied to a manufacturing business and then to a service business.

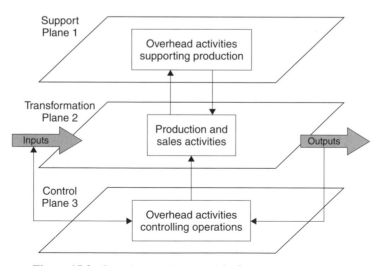

Figure 15.2. Generic operations model of any managed system.

Toyota Production

To see an example of the EVA of a manufacturing firm, we look at Toyota's automobile production system in the 1980s. Toyota purchased both manufactured components and materials for Toyota's own manufacturing processes from various supplying vendors. Next, materials went through various production processes to be formed into parts (such as forging, casting, machining, stamping, plastic molding). In addition, some of the components purchased went through further processing (such as heat treatment or additional machining) to be finished as components. Materials, components, and parts eventually were all used for three subassembly systems in fabricating the automobile:

1. *Power subsystems:* engine, axles, transmission, steering assembly, etc.
2. *Chassis subsystems:* frame, suspension, brakes, etc.
3. *Body subsystem:* body, seating, windows, doors, etc.

Various plating and painting processes prepared the power and chassis subsystems for final assembly, and the body was painted for final assembly. Then the three major fabrication subsystems were attached together as an automobile. After adjustments and inspection, the product emerged as a completed automobile.

Toyota purchased both materials and components used in fabricating automobiles: "During the early 1980s, a dozen assembly plants turned out Toyota automobiles at the combined rate of more than 800 per hour" (Cusumano, 1985, p. 262). At that rate of production, Toyota required close coordination with their suppliers in delivering parts and materials in time for production: "Toyota managers were responsible for coordinating deliveries of components and subassembly manufacturing with the schedules of final assembly lines, where workers quickly joined engines, transmissions, steering components and frames with body shells. . . . To manufacture a small car, from basic components (excluding raw-materials processing) through final assembly, Toyota and its subcontractors took approximately 120 labor hours . . ." (Cusumano, 1985, p. 262).

At that time, a major difference between Japanese automobile manufacturers and American manufacturers was a much lower vertical integration of the Japanese automobile industry: "Japan's 10 major automakers were more like a collection of manufacturing and assembly plants for bodies, engines, transmissions and other key components than they were comprehensive automobile producers. From the mid-1970s through the early 1980s, Nissan and Toyota accounted for only 30 percent of the manufacturing costs for each car sold under their nameplates; they paid the rest to outside contractors, including subsidiaries . . . loosely affiliated companies . . . and nonaffiliated suppliers" (Cusumano, 1985, p. 187).

In manufacturing, there usually is a choice of which components and parts to purchase from suppliers or to produce internally (in house). This is called the *degree of vertical integration* in the manufacturing. Cusumano (1985) presented figures that compared vertical integration among some companies in 1979: Nissan produced 26% in house, Toyota 29%, GM 43%, Ford 36%, and Chrysler 32%. The reason that Japanese managers choose low vertical integration was partly historical: "Subcontracting to subsidiaries or other firms reached these high levels in the Japanese automobile industry after demand expanded rapidly beginning around 1955. Managers decided that it was cheaper, safer and faster to recruit suppliers rather than to hire more employees or invest directly in additional equipment for making components" (Cusumano, 1985, p. 192).

BUSINESS INFORMATION ARCHITECTURE OF A MANUFACTURING ENTERPRISE

For manufacturing enterprises, the EVA operations model describes the value-added operations of the manufacturing business in the three planes of a manufacturing enterprise: transformation, support, and control (Figure 15.3).

Transformation Plane Activities

The continual operations that produce and sell the products (services) of the enterprise are performed as a sequence of product value-adding activities, as depicted in

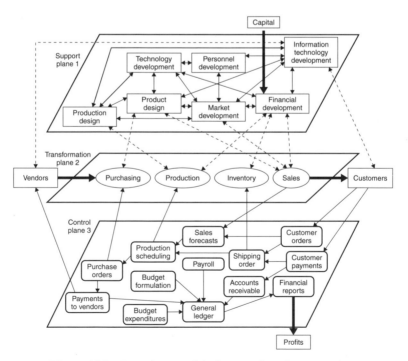

Figure 15.3. Operations model of a manufacturing enterprise.

the transformation plane of Figure 15.3. Products and services are produced and delivered by acquiring appropriate supplies, materials, and resources from vendors, then forming and assembling and using these to create a product or service sold to the customer. The activities in the transformation are organized in this sequence, and one can depict the direct value-adding activities in the enterprise value chain as purchasing, production, product inventory, and product sales. Purchasing activities are usually organized in a purchasing department, production in a production department, and sales in a sales and marketing department. Product inventory is stored in product inventory warehouses until shipped to dealers and customers.

Support Plane Activities

Projects in the support functions of a business are necessary to change operations for improvement. The types of improvements that are useful are in products, production, personnel, markets, finances, information technology, and other technologies. Accordingly, it is useful to have and depict the explicit project activities attending to these areas for improvement as noted in Figure 15.3 on the support plane: product design program, production design program, personnel development program, market development program, financial development program, information technology development program, and technology development program.

Strategic change in operations is planned and implemented in the form of specific projects: design projects, training projects (and programs), product lines and brand projects, financial analysis projects, information technology projects, and technology research and development projects. Since all businesses need some kind of change annually to continue to adapt to the future, some projects in some of these areas will be occurring each year. Long-term programs for change are usually organized within engineering departments, marketing departments, and research units and are performed as discrete projects by multifunctional teams. New product designs are performed by multifunctional design teams led by engineers in the engineering department. Production improvement projects are performed by multifunctional teams led by manufacturing engineers in the engineering department. Product line and brand market analysis projects are performed by multifunctional teams led by sales personnel in the marketing department. Cost analysis projects for product lines and new product lines are performed by multifunctional teams led by financial personnel in the finance department. Personnel development projects are performed by multifunctional teams led by personnel people in the human resources department. Technology development programs are performed in multidisciplinary teams led by scientists and engineers in the research and development (R&D) laboratories. Information technology development improvements are performed in projects by multifunctional teams led by personnel in the information technology department.

> **Strategic change in operations is planned and implemented as discrete change projects performed by multifunctional teams led by personnel in the appropriate organizational department.**

The double arrows within the support plane connecting the various areas of development indicate the kinds of informational interactions that occur between multifunctional projects in carrying out the projects for strategic change. For example, new technology developed in R&D projects might be used either in new product design or in new production design and in market and financial analyses and projections. Projects to improve information technology in the business can affect the processes of technology development, product design, production design, change in markets, and financial performance of the business. Information technology development projects can also improve communication and interactions with customers and with vendors.

Control Plane Activities

In a hard-goods business, the transformation plane describes materials flows as physical parts and materials are purchased from vendors and shaped and assembled into hardware products shipped to customers. The control plane describes the information flows that control the material flows of the transformation plane. For example, at Toyota, daily sales information controlled production scheduling.

Within the control plane, information systems process information on the operations and report performance of the operations. These systems must include the ability to:

- Receive customer orders
- Receive customer payments
- Create product shipping (or services) orders
- Forecast sales
- Schedule product production (or service delivery)
- Purchase supplies
- Pay vendors
- Formulate budget plans
- Control budget expenditures
- Pay personnel
- Control customer payables
- Account for finances
- Produce financial reports

The directional arrows in Figure 15.3 depict where the information system on the control plane directly affects the control of operational activities on the transformation plane. Customer orders control shipping orders and sales forecasts. Shipping orders control inventory depletion and co-control production scheduling. Actual sales control the sales forecasts, which co-control production scheduling, along with inventory depletion. Production scheduling controls the type and rate of production and the buildup of inventory. Purchasing orders control the types and rates of purchases from vendors and payments to vendors. The general ledger records customer accounts receipts and receivables, payments to vendors, and payroll and budget expenditures. Financial reports (daily, quarterly, and yearly) summarize financial performance of operations and calculate profits.

Information systems record and control the performance of operations.

CASE STUDY:

Ryanair

To see an example of the EVA of a service business, we look at Ryanair in Ireland in the 1990s. The airline industry around the world had been regulated until the 1970s, when deregulation of the industry in some countries began altering competitive conditions, as Lawton (1999) nicely summarized: "Beginning in North America and spreading more recently to western Europe, the airline passenger market has witnessed a growing intensity in price-based competition. This intensified competition has been facilitated by policy deregulation initiatives" (p. 573).

In the United States, a successful new airline begun then was Southwest Airlines, and in Europe, Ryanair: "The largest and most successful of Europe's low fare airlines is the Irish operator, Ryanair. It is also the longest established, having first commenced scheduled services in June 1985, operating a 15-seater aircraft between Ireland and England. The market entry of this independent, privately owned airline, symbolized the first real threat to the near monopoly which the state-owned Aer Lingus had on the routes within Ireland and between Ireland and the U.K." (Lawton, 1999, p. 574).

Ryanair provided a simplified service. No meals were served, seats were not reserved, and no restrictions were placed on tickets. To meet the competition, Aer Lingus eventually had to reduce their ticket price, and the lower prices increased the volume of passengers: "This 'egalitarian' scenario was in stark contrast to the more opulent, segregated and expensive service offered by . . . Aer Lingus. . . . Ryanair's arrival helped precipitate a growth in the total air travel market, particularly between Ireland and the United Kingdom. This growth occurred primarily in what has been described as the 'visiting friends and relatives' traffic" (Lawton, 1999, p. 574).

In the decade from 1985 to 1995, the number of air travelers between the UK and Ireland grew from 1.8 to 5.8 million annually. Ryanair's leadership in low costs allowed its low fares to provide excellent profits: "Costs have fallen faster than yields within Ryanair, allowing profits to rise consistently. . . . This [has] translated into steadily increasing operating profit margins . . . going from 10.3% in 1994 to 17.6% in 1997" (Lawton, 1999, p. 577). Ryanair's strategy had been not only to compete with low prices and low costs but also to open new routes. For example, it was first to offer services between Dublin and Bournemoth and between Dublin and Teeside.

In this case we see that the new-entry airline service competed against an established competitor with lower prices on existing routes, and it could do this through having lower costs. It expanded its markets by introducing new routes that had not had regular service previously. Since the cost of the major device, the airplane, that made the service possible was the same for all competitors, the lower-cost leadership of Ryanair had to be focused on other aspects of the service delivery system (e.g., costs of food service and seat reservations).

BUSINESS ARCHITECTURE OF A SERVICE ENTERPRISE

As in operations of a manufacturing enterprise, the operations in a service enterprise are complex. It is useful to construct a business architecture model of a service enterprise's operations. The operations of a service enterprise can also be modeled as three parallel planes of operations: (1) activities of support of the transforming activities, (2) activities of the functional transformation, and (3) activities of control of the transforming activities. To illustrate how this generic operations model of a managed system can be used to model airline services, we depict an airline service system operations model in Figure 15.4. Therein, the transformation plane shows the city-to-city routes of the air transport system of a territory served by an airline. Air travelers (and freight) are inputs into the service to be flown from one city's airport to another city's airport. The output of the transformation of the functional capability of flight is the transportation of travelers and freight from city to city.

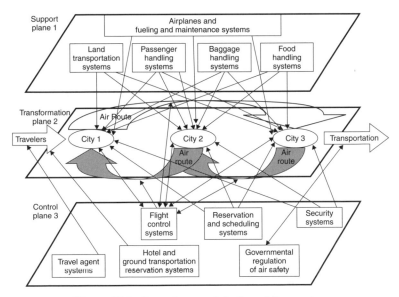

Figure 15.4. Operations model of an airline system.

Support activities for these flight transformations of geographic locale (air transportation) include:

- Airplanes and fueling and maintenance systems
- Land transportation systems for travel to and from airports
- Passenger-handling systems for passenger check-in and loading and unloading passengers onto airplanes at flight gates in airports
- Baggage-handling systems for accepting and loading and delivering baggage and freight to and from airplanes
- Food- and beverage-handling systems for refreshing passengers during flights

Control activities for the flight transformations include:

- Airport flight control systems to control air traffic
- Travel agent and reservation and sales systems for passengers to purchase air tickets and make flight reservations
- Airline reservation and scheduling systems
- Airport security systems
- Hotel and ground transportation reservation systems
- Governmental regulation of air safety

On the top support plane, support activities interact with producing activities on the transformation plane in a two-way interaction. For example, one support activity of Ryanair was a plane maintenance activity, which periodically takes a plane out of service and performs routine maintenance and/or repair on the plane. The number of hours of flying time of a plane would be information that would trigger the maintenance activity of that plane, after which the plane would be returned to flight service (or a problem regarding a plane would be reported by a pilot to repair personnel, who would in turn diagnose and repair the problem).

On the bottom, control plane, control activities acquire information from the service's outputs and use this information to control producing activities and respond to inputs). For example, one control activity of Ryanair was to receive reservations and ticket purchases from travelers for flights and then to schedule flights to serve these travelers.

> **Activities on all three planes are all operations essential to a service business, although not all activities may belong to a single business in the service.**

For example, in the airline service business of Ryanair, the airport flight control activities were essential to the taking off and landing of Ryanair flights (and all other flights to and from an airport), but these air control activities belonged to the respective airports, not to Ryanair.

Changes in any of the support, transformation, or control activities can improve service operations. Information technology, which can improve any of these activities, affects the business enterprise performance of service systems.

Dell

Next we look at how retail business architectures have been altered by the information innovation of the Internet in the 1990s. We look at Dell's adoption of business practices to the Internet. Back in the late 1980s during the early days of the personal computer industry, one entrepreneur who first saw the personal computer as a standardized product was Michael Dell. He understood that the two product features that individualized the personal computer, its operating system and its central processing unit, had been standardized under IBM's brand name entry into the personal computer market in 1985 but without IBM proprietary operating systems or central processing unit (CPU) chips; these were produced by Microsoft and Intel. Therefore, all PCs (except Apple's Macs) would look exactly alike, and the competitive situation in PC manufacturers depended on low-cost production and competitive pricing. Dell focused his business strategy on direct sales by telephone to large organizations and computer assembly by suppliers controlled by the Japanese manufacturing strategy using *just-in-time* processing.

By the mid-1990s, only three new PC companies had succeeded as big, dominant companies: Microsoft in PC operating system software, Intel in PC CPU chips, and Dell in PC computers. In fact, Dell's direct sales by telephone and low-cost production, minimizing product inventory, became a model that any business selling PCs had to emulate to be price competitive. Then as the Internet emerged as a major business channel, Dell rapidly adopted its use to Dell's business model: "Late in August [1999], 1,225 business men and women, ranging from information technology pros to buttoned-up CEOs, journeyed to Austin, Texas, like pilgrims flocking to Mecca. They filled half a dozen downtown hotels and endured three days of 101-degree heat, all for a chance to hear Michael Dell kick off Dell Computer's inaugural Direct-Connect Conference. Michael Dell, the oracle of Austin, would reveal how he'd made the Internet part of his company's success and explain how they could too" (Roth, 1999, p. 152).

Michael Dell had become famous in the business world in the 1990s from his successful operations in direct phone sales and his control of product inventory with just-in-time assembly by suppliers. Dell's profit margins were large because Dell had avoided the investment necessary to build production facilities when production gave no competitive edge. (PCs were assembled from standard purchased parts in part modules—cases, motherboards, disk drives, model cards, keyboards, monitors, etc.) The manufacturing challenge and competitive edge lay in the computer part production, not in the computer assembly. Therefore, Dell had outsourced production to minimize production costs and avoid production investments. Because the PC was such a standardized product (with "Intel inside" and Microsoft Windows installed), this strategy of outsourced, just-in-time assembly and shipping worked very well and made early Dell investors rich and Michael Dell famous.

Dell was relating his new Web business strategy to the world (and selling new Dell Web products). In 1999, Michael Dell was asked to speak on over 1700 occasions (accepting 35 of these). Each talk described how Dell was using the Internet to continue growing sales, as its online sales had grown to total 40% of total sales. Michael Dell had exploited the new market channel of the Internet to continue to grow his direct-sales mode, moving from phone sales to Web sales.

In 1994, Michael Dell had envisioned that the Internet could be another channel for Dell's mode of direct sales to customers. He formed a strategic project team to launch Dell's Web site, *Dell.com*. The site provided technical support information for Dell computers and price guides to help customers select the appropriate components in buying a Dell computer.

Dell continued to develop the Web site, and by 1996, Dell computers (PC desktops, laptops, and servers) were selling on the Web site at a rate of $1 million in daily sales. By 1999, the Web site was receiving 2 million daily visits and selling $30 million worth of products each day. (To appreciate this high level of retail sales in 1999, Amazon was receiving 11 million visits a day but selling only $3.5 million a day.)

Dell's successful high volume of sales on the Web generated intense business interest in hearing Michael Dell's talks. As Fred Buehler, then director of electronic business for Eastman Chemicals, commented: "Dell is clearly one of the top few companies that have really been successful on the Internet" (Roth, 1999, p. 154).

Buehler was so impressed with Dell's Web success that in 1997 he had replaced all Eastman's 10,000 PCs with Dell PCs. (Eastman Chemicals is a chemical producer in Tennessee with about $4.5 billion sales in 1999.) Next, Buehler and other managers of Eastman began traveling back and forth between Tennessee and Austin to study the business part of Dell's information strategy. In July 1999, Eastman started its own Web site, *Eastman.com*, using what they learned from Dell.

The strategy of informing customers and then selling products to customers to meet their needs worked well for Dell. In sales of servers to about 300 corporate IT users (surveyed in 1999), 30% bought their servers from Compaq but 15% bought from Dell. When the surveyor asked these companies who they would buy from in the future, the number that cited Dell grew: "Companies are confused; they don't know what the Internet means (says an analyst at one of Dell's biggest institutional holders). So when the CEOs get back [from Austin] to their offices, they call up their purchasing managers and say, 'Let's do business with these guys, they have their act together.' " (Roth, 1999, p. 156).

Dell also used the Web for supply chain management of its vendors. Dell's suppliers use a version of Dell's Web site called "Premier Pages" at the password-protected site on *Valuechain.Dell.com*. They used it to communicate with Dell about what orders they have shipped and about how Dell sees that they are measuring up to Dell's quality standards. In 2002, Dell was the largest online retailer, selling 22% of the total volume of retail sales on the Internet, with 12.5 million annual visitors to its site (Tedeschi, 2002).

BUSINESS ARCHITECTURE OF RETAIL E-COMMERCE

The Internet is a medium of communication and a channel of marketing. The information strategy of a business needs to use the particular characteristics of this medium in its activities that add value to the business's customers. The typical business architecture of *e-commerce retail businesses* in the late 1990s (such as Dell's) were structured as the transformation plane 2 in Figure 15.5. The starting point in the information architecture is the Web site on the Internet, which provides access by the customer to the e-commerce retail business. Within the business, the merchandising function supervises the appearance of the Web site, products presented, and interaction of the customer with the Web site. When the customer makes a purchase, the order-taking function records the order and informs the operations management function to direct product inventory and shipping to ship the products ordered. Operations management then reviews the products inventory and, as necessary, orders more products from product suppliers. When the customer purchases, the payment and security function bills the customer's charge card. Customer service records the customer's purchases, tracks, and is available for any subsequent interaction with the customer about the purchases. The accounting function records the purchases and operations expenses. The entire business is tied together through a system integration function. Normally, delivery is provided by an external business such as United Parcel Service or Federal Express. Product manufacturers provide products to the business or may even ship directly to the customer upon notification by the e-commerce business. What is especially important to note in the

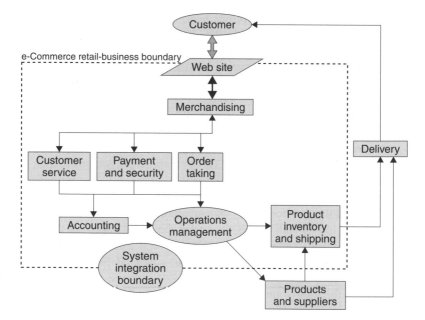

Figure 15.5. Information system architecture (plane 2) of a retail e-commerce business.

business operations architecture is that the boundary of the business is not defined by a physical space but by a virtual space, the software system integration function.

CASE STUDY:

Speed at Toyota

We next look at how the proper combination of business architecture and information architecture could alter the competitive conditions of an industry, particularly in changing the speed of business response. In the 1980s, Toyota was one of the most innovative manufacturing firms in the world. Toyota had a competitive advantage over other automobile manufacturers in the speed with which Toyota responded to its markets, by operating with fast and versatile business processes.

Bower and Hout (1988) described Toyota's strategy to use speed in manufacturing operations cycles as a competitive advantage: "Toyota [is] a classic fast-cycle company. . . . [The] heart of the auto business consists of four interrelated cycles: product development, ordering, plant scheduling, and production. Over the years, Toyota has designed its organization to speed information, decisions, and materials through each of these critical operating cycles, individually and as parts of the whole. The result is better organizational performance on the dimensions that matter to customers—cost, quality, responsiveness, innovation" (p. 111).

Toyota's product development business process used self-organizing and multifunctional teams focused on a particular model series. A product development team accepted full responsibility for the entire cycle of product development for a model—making the style, performance, and cost decisions and establishing schedules and reviews. In addition, the product development teams selected and managed the supplier input, bringing suppliers into the design process early. As a result, in 1988 Toyota was capable of a three-year product development cycle. (At that time the average car development cycle of U.S. automobile manufacturers was five years.)

A fast response time was used not only for product development but also for production control in Toyota. Toyota dealers in Japan were connected online to Toyota's factory scheduling system. As soon as an order was taken, the information on the model and options selected was entered immediately into the scheduling function. Toyota's purpose in integrating sales information with production scheduling in real time was to minimize sharp fluctuations in the daily volume of production while also minimizing inventories. Toyota could produce on each production line a full mix of models in the same assembly system, utilizing flexible manufacturing cells.

Bower and Hout (1988) saw that Toyota's attention to responsiveness pervaded its organization: "Much of Toyota's competitive success is directly attributable to fast-cycle capability it has built into its product development, ordering scheduling, and production processes. By coming up with new products faster than competitors do, it puts other manufacturers on the marketing defensive. . . . By

continuously bringing out a variety of fresh products and observing what consumers buy or don't buy, it stays current with their changing needs and gives product development an edge market research cannot match" (p. 112).

INFORMATION TECHNOLOGY AND COMPETITIVE SPEED

One of the impacts of information technology that was pervasive in business enterprises in the late twentieth century was to add as a competitive factor the idea of
speed in market responsiveness. Then many business writers (such as Bower and
Hout, 1988) urged businesses to use time as a competitive advantage; they called
such businesses *fast-cycle companies.* Speeding up the response time of companies
to changes in customer needs and the economic environment required more than
simply working faster. Speed also required working "correctly," thinking about why
it takes time to respond, whether responses are correct, and how to respond more
quickly and correctly. They argued that a sustainable competitive advantage from
attending to time is through better satisfying customers more quickly.

Innovations in the information technologies in the second half of the twentieth
century in manufacturing firms provided capabilities for faster response by enabling
(1) rapid product development cycles, (2) manufacturing science bases in unit production processes, (3) intelligent control in operating systems, and (4) teamwork
and communications that integrate the business enterprise. Progress in information
technologies enabled a fast product development cycle for correcting product mistakes, refining product successes, and emulating competitors' product successes.

Accordingly, the information function in a business is responsible for developing and maintaining effectively linked technical and business architectures as a
fast-cycle company. The information function enables information capabilities in the
other functions of the business:

- *The marketing function:* uses information systems to find and communicate
  with and inform customers and to make sales
- *The production function:* uses information systems to obtain supplies and to
  schedule and control production and inventories
- *The finance function:* uses information systems to record and control expenditures and revenues and to plan investments
- *The administration function:* uses information systems to track, remunerate,
  and train personnel
- *The research function:* uses information systems to plan, perform, monitor, and
  transfer technology progress
- *The engineering function:* uses information systems to design and develop new
  products and services and production processes

The information function is needed to provide all these kinds of information
capabilities within the different business functions. But in so doing, there often

arose *islands of information automation,* subsystems of information in the vari-uos functions that do not communicate with each other. The information function also needed to integrate the communication of information across all the business functions, avoiding isolated islands of information automation. The information function also needed to upgrade the information technologies of a business peri-odically to match the progress being made in information technology.

> **The contribution to productivity and competitiveness of the information function lies in its direct impact on the *speed* of business operations for proper responses to market opportunities.**

SUMMARY AND PRACTICAL IDEAS

In case studies on

- Information technology and productivity in the 1990s
- Toyota production
- Ryanair
- Dell
- Speed at Toyota

we have examined the following key ideas:

- Technical and business information architectures
- Economic value added
- Enterprise systems
- Business architecture for manufacturing
- Business architecture for service
- Business architecture for retail e-commerce
- Speed in competition

The practical use of these ideas provides understanding of the information busi-ness function and how business architectures are made competitive as fast-response businesses through information technology.

FOR REFLECTION

Locate articles and histories of IBM, and trace IBM's business transformations from a cash register business to a punch-card sorter business to a mainframe computer business to a Web services business. What key roles did CEOs play (or not play) in the major transformations of IBM's businesses?

16

HIGH-TECH PRODUCTION

INTRODUCTION

As we saw in the EVA model of a manufacturing business's architecture, the production function of a business provides the transformational activity. Products and services must be produced and sold for a business to make profits. Also, we saw earlier (in the life cycle of an industry) that technological innovation must occur not only in new products but also in new production processes—to produce products quickly, cheaply, and at high quality. Over time, the number of innovations in production exceeds the number of innovations in products. Accordingly, innovation strategy should focus not only on high-tech products but also on high-tech production.

Production has been called by different names in different industries. In the hardware industries, production is called *manufacturing*. In the civil structures industry, production is called *construction*. In service industries, production is called *service delivery*. We will use the term *production* generically. Software production is either in a form of hard-goods production (e.g., distributing software by means of disks) or in the form of delivery of information (e.g., electronic format). The creation of new software is a product design problem. Therefore, software production can be covered either by production systems for hard goods or by service delivery systems. In this chapter we review the forms of production in products and services, focusing on technological innovation in production and delivery; reviewing the key ideas of *production systems, unit processes, learning curves, quality,* and *technology audits.*

CASE STUDY:

Biotechnology Innovation at Bristol–Meyers Squibb

This case study illustrates how in the early biotechnology industry, innovation of product and production was tightly linked. The biotechnology industry could not innovate new products without also innovating new production processes. The historical setting of the case was in the 1990s, when the biotechnology industry was pursuing new therapeutic protein products and each new product required a new unit production process. Cells had to be modified genetically to produce any particular product.

In 1997, Thorne presented an overview of the early phases of process development at Bristol–Meyers Squibb Co., describing the general stages in biotechnology drug innovation (particularly for creating drugs based on antibodies):

1. Identify the therapeutic target.
2. Obtain and characterize therapeutic protein (such as an antibody).
3. Generate a recombinant protein (e.g., antibody) in recombinant-engineered host cell production.
4. Evaluate recombinant antibody as a drug:
   a. Suitability of antibody for curative target.
   b. Side effects of humanization, such as the loss of affinity as an antibody of the recombinant-DNA-engineered antibody or harmful effects.
5. Reengineer and reevaluate:
   a. Redesign recombinant-engineered antibody for affinity and safety.
6. Develop into a therapeutic product (quantity and purity).

Since recombinant-engineered cells produce the desired therapeutic protein (such as an antibody) in tiny quantities, the first problem in drug discovery is to produce enough of the engineered protein for therapeutic testing as to effectiveness and safety. Since each engineered protein must be produced by modifying a group of cells, the speed of development of the production process (batches of modified cells) is critical to the speed of product development. All new drugs require extensive testing in the United States for effectiveness and safety before being allowed on the market by the federal government.

For this reason, biotechnology-based drug firms had to perform process research as well as product research. Thorne sketched how the two research procedures for product and production interacted on a time line (Figure 16.1). The time line of new product development requires two sides: product development and process development.

The time line begins on the product development side with the choosing of a candidate protein (such as an antibody) targeted for a therapeutic application. Next, a DNA construct must be made which inserted in a cell will produce the target protein. This construct is then handed from the product research side of the innovation procedure to the process research side.

Then process research takes the DNA construct and puts it into a host cell (transfection). The host cell is a mammalian cell line which then produces and secretes the targeted protein. Both quantity and quality of DNA-constructed protein produced are goals of the process research. When the DNA is inserted into host cells, it may be incorporated randomly into the cell's genome. Some of these locations will affect the survivability and growth of the cell line and the productivity of the cell line in producing and secreting the target protein. Process

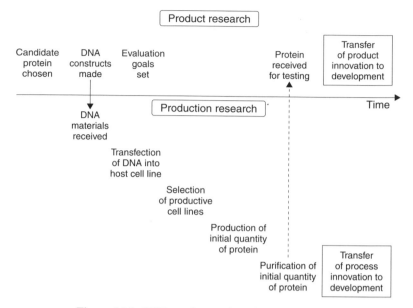

Figure 16.1. DNA product and production time line.

research then uses techniques for selecting and screening productive cell lines for the target protein. Once this is accomplished, an initial quantity of the protein is produced and purified for handing back to the product research side.

On the product research side, personnel must begin testing the protein for efficacy. Then the new product and new processes are transferred to the development personnel for drug testing to meet federal standards. The speed with which the initial production process is created affects the product research side, since they must depend on process research to produce sufficient quantities to begin drug testing.

This case study illustrates the need for rapid process development to complement new product development for the biotechnology products of a pharmaceutical firm. Development and innovation of new products in the biotechnology industry produces new sources of sales. The ability to innovate a variety of new products for a biotechnology company created an *economy of scope* in a biotech firm. Thus process research was aimed at facilitating an economy of scope in the firm. Later, after a new biotechnology product was innovated, further process innovation would be aimed at reducing the unit cost of production to create an *economy of scale*.

PRODUCTION SYSTEMS

The transformational logic of *production* consists of a *production system* containing *unit production processes*. A production system is an ordered set of activities that transforms material, energy, and information resources into products. The logic of a production system differs according to whether the product is a hard good, software, or service. Unit production processes are discrete steps in the transformations needed to produce a product from resources. Production systems and unit production processes differ by industry.

Innovation in production systems can provide competitive edges of two economic types: *economies of scale* or *economies of scope*. An economy of scale is a unit-cost advantage in production that arises from technical efficiencies in the volume of the production scale. For example, in the materials processing industries (such as steel or chemicals), a larger scale of production can often produce a unit volume of material with lower energy costs and lower waste. An economy of scope is a production flexibility that allows a producer to market a broad range of products. For example, in food processing or retailing industries, an ability to distribute a wider variety of products often creates a larger volume of total sales than those of smaller competitors. When innovating new technology products, economies of scale are important in building large distribution capability, creating brand recognition, and establishing a dominant market position. When competitors' production capacities begin to saturate a market and product technologies mature, economies of scope are important in maintaining market dominance and in maximizing returns on investment in production capability. For example, Suarez and co-workers (1995) studied flexible production in the printed-circuit-board industry and emphasized that manufacturing flexibility can provide flexibility either in product mix or in volume production.

HARD-GOODS PRODUCTION SYSTEMS

Unit production processes differ by industry. Chemical production uses a set of unit processes, such as reacting, distillation, and filtering. Mechanical production uses a different set of unit processes, such as swaging, milling, machining, molding, forging, assembly, and painting. The electronic industry uses a different set, such as wafer growth, lithography, etching, metallization, component insertion, soldering, and assembly. Construction uses a different set, such as pouring, footing, framing, erection, siding, roofing, insulation, wiring, and plumbing.

Unit production processes are the discrete transformation steps in taking purchased material into fabricated and assembled products. Material unit production processes include:

- *Forming processes* (such as molding, stamping, machining, forging, reacting, fermenting, lithography, etc)
- *Separation processes* (such as distillation, shredding, combing, slurring, masking, layering, etc.)
- *Finishing processes* (such as grinding, polishing, painting, coating, filtering, etc.)
- *Assembly processes* (such as joining, welding, gluing, assembly, fabrication, construction, mixing, etc.)
- *Packaging processes* (such as boxing, coating, packaging, wrapping, etc.)

Each material unit process embodies a core technology system. Improvement in the technical capability of a material unit production process can come from improved understanding of the phenomena of the technology or from control of the process.

In addition to the set of unit processes, materials need to be moved into and between the unit processes for processing in a production system until the final material parts are assembled into the product. In the chemical industry, chemical materials are moved through pipes. In the mechanical industry, materials are transported by conveyer belts or carts. In the electronic industry, materials are transported by belts, carts, or robots. In the construction industry, materials are transported by trucks, cranes, and people.

Materials must thus be transported to the production site, moved to and between unit production processes, packaged, stored in finished-product inventory, and shipped into distribution channels. The system of coordinating all this is called the *production system*. Control in the production system operates at two levels: scheduling of production and control of work at and between unit processes.

Accordingly, the production system of a firm is a series of process subsystems of unit production processes embedded within a sociotechnical system of the organization of production. Technological change in a production system can therefore occur in any aspect of the production system:

1. At the boundaries of the production system
2. In unit production processes within the production system
3. At connections between unit processes, such as materials handling technologies
4. In production system organization, communication, and control

Thus, technical innovation may occur in any part of a production system:

- New or improved unit processes of production
- New or improved tools or equipment for unit processes
- New or improved control of unit processes
- New or improved materials-handling subsystem for moving workpieces from unit process to unit process
- New or improved tools and procedures for production scheduling
- New or improved tools and software for integrating information for design and manufacturing

UNIT PRODUCTION PROCESS INNOVATION

Technological innovation can be implemented in both the unit processes and the control of the unit processes of a manufacturing system. Innovation of unit processes can occur as:

1. New unit processes
2. Improvements of existing unit processes

3. Improved understanding and modeling of the physics and chemistry of unit processes

4. Improved control of unit processes

Unit process improvement may require invention, and the invention of new unit processes may be required by:

- A need to process new kinds of materials
- A need to process with improved precision (improved purity or at greatly reduced tolerances)
- A need to process with reduced energy consumption
- A need to speed up throughput of production or to scale up the volume of production

Research and technological innovation can improve unit process control in two ways: (1) by real-time control through intelligent sensing and control, and (2) through experimental design for processes that one cannot presently model. Real-time control of unit processes requires:

1. Sensors that can observe the important physical variables in the unit manufacturing process

2. Physical models and decision algorithms that compare sensed data to desired physical performance and prescribe corrective action

3. Physical actuators that alter controllable variables in the physical processes to control the manufacturing process

Technological innovations in unit processes should be introduced into the production system as discrete modules, previously debugged, into an existing production system.

PRODUCTION LEARNING CURVE

Production innovation is critical to the beginning of the production of any new product, as improvements in production to improve product quality and lower product costs are essential to competition. When a new production system is innovated, the initial production of the system will always be more costly and of lower quality and less safe than later production. Learning better how to produce is the common experience of successful producers. This has been expressed in a form called the *production learning curve* (Figure 16.2). Plotting the unit cost of production over time shows how unit production costs can decline as technologies and skills of production are improved.

There are several mathematical forms that can be used to model production learning curves, and all forms result in a declining curve on semilog scales. A recent

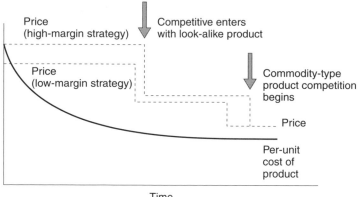

Figure 16.2. Production learning curve and pricing cure.

summary of mathematical forms appears in an article by Badiru (1992). Although these different models are of intellectual interest, in practice exactly which model is used is less important than how the curve is used as a management tool. It is important to use the production learning curve as a planning tool to set *targets* for production improvement. In starting up a new production process, the production team should set target goals for production improvement and then try to meet them, as shown in Figure 16.2. A high-margin price strategy maintained for too long invites competition to enter with lookalike products. Pricing with a lower margin can discourage entry, and price cuts before entry can hurt new competitors. Finally, after the product no longer includes innovation and becomes a commodity-type product, margins will be forced low because of competition. Then the lowest-cost producer survives.

Setting target goals for production improvement is particularly important when innovating a new high-tech product for which new production processes may also need to be innovated. New high-tech products soon meet price competition from competitors, and the innovator should not only be improving the product but moving down the production learning curve faster than competitors. For example, Leonard-Barton (1992) suggested that manufacturing be treated as a kind of learning laboratory, emphasizing the production improvement processes: problem solving, continuous experimentation, and aggressive acquisition of new process knowledge.

CASE STUDY:

General Electric's Refrigerator War in the 1980s

The design of products and of production processes are interactive challenges. We recall that continuing improvements to production are essential to continuing in business, to remaining competitive with higher-quality and lower-cost products.

We look at a case of innovation in manufacturing technology aimed at maintaining a product line against competition which occurred in the last quarter of the twentieth century. U.S. manufacturers had slipped seriously behind Japanese manufacturers in both the design and manufacture of physical products. This case examines the strategic changes needed in one of GE's manufacturing businesses in the 1980s to survive the strong Asian manufacturing competition of the time. In this case we will see a historic example of a large firm redesigning one of its major businesses, product and production, to improve competitiveness and remain in business—GE's large consumer appliance business.

One result of World War II was that the undamaged industry of the United States enabled its manufacturers to dominate world markets for 20 years after the war. During this time, European and Asian firms rebuilt capabilities. In the 1970s, U.S. manufacturers found themselves under competitive assault as one North American manufacturing business after another again faced competent foreign competitors. Moreover, the modernized rebuilt industries of Europe and Asia often were producing at lower cost and higher quality. As GE's consumer appliance business faced new competition, GE decided to quit the small appliance business but stay in the large appliance business. Magaziner and Patinkin (1989) described the challenge GE then faced: "Fifty miles south of Nashville . . . is one of the world's most automated factories. Had it not been built, U.S. households might soon have had yet another product—the refrigerator—stamped 'Made in Japan.' Instead here in the heartland, General Electric found a way to build products better and cheaper than those made by foreign workers paid one-tenth American wages" (p. 114).

The GE manufacturing manager who helped manage the production improvement was Tom Blunt: "Tom Blunt still remembers the day in 1979 that he first stepped into Building 4, the plant in Louisville, Kentucky where compressors for GE's refrigerators were made. . . . The plant was a loud, dirty operation built with 1950s technology: old grinders, old furnaces, too many people. Finishing a single piston took 220 steps. Even the simplest functions had to be done by hand. Workers loaded machines, unloaded machines, carried parts from one machine to the next. The scrap rate was ten times higher than it should have been, 30% of everything the plant made was thrown out" (Magaziner and Patinkin, 1989, p. 114).

In 1979, this sort of practice was common in U.S. manufacturing. U.S. industrial capacity had aged after World War II. In the 1970s, the OPEC oil cartel's rise in energy prices had pointed out that obsolescence, while the accompanying inflation reduced incentives to make investments necessary to improve manufacturing technology. But while the U.S. manufacturers hesitated, the Japanese had not hesitated. By 1980, the United States had lost manufacturing superiority and was losing manufacturing bases in many industries.

Blunt, a manufacturing engineer, had recently joined GE's Major Appliance Business Group (MABG) as chief manufacturing engineer for ranges, and he was worried about the manufacturing capability. He saw that manufacturing had to be improved at GE's Louisville plant. He found that other managers in GE were worrying about the poor state of GE's manufacturing. The major appliance business group's (MABG) profits and market share were declining. One competitor, Matsushita, was manufacturing better and cheaper compressors in Singapore and was selling them to the GE appliance subsidiary in Canada. Matsushita was also experimenting with new rotary compressor technology. This was a technology GE had invented but used only in its air conditioners. Moreover, Whirlpool was moving its compressor manufacturing to Brazil to lower labor costs. In the fall of 1981 several manufacturers approached GE offering lower-priced compressors than GE could produce itself. GE's managers began talking about the strategy of sourcing rather than of producing the compressor.

As a manufacturing engineer, Blunt didn't like that strategy. The compressor was the core technology of the refrigerator, important to the performance and energy efficiency of the refrigerator. Blunt was promoted to head of advanced manufacturing for refrigerators. Since the compressor was the key component of refrigerator, Blunt decided that GE had to innovate its compressors to keep its refrigerator business. For a product, some aspect may or may not provide a competitive advantage. If an aspect does provide a competitive advantage for differentiating the product, one should not outsource that aspect. In refrigerators, energy efficiency is one differentiating competitive factor; so it would have been poor competitive strategy to outsource the compressor.

In 1981, Ira Magaziner was then a manufacturing consultant hired by GE to provide information on their competitor's refrigerator manufacturing costs. He visited all the compressor manufacturers in the world. When he returned to Louisville and gave his report to GE's MABG, he had documented a major cost problem for GE. It cost GE's MABG more than $48 to make a compressor, whereas the costs at Necchi and Mitsubishi for the same compressor were $32 and $38, respectively. In addition, other competitors were designing new plants that would down further drive costs, with Hitachi and Toshiba aiming at a cost of $30 per compressor and Matsushita and Embraco building new plants to produce compressors at $24 apiece.

One reason was for their competitor's cost advantage was cheaper labor costs: $1.70 per labor hour in Singapore and $1.40 per hour in Brazil, compared with $17 an hour with benefits in Louisville. Another reason was productivity. It took GE 65 minutes of labor to make a compressor, whereas the cheaper labor was also faster. Singapore built one in 48 minutes, Brazil in 35 minutes, and Japan in 25 minutes. Their new plants and designs were more efficient.

It was obvious to Magaziner that GE's only hope was in a new design of a rotary compressor because it had fewer parts and could therefore be made more cheaply than a reciprocating compressor. GE had invented the rotary compressor for air conditioners but hadn't bothered to use it in refrigerators, where it would have to stand up to harder use. Magaziner next went to see Tom Blunt and asked him if he could design a plan to produce the rotary. Blunt replied that that's what he did for a living, if he got the chance. And Blunt would get his chance. Truscott, MABG's chief engineer, assembled a team of product design engineers and came up with a model they thought could be made cheaply. But the design required a precision in working parts to fifty millionths of an inch—more precise a tolerance than that used by any mass-produced consumer equipment.

Next, Truscott told Blunt to assemble a team to design a factory to produce it. Blunt assembled a team of about 40 people, including engineering product design people and manufacturing people (a concurrent approach). He put the product and manufacturing engineers in offices across the hall from one another. Together they refined the design of the rotary pump to the most automated model, with fewer than 20 parts.

But there was a major technical bottleneck in the production system. Existing production equipment at that time normally could not meet the close tolerances required of the new product. Blunt and his engineers also had to improve the equipment. One of Blunt's engineers was Dave Heimedinger, who negotiated with suppliers of grinding and gauging machines for equipment with the needed processing accuracy. This was much higher than was traditionally done, and Heimedinger and his team had to develop combinations of equipment that could produce the parts at the required low tolerances. Often, existing production processes need to be improved to produce innovative products.

Finally, Truscott and Blunt had a design for an inexpensive rotary compressor and an automated factory to produce it. Should GE build it? The investment was high, at least $120 million. Millions more would have to be spent on redesigning GE's refrigerators to fit around the new compressor. The GE engineers would have to sell the project to the finance people in GE's headquarters in Connecticut, who were wary of large capital investments in appliances, since recently they had lost a lot of money in a new washing machine plant that had failed. To prepare for their visit to headquarters, Truscott and Blunt asked Magaziner (who was still consulting for them) to check the proposal for a new plant. Magaziner advised them that it was a close call, but he would recommend the investment. Truscott sent the proposal to Roger Schipke, then head of MABG.

Schipke, Truscott, and Blunt flew from Kentucky to Connecticut to present a proposal to the CEO of GE, Jack Welch. Welch listened to their presentations and agreed that it was strategically important to keep major appliances as a core business. He understood that it would require a major investment to improve manufacturing. The new plant would be built in Columbia, 200 miles from Louisville.

There was a major business risk depending on a technological risk. The compressor technology worked. Could the manufacturing technology produce it within the tolerances required? In addition to the technical risks, there were also people risks, since even the most automated production process is still a sociotechnical system. GE would have to train its manufacturing people in new processes. Accordingly, management developed a training plan for its workforce to staff the new factory and asked the workers to undergo extensive training. The workers were eager to get training, as they knew that jobs were lost when manufacturing companies became technically obsolete and hence uncompetitive. The new plant was built, workers trained, and production of the new, efficient rotary compressor refrigerators began.

DESIGN FOR MANUFACTURING

We recall that competitive strategies require a focus on product differentiation or on manufacturing leadership in low-cost, high-quality production. Yet these two strategies need not be alternatives.

> **Innovation strategy should include improvement not only of product performance but also of product quality and costs.**

Since all products eventually become commodity-type products, product redesign must include not only improvements in the product model but also improvements in the manufacturability of the product. Considerations in the design of products to lower manufacturing costs and improve manufacturing quality have been termed *design for manufacture* (DFM). For example, Crow (1989) summarized the goals of DFM as (1) the improvement of product quality, (2) the increase of productivity and capital utilization, (3) the reduction of leadtime, and (4) the product flexibility to adapt to market changes. DFM concepts focus on design of parts with appropriate tolerances and producibility rules, simplifying parts and numbers of parts, and simplifying fabrication and assembly. Crow listed some rules of thumb that DFM practitioners have collected:

1. Reduce the number of parts, because for each part there is an opportunity for a defective part and an assembly error.
2. Make the assembly design foolproof so that the assembly process is unambiguous. Components should be designed so that they can be assembled in only one way.
3. Design verifiability in the product and its components. For example, for mechanical products, verifiability can be achieved with simple go/no-go tools in the form of notches or natural stopping points.
4. Avoid tight tolerances beyond the natural capability of the manufacturing processes and avoid tight tolerances on multiply connected parts. Tolerances on connected parts will stack up, making maintenance of overall product tolerance difficult.
5. Design robustness into products to compensate for uncertainty in a product's manufacturing, testing, and use.
6. Design for parts orientation and handling to minimize non-value-added manual effort and ambiguity in orienting and merging parts. Parts must be designed to orient themselves consistently when fed into a process.
7. Design for ease of assembly by utilizing simple patterns of movement and minimizing fastening steps. Complex orientation and assembly movements in various directions should be avoided.
8. Utilize common parts and materials to facilitate design activities, to minimize the amount of inventory in the system, and to standardize handling and assembly operations.

9. Design module products to facilitate assembly with building block components and subassemblies.

10. Design for ease of servicing the product. Easy access should be provided to parts that can fail or may need replacement or maintenance.

CASE STUDY CONTINUED:

General Electric's Refrigerator War in the 1980s

One may hope for a happy, fairytale ending to brave cases of business innovation such as GE's refrigerator case—a successful ending without problems. But in real life and in real business, there are always problems. This story was no exception: "The celebration was short-lived. In January 1988, twenty-two months after the first compressor rolled out of the new factory, a problem surfaced. Some of the larger compressors—those in GE's bigger refrigerators—began to fail" (Magaziner and Patinkin, 1989, p. 121).

Although the failures constituted only a small percentage of the plant's total production, a few failures could destroy GE's quality reputation with customers. Management and engineers turned to the problem. Schipke formed a project team to analyze the problem. They worked incessantly, often through the night, for several weeks. It was a difficult problem because only a small portion of the compressors had yet failed in service. But they finally found the source of the problem: Lubrication had not been properly designed and was allowing one of the compressor's small parts to wear more quickly than the designers had calculated. This is an example of what we earlier discussed as design bugs. Design bugs will probably occur in the most complex part of a product system. In engineering, defining the problem is halfway to the solution. Truscott and other GE engineers solved the compressor problem by improving lubrication.

Still, the management problem was not over. Schipke first approved a customer service plan to replace immediately at no cost to the customer any compressor that broke down in the refrigerators that GE had already produced and sold. But it would still take several months to redesign the compressor for manufacturing. This meant that production would continue during that time to make a product that would prematurely fail—or Schipke could stop production until the redesign was ready. Either way, it would cost a lot of money and lose customers.

Schipke cut through the Gordian knot of his manufacturing dilemma a hard, direct way. He didn't wish to lose GE's reputation for quality by shipping refrigerators that might develop problems. He decided to purchase older reciprocating compressors from abroad while engineering fixed the problem. It was a tough decision, but it was the only way that GE could keep major appliances as its core business. The final responsibility of the quality and cost of a product is manufacturing's responsibility. GE paid the price, and when the twenty-first century began, GE was still in the major appliance business.

This case illustrates an important point about innovation in manufacturing. To maintain competitiveness in hard-goods products, both products and production must be improved periodically through innovation. Product systems and production systems of an enterprise are interrelated issues. The choice of the technology for rotary compressors was primarily that they were simpler in numbers of parts and could therefore be manufactured more cheaply than other designs. Since refrigerators were commodity-type products, cost was the primary competitive factor. Redesign of the compressor required much tighter manufacturing tolerances than existing equipment by vendors could certify. Much manufacturing innovation had to be made by GE engineers in adding sensors and online control to machining equipment to make the machines capable of producing the tighter tolerances.

QUALITY IN MANUFACTURING

In the GE case we see the importance to competitiveness of both product cost and product quality. We review briefly the several meanings of *product quality* as used in business production:

- *Quality of product performance*
  —How well does it do the job for the customer?
- *Quality of product dependability*
  —Does it do the job dependably?
- *Quality of product variability*
  —Does it produce the product in volume without defective copies?

Product performance is often primarily a focus of engineering design. However, manufacturing, as well as design, is important to the second and third meanings of quality (dependability and variability). In U.S. manufacturing up to 1980, traditional manufacturing quality control focused primarily on the third notion of quality— product variability. Then the standard technique was to control quality after production by sampling batches of products to determine if product variation was within acceptable specifications. However, newer approaches to quality have emphasized the importance of considering product dependability and product variability together. In the 1980s, after witnessing Japanese manufacturing leadership in quality, U.S. manufacturing began paying more attention to the issues of manufacturing quality. For example, Taguchi has emphasized the total costs of poor quality: "When a product fails, you must replace it or fix it. In either case, you must track it, transport it, and apologize for it. Losses will be much greater than the costs of manufacture, and none of this expense will necessarily recoup the loss to your reputation . . ." (Taguchi and Clausing, 1990, p. 65).

Taguchi argued that quality should be seen from the perspective of the customer. From the customer's perspective, the important aspect of quality in a product is whether it works immediately on taking it out of the box and into the field. Too often the quality engineer has been overly concerned with staying "in spec," forgetting whether simply staying within manufacturing specifications meant real quality to the customer. To correctly use the notion of manufacturing specifications with the customer's perspective, Taguchi has emphasized setting the specifications in the design of a product so as have product performance be to the field conditions of product use: *product robustness*. Technological innovation in products improves the quality of performance when the product functions more effectively in the hands of the customer.

Technological innovation in production improves product quality when production reduces deviation from target design specifications. When all parts of a product are on target specifications, the product's performance in field use becomes more robust and independent of field conditions. Manufacturing requires *repeated* acts of production with *each and every* detail under *tight* control. Tightly controlled repetition of details means that implementing new technology into manufacturing will disturb a process whose main goal is to *avoid* disturbances.

> **Technological innovation can improve product quality in three ways: through improving performance or dependability or through lowering product variability.**

EVALUATING INNOVATIONS IN PRODUCTION

The economic benefits of implementing technical improvements in the production system should be evaluated in terms of improving:

- *Production quality:* reducing rejects and improving production accuracy
- *Production efficiency:* reducing material wastage and/or energy use
- *Production effectiveness:* increasing flexibility and variability of producible products
- *Production capacity and throughput:* increasing the volume of production per unit time
- *Production responsiveness:* decreasing the time of changeover for product variability
- *Production cost:* decreasing the overall cost of a unit product

Accordingly, evaluating the benefits of innovation in manufacturing (and selling those ideas to senior management) requires the research unit to express the benefits of the improvements in terms of one or more of the following criteria:

1. Improvement in production precision
2. Reduction of material waste
3. Improvement in production flexibility
4. Reduction of throughput time

5. Improvement of production control and scheduling
6. Reduction of work-in-process inventories
7. Reduction of indirect product costs

Financial evaluation of technical innovation in manufacturing *solely* on a direct-cost return-on-investment criterion misses the impact of quality on competitiveness and profitability. Investments in manufacturing should include considerations of the opportunity cost/benefits and the cost of lost opportunities. For example, Drucker (1990) emphasized that manufacturing should be viewed in its total enterprize setting. He argued that not only should innovations be measured as to their impact on *productivity* but also as to their impact on *quality*.

> **Implementing new technology in manufacturing should have the dual goals of improving productivity for the manufacturer and improving the economic value for the customer.**

HARD-GOODS PRODUCTION TECHNOLOGY AUDIT

To plan production innovation, it is important to have a complete list of all the technologies involved in production. Twiss (1968) called a systematic identification and listing of all technologies in a firm a *technology audit*. It is particularly useful for innovation in production systems because of the need to be systematic to avoid production downtime. We will call a systematic list of all technologies involved in production a *production technology audit*.

A production technology audit should be oriented to the value-adding transformations of the firm producing the product, and may be extended downstream and upstream in the industrial chains of the firm's businesses. The technologies for production of hard goods include (1) unit-production processes, (2) materials-handling processes, (3) production-control processes, (4) environmental-control processes, and (5) product-distribution processes. In each of these processes, there will be physical equipment and phenomena and control equipment and algorithms. Accordingly, one can categorize a production technology audit for a hard-goods producer in a matrix as in Figure 16.3.

SERVICE DELIVERY TECHNOLOGY AUDIT

In contrast to hard-goods production systems, the logic of service delivery consists of different stages:

- *Service referral.* A customer must arrange for a service by contacting the service deliverer, as for example, selecting a doctor or opening a bank account.
- *Service transaction.* Service delivery must be scheduled, as for example, making a appointment and visiting the doctor's office or writing a check or making a deposit or withdrawal from the account.

	Physical Equipment and Phenomena	Control Equipment and Algorithms
Unit production processes		
Materials-handling processes		
Production control processes		
Environmental control processes		
Product distribution processes		

Figure 16.3. Hardware production technology audit.

- *Selection of service application.* The appropriate application in the service must be selected, as for example, diagnosis by a doctor of the patient's illness or recording and accounting a banking fund transaction.
- *Service application.* The application selected in the service must be provided, as for example, prescribing a drug, performing surgery, or transferring bank funds electronically or in cash.
- *Payment for service.* Payment must be received by the service provider from the customer, as for example, billing a patient's insurance company or billing a client's bank account.

Technologies are used in the various stages of the service delivery: such as devices and techniques, information and communication technologies, and professional knowledge base technologies. Evaluating software for service delivery is a particularly important process in improving service delivery. For example, Bard and co-workers (1997) examined software selection procedures for the Canada Post Corporation. They evaluated competing software programs on the criteria of modeling features, debugging ease, animation, report generation capability, documentation, and performance.

Service industries require constructing some kind of physical infrastructure, and once this is in place, service delivery requires the coordination of several subchains of value-adding sectors. Technologies for a service firm group into:

1. Technologies for devices used in service delivery
2. Technologies for supply and maintenance of devices used in service delivery
3. Technologies for services delivery
4. Technologies for services development

	Physical Equipment and Phenomena	Control Equipment and Algorithms
Necessary Infrastructure		
Major device system		
Facility system		
Sales system		
Parallel Operating Systems		
Professional personnel system		
Support systems		
Scheduling systems		
Maintenance/resource systems		

Figure 16.4. Service technology audit.

5. Technologies for assisting the customers' applications of services
6. Technologies for communicating and conducting transactions with customers and suppliers
7. Technologies for controlling the activities of the service firm

A service delivery technology audit can be categorized, as shown in Figure 16.4, in terms of the necessary infrastructure and the parallel operating systems. Within these, as in the manufacturing technology audit, identify the physical equipment and phenomena and the control equipment and algorithms.

SUMMARY AND PRACTICAL IDEAS

In case studies on

- Biotechnology innovation at Bristol–Meyers Squibb
- GE's refrigerator war in the 1980s

we have examined the following key ideas:

- Production system
- Hard good production systems
- Unit process innovation
- Production learning curve
- Design for manufacturing
- Quality in manufacturing
- Evaluating innovations in production
- Technology audits

These theoretical ideas can be used practically for planning innovation in production systems and for innovations necessary to improve economies of scale and scope.

FOR REFLECTION

Select an industry and plot the reduction of unit production costs for one of its product lines over time. What kinds of production technology innovations contributed to cost reductions and/or quality improvement?

17

HIGH-TECH MARKETING

INTRODUCTION

In addition to production, all businesses must have a marketing function. But a high-tech business must have a marketing capability to sell new products of high performance. After a new high-tech product/service is produced in volume, the new product must be sold. The more radical the new technology of the product, the more existing markets must change to accept it. Marketing high-tech products (or services) is a challenge because change in technology creates change in markets. We examine the special problems of high-tech marketing in the ideas of *innovation sources, new market creation, completing applications, proprietary knowledge,* and *market experimentation.*

CASE STUDY:

Personal Computer Industry

From the case of the Commodore 64 personal computer in Chapter 6 we saw how it made an impact in the early personal computer market, due to its lower price. Let us return to that era and look more carefully at the origin of the personal computer industry. The growth of the early personal computer industry provides a clear example of how a new high-tech product creates a new market. Then the biggest challenge of the new personal computer industry was just that—what was to be the market for personal computers?

The personal computer (PC) began as kind of a hobbyist market (a technology innovation looking for market applications). Yet it grew to become one of the great market changes in modern society. Later, the PC would become essential to the Internet and electronic commerce. But at the end of the 1970s, the personal computer was seen as only a very low end product (barely even a computer). It was made possible by technical advances in semiconductor chip technology at the stage of large-scale integration (LSI). We recall that LSI was a next-generation technology with tens of thousands of transistors on a chip which had made possible a computer-on-a-chip [the central processing unit (CPU) of 8-bit word length] and a memory-on-a-chip [the 16K dynamic random access memory (DRAM)]. Putting a CPU chip and a memory chip together into a box with a data bus chip and other parts made it possible to build a very small computer for personal use. Yet this new product line of the personal computer was innovated not by existing computer companies but by individual entrepreneurs. Entrepreneurs started new firms to exploit the new technological opportunities. By the middle of the 1970s, computers were being produced in three product lines: super, mainframe, and mini computers. The personal computer added a fourth low-end product line.

Because of the existence of the bigger computers, many people with technical backgrounds were wishing that they had their own computer. But computers were expensive and then purchased only by large organizations. One of the first hobbyist groups, the Amateur Computer Society (ACS), was started in 1966 by Stephen B. Gray. ACS published a bimonthly newsletter containing problems. answers, and information about where to get parts, schematics. and cheap integrated circuits. Still, trying to put a computer together was a massive job, beyond the range of most people (Gray, 1984).

The situation for amateur computer makers remained this way from the late 1960s to 1973—when Frederick Faggin, at Intel, designed a microprocessor on a chip. Intel produced and sold this as the Intel 8008 chip. Nat Wadsworth, a design engineer at General DataComm in Danbury, Connecticut, attended a seminar given by Intel about its new 8008 chip. The chip used an 8-bit word length and could perform all the mathematical and logical functions of a big computer. Wadsworth suggested to his management that they use the chip to simplify their products, but they were not interested. Then Wadsworth talked to his friends: "Why don't we design a nice little computer and each build our own to use at home?" (Gray, 1984, p. 12).

Two friends joined Wadsworth, and they designed a small computer. They laid out printed-circuit boards and ordered several constructed by a company. Then Wadsworth decided to manufacture a small computer, and in the summer of 1973, Wadsworth quit his job, incorporating the Scelbi (Scientific, Electronic, Biological) Company. In March 1974, the first advertisement for a personal computer based on a microprocessor appeared in QST, an amateur radio magazine. Unfortunately, Wadsworth had a heart attack in November 1973 at the age of 30. Then he had a second attack in May 1974. He sold about 200 computers, half assembled and half in kits. While in the hospital, he began writing a book, *Machine Language Programming for the 8008,* which sold well.

TECHNOLOGY-PUSH AND MARKET-PULL STRATEGIES

Technological innovations have been motivated either by conceiving a technological opportunity or by perceiving a market need, called, respectively, *technology-push strategy* or *market-pull strategy*. Technology-push innovation strategy occurs in cases of technological innovation motivated primarily by exploring nature and its manipulation. Market-pull innovation strategy occurs in cases of technological innovation motivated primarily by market need. Both sources of innovation have been important in the history of technology. Yet in the early studies of technological innovation, there used to be arguments about which was best.

For example, in the 1970s, von Hippel studied samples of innovations in the scientific instrumentation industry: gas chromatography, nuclear magnetic resonance spectrometry, ultraviolet absorption spectrophotometry, and transmission electron microscopy (von Hippel, 1976). He found that the user scientists of the instruments made 80% of the innovations in his sample and that manufacturers of the instruments made 20%. From these and other studies (e.g., von Hippel, 1976, 1982), von Hippel argued that market pull was a more important source of innovation than technology push. One notes that his conclusion about the importance of market pull as a source of innovation followed from his selection of cases—wherein the user of the innovation was technically sophisticated and had research capability. (Historically, research instrumentation has been invented by scientists.)

A contrasting study was performed by Knight (1963), who looked at innovations in computers and found that manufacturers (rather than users) dominated innovation from 1944 to 1962. In computers during this period, industry was more technologically advanced than were most computer users, and in this case, technology push was a more important source of innovation than market pull. Other studies, such as those by Berger (1975) and Boyden (1976) of plastics innovation wherein manufacturers were technologically sophisticated and had research capability, also documented the importance of technology push.

The conclusion is that both manufacturers and users can be sources of innovation but only when either one is technologically sophisticated and a research performer. It is the *locus of sophisticated technical performance* in an industrial value chain that determines whether market pull or technology push for any sector will be most important for innovation. But whatever the source, for technological innovation to be commercially successful, one must match technology to markets. As Freeman (1974) nicely summarized: "Perhaps the highest level generalization that it is safe to make about technological innovation is that it must involve synthesis of some kind of [market] need with some kind of technical possibility" (p. 193). In the history of major new innovations, technology-push strategy has always preceded any market-pull strategy. However, once a basic innovation has occurred, good market-pull strategies can be formulated.

Market-pull innovations most often create incremental innovations; and technology-push innovations most often create radical innovations.

CASE STUDY CONTINUED:

Personal Computer Industry

It was technology push that motivated the innovation of the first microprocessor by Intel. It was a kind of market-pull desire that motivated computer buffs such as

Wadsworth to create a personal computer. The computer industry was producing several computer product lines by the middle of the 1970s—mainframe computers, supercomputers, and minicomputers—but all too expensive for personal use. It was this personal use vision that was a market-pull motivation to produce a personal computer. But it was technology progress in IC chips, the computer-on-a-chip, that was providing the technology push for a personal computer. Moreover, it was not clear what the market for personal computers would be beyond that of the computer hobbyist and buffs who were creating the new personal computers.

About the time of Wadsworth's first computer efforts, H. Edward Roberts (another entrepreneur) was also building one of the first personal computers, using the new Intel 8008 chip. Roberts was working at the Air Force Base in Albuquerque, New Mexico. Earlier in 1971, he had formed a company with a partner, Forrest Mims, called Micro Instrumentation Telemetry Systems (MITS). They manufactured electronic calculators, and Roberts bought out his partners. But by 1974, big firms had moved into the electronic calculator business with better products and cheaper prices. MITS was $200,000 in debt, and Roberts decided to make computers with the new Intel 8008 chip (Mims, 1984). At the same time, Arthur Salsberg, an editor at the magazine *Popular Electronics,* was looking for a computer project using the new Intel chip and learned of Roberts's project. Salsberg called Roberts and told him if he could deliver an article in time for the January 1975 issue, it would be published. Roberts called his computer the Altair, and it was featured on the front cover of *Popular Electronics:* "Project Breakthrough! World's First Minicomputer Kit to Rival Commercial Models." Salsberg titled his editorial in that January 1975 issue: "The Home Computer Is Here!" (Mims, 1984, p. 27).

The publicity worked, and orders poured into the company. But the product was awkward and difficult to use. In 1975, Altair cost $429 in kit form and came without memory or interfaces. It had no keyboard and had to be programmed directly in binary by setting switches. But it was a computer.

The first programming language to be adapted to the Altair was Basic, written by a Dartmouth professor as a programming language easier to learn than Fortran. In 1975, Bill Gates and Paul Allen wrote a Basic language interpreter for the Altair (Gates, 1984): "Ed Roberts still recalls the day in late 1975 when William H. Gates, the co-founder of Microsoft, walked into his office, anxious and depressed. A rail-thin young man with a voice that cracked when he was excited. Mr. Gates had just turned 20. People were stealing his software, he complained, and the prospects for his fledgling company appeared dim. 'I'm going to quit,' Mr. Gates said, according to Mr. Roberts. 'There's no way I'll make money on software.' To help tide a needed supplier through difficult times, Roberts took Mr. Gates on as a $10-an-hour contract worker. Mr. Gates has a different recollection. . . . The reason he did hourly contract work for about four months, he said, was to write some special test software to catch defects in computers made by MITS" (Lohr, 2001, p. 1).

Roberts used a computer bus (the S-100 bus) that was to became a first standard for the new personal computer industry: "For a couple of years, MITS was the dominant company in an industry that was just taking shape. Its starring role was shortlived, as Mr. Roberts sold MITS in 1977 amid a legal dispute with Microsoft that left him embittered for years. But during its brief run as the leader, MITS pioneered innovations and set the pattern that the personal computer industry still follows. The MITS–Microsoft partnership established software as a separate business in personal computing" (Lohr, 2001, p. 12).

After Roberts sold MITS, he went to medical school and became a doctor and practiced for the next 20 years. In 2001, he was still practicing medicine in Cochran, Georgia, with modern personal computers in each examination room for taking notes and analyzing tests (Lohr, 2001).

NEW MARKETS AND NEW MAJOR DEVICES

Earlier we saw how in a new industry an industrial value chain emerges to provide the high-tech product for the consumer. In the computer industry value chain, the computer is the major device, and the CPU chips and memory chips are the parts providing the core technology of computation. The major device manufacturers of personnal computers were new businesses such as MITS and later Apple and Commodore. It has often happened that an old-line industry fails to innovates a new basic innovation; so the older mainframe computer industry did not innovate the personal computer industry.

Marketing continues to be a dynamic activity in the early growth of the new high-tech industry, because product performance changes rapidly, depending on many improvements in the technological systems embodied in the product system. In the early growth of the personal computer industry, the market rapidly grew from computer hobbyist customers to kids and then to young business managers.

> **As technical performance in a new high-tech product rapidly advances, new customers come into the market and the market changes.**

The first products in any new technical area are never really easy to use, are limited in performance, and are costly. Accordingly, the room for improvement is large, and competition begins immediately. One mustn't be too hard on first products in a basic new technology (such as Roberts's Altair). The early products show the general shape of things to come and demonstrate the market, and this is very important to innovation and to beginning the market. Then it is a rapid series of improvements in technologies within and around the product system that advance the product performance and growth of the market.

CASE STUDY CONTINUED:

Personal Computer Industry

Important advances continued in components of the personal computer. Frederick Faggin, the designer of the 8008 chip, left Intel to start a new company, Zilog, producing the Z80 chip, Chuck Peddle, at MOS Technology, created a new 8-bit microprocessor chip with an extended instruction set, the 6502. These two chips, the Zilog Z80 and the MOS 6502, were to become the basis of the first expansion of personal computers.

The knowledge infrastructure of the mainframe computer industry had created a latent market for personal computers, and the first customers were all computer hobbyists. Amateur groups had formed and hobbyist magazines were looking for the right product. Next, marketing infrastructure to sell personal computers had to be created. In 1975, Dick Heiser started the first retail computer store in Los Angeles, California. Upstate, Paul Terrell opened a retail store in Mountain View, the Byte Shop (Ahl, 1984).

Computer hobbyists needed the entire application system for personal computers. The first marketing challenge of the new industry was not only to provide the major device but also peripheral devices and supplies to enable the personal computer to perform computational applications. The market for the first personal computers therefore evolved as a technical market, with high-tech products needing to be completed as an application system of personal computation. The first high-tech products were created by entrepreneurial engineers, who left employment in large companies and government labs to start their own companies. Publicity for new products in hobbyist journals was essential to the commercial success of a new consumer technical product.

Another important component for the personal computer was the programming language; and we described earlier, Bill Gates and Paul Allen had written a Basic language interpreter for the Altair (Gates, 1984). The early personal computers used Basic as a programming language.

COMPLETING HIGH-TECH PRODUCTS FOR APPLICATIONS

New high-tech products usually begin in a form incomplete for any real application; and we recall that applications are systems. We can see in the case of the first personal computers that the first marketing problem in a new high-tech product was getting the product complete enough for a customer's application. Until a customer can apply the product to an application, a customer can gain no value from the product, no matter how technically clever the product.

> **For a market to emerge, a new high-tech product must be completed for an application.**

We recall that technology, product, and application are different systems. A technology is a system of knowledge about how to manipulate nature in a logical scheme to achieve a functional transformation. A product is a system of a device or process that embodies a technology to perform the functional transformation. An application is a system of tasks requiring a transformative device. Because they are different systems, it usually takes more technological systems than the core technology to complete a product system, and it usually takes more products than the major device project to complete an application system.

In the design of the major device of the personal computer as a product system, it took more technology systems than the core technology of the CPU chip. It also required the technology systems in the data bus, internal memory of the memory chip, software technology of the operating system, and so on. To use the personal computer in an application for computation, the customer required additional peripheral devices, such as a keyboard for data input, a monitor for data output, a recorder or floppy disk for data storage, a programming language such as Basic with which to program the computer, and instructions for understanding how to use the computer.

> **The first marketing problem for a new high-tech product is to determine what, how, where, and when it can be completed by additional products for a customer's application.**

The first market of the personal computer industry was a market of computer hobbyists who could build applications of the personal computer. Bill Gates was one of the first customers for Roberts's Altair computer, to adapt the programming language of Basic to run on the computer. Gates's purpose was to become a personal computer entrepreneur, as was the purpose of many of the first customers of the new high-tech personal computer product.

> **The value added to a customer of a high-tech product is as a device in the customer's application system.**

Although illustrated by the personal computer, these lessons are general—technical entrepreneurs selling first products for a hobbyist or scientific or industrial or business market. Technical education and communications and publicity play an essential role in the commercial success of the early marketing of high-tech products. Thus both major device and peripherals and supplies and a technical infrastructure around a new application system are all necessary to come together to begin an high-tech industry.

> **Marketing of new high-tech products is not only an advertizing challenge but also but an application-system-completion challenge.**

CASE STUDY CONTINUED:

Personal Computer Industry

We return to the growth of the new personal computer industry in the late 1970s. The personal computer that finally ignited the mass market was the Apple computer. In January 1976, at a west coast electronics trade show (Wescon), Chuck Peddle decided to sell the new MOS 6502 chips at $20 each. (The first Intel 8008 chip had cost Roberts $100 a chip.) One of the first customers for Peddle's 6052 chip was Steve Wozniak, who was then a technician at Hewlett-Packard. Wozniak took the chip home and wrote a Basic language interpreter for it. Then he made a computer, calling it Apple I. He showed it in the spring of 1976 to the Homebrew Computer Club in Palo Alto, California. Again, it was only partially complete, without a keyboard, case, or power supply. Yet two friends, Steve Jobs and Paul Terrell, were impressed. Jobs formed a company to sell the computer. Terrell ordered 50 units as assembled machines for his Byte Shop. He did not want to sell kits (Ahl, 1984).

With orders in hand and needing cash for production, Jobs sold his Volkswagen and Wozniak sold his two HP calculators. Wozniak stayed on his job at HP, while Jobs hired his sister and a student to assemble the units. In 29 days, they delivered their first 50 units. Then the garage-based company sold another 150 computer for $666 each. By the end of the summer, Wozniak was designing the successor product, which was to have a keyboard, power supply, and plug-in slots for the S-100 bus. The entrepreneurs understood that they had to improve their.first design for better marketability.

Jobs and Wozniak almost had it together as a new venture: with engineer, manager, first products sold, and an improved model. What was missing was business experience and capital. The next member of the Apple team to be added was A. C. "Mike" Markkula. Markkula was trained as an engineer and had worked for Intel and Fairchild during their meteoric growth in the days of the first integrated circuits. Intel stock options had made Markkula a millionaire, and he retired at age 34. Markkula visited the garage of Jobs and Wozniak and was impressed. He invested $91,000 of his own money and began an active role in planning and management. He hired Mike Scott as president. The four of them, Wozniak, Jobs, Markkula, and Scott, set out to make Apple a Fortune 500 company (and for a time, it became one).

Apple was one of the four major competitors for the major device product in the initial stage of industrial growth of the personal computer. The other early competitors were Radio Shack, Commodore, Texas Instruments, and Atari. (We saw earlier saw that the low price of the Commodore 64 drove Texas Instruments and Atari out of personal computers.) Later, the IBM PC would drive Commodore out of the industry.

The next steps in innovation in the personnal computer industry were in peripheral devices and in applications. The tape cassette had early been used to store programs, but it was slow. In the middle of 1978, two companies, Apple and Radio Shack, introduced floppy disk drives for their computers. This facilitated software applications.

Earlier in 1975, Michael Shrayer had purchased an MITS Altair, and he had written a *text editor* routine for it. A text editor is a software tool that makes it easier to write and alter programs. On completing the routine, he wrote a second piece of software to help him document the software, which he called the Electric Pencil. This was the first word-processing software for the personal computer, which he began selling in 1976. Word processing became one of the major applications of personal computers (Shrayer, 1984).

Later, a second major application was the spreadsheet. In 1978, Dan Bricklin was a student in the Harvard Business School. He and his friend Bob Frankston had been working on a spreadsheet program for the Apple. Spreadsheets were just what business students needed. Bricklin had an assignment in his consumer marketing class to analyze a Pepsi Challenge campaign. Bricklin used his new spreadsheet routine, projecting the financial results of the Pepsi campaign five years out (instead of the two years projected by all his other classmates). His professor liked it. Bricklin and Frankston knew that they had a useful business tool, VisiCalc. They introduced the new product at the National Computer Conference in New York in 1979 (Bricklin and Frankston, 1984). With the spreadsheet application, Apple sales really took off. The personal computer had become a useful personal business tool for the business manager.

By 1980, the major device, parts, relationships, and infrastructure of the personal computer system were developed, and the industry began extraordinary growth.

MASS MARKETS

Traditionally, markets provide a structure for exchanging goods and services between sellers and buyers. Markets are ancient forms in all civilizations, and the form of the traditional market still exists [as described by an economic historian, Fernand Braudel (1979)]: "In their elementary form, [traditional] markets still exist today. Survivals of the past, they are held on fixed days. And we can see them with our own eyes on our local marketplaces, with all the bustle and mess, the cries, strong smells and fresh produce. In the past, they were recognizably the same: a few trestles, a canopy to keep off the rain; stall holders, each with a numbered place . . . a crowd of buyers and a multitude of petty traders— . . . bakers selling coarse bread . . . butchers with displays of meat . . . wholesalers selling fish, butter and cheese in large quantities . . . straw, hay, wood, wool, hemp, flax . . ." (p. 29).

The traditional market developed in agricultural societies as a way of getting the surplus produce from the country to the city—exchanging these for crafted products produced by the artisans and merchants of the city: "Markets in towns were generally held once or twice a week. In order to supply them, the surrounding countryside needed time to produce goods and to collect them; and it had to be able to divert a section of the labor force to selling the produce" (Braudel, 1979, p. 29).

This ancient *country-to-city market* is the historical origin of all markets, bartering goods and services. When money was introduced into a market as the basic unit of exchange, the reach of the market was extended over space and in time. Money allowed exchange of goods and services without an immediate barter. So it is that markets go back to the dawn of civilization, to the beginning of agriculture, to the origin of cities.

All forms of markets are basically ways for society to trade economic uses of nature.

Markets can be stable, or they can change. The major force for change in markets has been technological innovation. Moreover, in business history, innovation in markets has always fostered a linear kind of change—toward a new form of market that never returns to an older form. Innovations have changed market structures from local bartering-markets to money-exchange markets to distant markets to global markets. In the 1990s, the innovation of the Internet in the 1990s added another new market structure to the world—electronic marketplaces. Electronic markets are a new way to perform some of the age-old matching of supply to demand.

Markets change (1) when the kinds of goods and services change and/or (2) when the ways of exchanging goods and services change in an economy.

Major devices with radical or next-generation technology do not market initially in mass consumer markets, because of the technical difficulty and price of the new devices. Marketing radically new innovations usually begins with the niche markets, which are receptive to new technologies: the military, scientific, hobbyist, and technically oriented industrial markets. In the twentieth century, the military market became highly organized to fund new technology and new applications. The scientific market is trained to create and improve new instruments. The hobbyist market enjoys learning and views technical tools as a pleasurable avocation. Technically oriented sections of industrial markets (such as engineering design) have technically trained personnel to adopt and improve technical applications. These markets are the first to adopt new high-technology products. They are the traditional markets for small high-tech companies. For a mass market, the cost of the high-tech product must be reduced to a low price.

To move high-technology products into the mass markets of consumers or of general business, a company must be both a technology leader and a marketing leader. The reason is that mass markets are sensitive to both education and price. They require substantial training and service to accept an innovative product. They also require a relatively low price, high level of safety, and ease of use. In a mass market, a technology leader runs the risk of demonstrating a new market, which can later be captured by a technology follower with stronger marketing and lower-cost production capability—the "me-too" competitor. The advantage of such a company is that the potential market factors for success has been explored by the technology leader and thereby eliminated. The competitor can then improve its product, target and price the market more carefully, and provide features and service which together may cancel the lead-time advantage of the technology leader.

CASE STUDY CONTINUED:

Personal Computer Industry

In computers, marketing leadership has gone to the company whose hardware performance is almost as good as anyone else's but provided superior software for applications. With this principle, IBM dominated the mainframe computer market from the 1960s through the 1980s. IBM was watching the new personal computer market and entered it in 1981 with an IBM PC. At the time, the IBM PC had the superior technology of a new Intel chip with 16-bit word length, but it was not IBM's chip. IBM claimed 7% of the market that year. In 1982, it matched Apple's share at 27%. In 1983, IBM emerged as the clear and dominant leader with 36%, and Apple slipped to 24%. IBM's brand name justified the new product to the business market as not just a toy but a true computer. IBM alone had established the industrial standard for personal computers.

IBM had played a "close-follower" technology strategy in 1981 in entering the personal computer market. It used just the right balance of technology leadership and marketing leadership—a little technology leadership and a lot of marketing leadership. The IBM PC had a 16-bit central processing and chip, superior to the 8-bit chip of the Apple and competitors in the personal computer market in 1981 (superior in the sense that it allowed a larger memory address). Yet since the IBM PC used an 8-bit bus, the combination made it easy for software applications to be written for the IBM PC (using the larger memory that the larger addressing capacity of the 16-bit processor made possible). Therefore, the principle of IBM's entry into the personal computer market in 1981 was just a little technology leadership, but not too much.

One marketing aspect of IBM's successful entry into the personal computer market in 1981 was that IBM focused on the business market, pricing their PC for business and marketing it through IBM's own sales personnel, opening new retail outlets, and distributing it through the new personal computer dealer networks. The important marketing edge was the IBM name, with the installed base of IBM business customers and IBM's reputation for service. At first, this combination of marketing and technology leadership worked well for IBM. There was an IBM computer for small businesses and for the desks of sharp managers. Another factor in IBM's early success in personal computers was the choice of an open architecture and operating system (imitating Apple), which facilitated the rapid transfer and development of software applications for the new IBM personal computer.

But in the end, the story did not turn out happily for IBM. When IBM introduced its first personal computer, the product manager responsible for introduction of the IBM PC failed to look at the long-term marketing consequence of his product design. Instead of drawing on IBM's technical capabilities and developing a proprietary operating system, he chose to license an operating system from Bill Gates's fledgling company, Microsoft. Also, instead of having IBM develop a proprietary microprocessor chip for the computer, he chose to purchase CPU chips from the semiconductor manufacturing firm Intel. Both choices turned out to be fatal for the long-term position of IBM in personal computers.

IBM had mistakenly introduced a high-tech product without significant proprietary technology.

This was IBM's big favor to Intel and Microsoft. During the 1980s and 1990s, Intel and Microsoft prospered while IBM's market share in personal computers was destroyed by lower-cost clone PC manufacturers (such as Compac and Dell). IBM dropped from a high of 37% of the personal computer market in the middle 1980s to a low of 8% in the 1990s.

PROPRIETARY TECHNOLOGY

Technology is implemented into products, production, service, and operations. Some of the technology implemented can be *generic* (publicly known) and some *proprietary* (known only by the company). Both generic and proprietary technology are important to competitiveness, but in a different way. Generic technology is necessary but not sufficient to competitiveness, whereas proprietary technology is not necessary but sufficient. Any competitor's technology must be up to the same standards as the best competitor's generic technology, for these are the publicly available standards of engineering. Products, production, and services that do not embody the most advanced generic technology are not globally competitive. In contrast, proprietary technology consists of the additional details of making generic technology work completely in a product, production, service, or operation. Proprietary technology consists of the details that make the competitive differences between competitors, who are all sharing the best generic technology available.

> **Products without proprietary technology simply invite imitators into its market.**

Porter (1985) emphasized that competitors within a market can gain competitive advantages either through differentiation of their products or through cost advantages. Product differentiation can be either on quality or on market focus. Cost advantages may be through lower prices or higher profit margins (allowing greater expenditures for marketing, research, etc.). Thus the relationships within a market between competitors can be changed by technology that either differentiates products or lowers prices. Technology may also affect the barriers for entry of new competitors into a market. For example, the microprocessor on a chip dramatically simplified the technical problems of making a computer and lowered the costs of producing a computer. This allowed a rash of new competitors to enter the computer market, challenging established computer makers. Thus Apple went from a small business to a major competitor by creating a market for low-end, low-cost personal computers.

New technology can also affect a market through product substitution into an industry's market by products from another industry. The lowering of computational circuit prices by microprocessor chips created market opportunity for product substitution. Microprocessors were first used in minicomputers to take the lower part of the market for computers and then later in personal computers for an even lower-performance end of the personal computer market. Personal computers took over the market of "dumb" terminals on time-sharing mainframe computers. New technology can create surprising market changes.

A brand name is always important in marketing because it uses name recognition to indicate the reputation of the company. In commodity products, only brand name and price features distinguish the products of competitors. But in high-tech products, an additional distinguishing factor is performance. To establish a new brand by a new company, technical performance is essential. Proprietary technology and quality are important to protect a brand time against competition over time.

> **A brand name cannot maintain a premium when customers learn that the brand does not a provide higher quality than that provided by competing products.**

CASE STUDY CONTINUED:

Personal Computer Industry

After IBM's successful establishment of the personal computer market in 1981, the competition in personal computers continued. If Apple was to survive IBM's successful attack on the new personal computer market, Apple would need a competitive advantage: advanced and proprietary technology. A second generation of technology in personal computers was introduced by Apple in about 1984. We recall that Steve Jobs visited Xerox PARC and there saw a technical vision of the new personal computer. This vision was first implemented in Apple in the Lisa project. In 1983, Apple introduced the Lisa to compete with IBM but met with disappointment. Although recognized as a technology leader, Lisa had been priced too high, at $10,000, almost twice the cost of the IBM PC XT (Morrison, 1984). Then Jobs at Apple responded in 1984 by introducing the Macintosh, a smaller version of the Lisa and priced in the IBM PC range: $2000 to $3000 (*Business Week*, 1984a). This model sold well and kept Apple as a major competitor to IBM in the personal computer market in the middle 1980s. Apple had made the strategic choice of playing technology leader against IBM's market leadership. Without IBM's prestige mainframe business and extensive business installations, Apple could not then match IBM's marketing strength. The Apple Mac continued to lead technology over the IBM PC clones for over a decade, until these finally caught up with Apple in technological performance with Microsoft's Windows 95. By the year 2000, Microsoft and Intel dominated the parts market for the core technologies of the personal computer (microprocessor and operating system software), but no single firm dominated the personal comuter major device market.

MARKET POSITIONING

A market's evaluation of the potential of new technology in a product can be examined through several criteria:

- Quality   through product superiority
- Value   through correct market focus
- Price   through efficient production
- Opportunity   through timing of innovation
- Profitability through investment return

The quality of a product to be delivered depends on its embedded technologies. Will new technology improve the quality of existing products or provide new quality products? The value to a customer depends on the application to which the customer applies the product. Will new technology improve the value or provide new value for customer applications? The price of the product depends not only on the technologies in the product but also on the technologies in the production of the product. Will new technology reduce the cost of the product?

New technology will provide a competitive advantage to the innovator only when its innovation occurs before competitor's use of the new technology, and thus timing is important to consider in evaluating new technology. Is there a window of opportunity to gain competitive advantage or, conversely, to catch up with and defend against a competitor's innovation? Finally, new technology must be evaluated on the capital required for innovation. All innovations cost money to develop and commercialize. What capital is required? Can that be recovered in a timely manner? What contributions to profitability will arise from the technological innovation?

As new technology develops, new applications are often discovered as well as new relationships to existing technologies and applications. Successful product innovation of the new-generation technology requires that corporate management recognize its significance and correctly focus marketing of the new product lines. However, this is the most difficult thing to do, since the new technologies will probably affect new markets in ways not envisioned. Therefore, it is important that technological development in a corporation be accomplished in a manner that strongly encourages a shared vision of the directions and rates of technological advance between the research laboratories and production divisions. The research personnel will be most creative about technological opportunity; the production and sales divisions will be most sensitive to market needs. Technology leadership thus must be aimed at the right market at the right time and at the right price, or the technology leader may end up showing the way to competitors without capturing a dominating lead. Since this may not happen in the first pass, it is important for the innovating corporation to be flexible and quick at adapting the new product to evolving market situations.

Marketing of new high-tech products is difficult, as the radically new technology of the product and its applications must be developed together—simultaneously and interactively.

To emphasize this challenge in marketing high-tech products, Shanklin (1983) called the making of new markets by a new technology a kind of "supply-side marketing—any instance when a product can create a market." Traditional market analysis for new products approaches the task by market segmentation, identifying the customer group (industry or consumer) upon which to focus the new product or service. This can be done because earlier or similar products have defined the function and applications of the product. Incrementally innovative new products can target a segmentable market. In radically new products, however, the largest market has often turned out to be different from the market envisioned initially. For example, Kodak's plastic photographic film was intended for a professional photographer's market, which at that time used glass plates, but it quickly found a larger market in the amateur photography market, which the film created.

Technology strategy should emphasize product flexibility, improvement in performance, and lowering of cost in order to seek new applications and create new markets. The more radically new an innovative product, the more flexible should be the view of the potential market. Critical points of market acceptance are when the application, performance, safety, ease of use, and price match a new market group of customers. Ryans and Shanklin (1984) called this kind of marketing for innovative products *postioning* the product: "Stated simply, market segmentation is a too narrowly defined term to describe the target marketing activities that need to be employed by the high-tech company. Rather, positioning seems to best describe the steps that the high-tech marketer needs to follow if it is to identify correctly the firm's target markets and to place them in priorities" (p. 29). In positioning, Ryans and Shanklin suggested that the marketer first identify a broad range of potential users, listing applications, performance, ease of use, and price required for each group. Next, the marketer should prioritize these groups in terms of preferred market for the product. In the product design, flexibility that covers a wider range of groups, or is easily adapted from group to group, will increase the likelihood of commercial success.

Ward (1981) reviewed the steps in market positioning:

1. Focus on the range of applications possible with the new technology.
2. Project the size and structure of corresponding markets for the applications.
3. Judge the optimal balance of performance, features, and costs to position for the markets.
4. Consider alternative ways to satisfy these markets.
5. Analyze the nature of the competition.
6. Consider the modes of distribution and marketing approaches.

Schmitt (1985) also emphasized the importance of marketing personnel working closely with research personnel to facilitate the linking of technological opportunity with market need: "These [marketing] experts should not, however, give blind allegiance to the latest analytical techniques or the dogma of marketing supremacy. Rather, they should have the temperament of a research experimentalist, putting forth hypotheses about the market and devising economical and efficient market experiments" (p. 126).

Positioning in a new market through experimentation is the key to improving an understanding of the eventual market implications of new technologies.

> **The more radical the innovation, the more experimental must be the marketing approach.**

SUMMARY AND PRACTICAL IDEAS

In a case study on

- Personal computer industry

we have examined the following key ideas:

- Technology push and market pull
- New markets and new major devices
- Completing high-tech products for applications
- Market change
- Proprietary technology
- Brands and performance
- Marketing experimentation

These theoretical ideas can be used practically in understanding the challenges and dynamics of developing new markets with new high-tech products, services, or processes. The new market really depends on the value-adding capability of the new product/service as performance in a customer application.

FOR REFLECTION

Identify a new radical technological innovation. What were its first market applications? How did the market grow and change as performance increased and cost decreased for products and services based on the new technology?

18

HIGH-TECH FINANCE

INTRODUCTION

We have examined some of the business functions of an innovative firm: engineering, research, information, production. We also looked at the special challenges of the marketing function for high-tech products. Now we look at the challenges of the finance function for high-tech firms. Significant investments are necessary to develop new products and markets simultaneously. We look at these investment requirements in terms of (1) *company life stages* and (2) *financial fundamentals*.

Cisco Systems

We look at the case of Cisco Systems, one of the few successful new companies of the Internet in the 1990s. Cisco's founders, Sandy Lerner and Leonard Bosack, met in 1977. She was a graduate student in statistics and computer science at Stanford University, and he was teaching in the computer science department. Stanford was one of the research universities that contributed significantly to information technology progress in the second half of the twentieth century.

We recall that the computer was invented at the University of Pennsylvania. We also recall that ARPAnet, the predecessor to the Internet, was researched and begun at U.S. universities. By 1979, ARPAnet was networking computers across university research universities. Stanford had about 5000 computers on its campus, which could communicate only by going outside through ARPAnet. Xerox's PARC gave Stanford a copy of its Altos computer network [and we recall that Altos had innovated a local area network (LAN) through an Ethernet connectivity]. Using PARC's Ethernet LAN technology, Stanford's medical school and computer science department installed separate LAN networks. Stanford's engineers began working to connect these LANs.

The idea of a *router* arose to route messages from network to network (LAN to LAN to ARPAnet). Bill Yeager, an electrical engineer working in Stanford's medical school, began designing routers for the school's networks. In 1980, he developed a prototype of a router using a DEC minicomputer and connected the medical school and computer science department networks. From 1980 to 1982, efforts continued at Stanford to construct networks across the entire campus, and the project was named Stanford University Network (SUN). But the project had not used the router concept and succeeded in connecting only six computers.

Sandy Bosack had graduated and was director of computer facilities in Stanford's business school. Leonard Bosack was director of Stanford's computer science department. They liked Yeager's router technology and decided, with some other colleagues, to "bootleg" the router idea into practice. Without sanction from higher university authorities, they ran coaxial cables from one building to another across Stanford's campus and installed routers and servers to communicate betweens LANs and ARPAnet. Yeager added more code to the routers to coordinate the network, and others added more to provide additional network services: "The project was a success. The router enabled the connection of normally incompatible individual networks. . . . Soon enough, the bootleg system became the official Stanford University Network" (Bunnell, 2000, p. 6).

Sandy and Leonard Bosack next went to Stanford's administration with a proposal to build routers for sale under the school's structure: "Although [Stanford's] Office of Technology Licensing was cognizant of the opportunity the couple had offered, they were unable to take any action to support the development of Len and Sandy's router business in any time less than a year or two. . . . The decision makers did not give Len and Sandy permission to continue their business on campus or to use school resources for making routers for colleagues at Xerox Labs and Hewlett-Packard. Livid, the couple decided to gather up their technology, quit their jobs, and leave Stanford to start their own business" (Bunnell, 2000, p. 7).

In 1984, they started their own router business, financed with their credit cards and a mortgage on their home, naming the new company Cisco Systems. Sandy Bosack took a daytime job at Schlumberger to support them while the new business was started. For a year and a half, they worked out of their home (along with colleagues Kirk Lougheed, Greg Satz, Richard Troiano, and others) to write code, assemble computer hardware, and test new prototypes of routers. They sold routers by word of mouth and e-mail to universities and big corporations that knew they needed to hook their networks together.

They priced routers between $7000 and $50,000. In the fiscal year ending July 1987, they had a profit of $83,000 on $1.5 million sales. They moved Cisco to a business building in Menlo Park, California. The timing of Cisco was opportune, in that by 1985 university and corporate demand for computer networking was exploding. Companies would pay up front for the unique product. Cisco was therefore able to grow on cash flow, without expensive debts for production expansion: "[Cisco was] one of those rare companies that was started at a moment in time where the problem was so vital that customers would pay in advance. . . . Cisco in 1987 filled a desperate need. Customers were tearing the hinges off the door to get the products" (Valentine in Cringley, 1997, p. 298).

The Bosacks saw that Cisco Systems needed to grow very rapidly to dominate the market. They decided to seek capital and professional management to assist in expansion. They pitched their company to about 75 venture capital firms without success until they found Don Valentine, founder and general partner at Sequoia Capital. But Valentine's terms were tough: "Did Valentine cut an almost obscene deal for himself? Of course he did; he's a venture capitalist. In return for just $2.5 million—money that Cisco never even spent, as its own revenues continued to grow—Valentine's venture fund . . . received close to one-third of the company's stock . . . [Learner and Bosack retained 35% of the stock]. In addition to stock, Valentine stipulated that he be able to shape the company's management team" (Nocera, 1995, p. 117).

The first thing Valentine did was to hire John P. Morgridge as CEO of Cisco, moving the Bosacks from top control: "[Valentine] wanted someone who was an industry veteran, a proven leader, a fiscal conservative, and a grown-up presence. Ironically, the best place to find such executives is often at a company that is failing. Companies in trouble shed their best executives first" (Bunnell, 2000, p. 16).

In 1988, John P. Morgridge was president of a failing personal computer company called Grid Systems. The company was selling a top-of-the-line and high-priced portable computer without any major proprietary performance advantages and thus was losing out to competitors (and was then being sold to another company). Morgridge was 54 years old and accepted Valentine's offer to run Cisco, receiving a stock option of 6% of the company. Previous to Morgridge's two-year stint at Grid, he had spent many years as a salesman at Honeywell. Conflict began right away between the founders and the new management: "Morgridge, as the new boss, found himself sparring with the company's founders every step of the way. Cisco was becoming a real corporation, and Sandy in particular was not the corporate type. Sandy had always been a self-proclaimed rebel and iconoclast" (Bunnell, 2000, p. 17).

In 1989, Cisco had $4.2 million in profits and growing. On February 16, 1990, Morgridge took Cisco public with stock price closing that day at $22.50. But management tensions continued at Cisco: "The IPO had made [Sandy Learner Bosack] rich, but with her role diminished and her company moving further away from its roots, she was deeply unhappy. Increasingly, she began lashing out. This she largely concedes—'I was screaming about a lot of things,' she says—but her view is that there were a lot of things going wrong at Cicsco, and her screaming was necessary. 'Yes,' she adds, 'I'm guilty of standing in someone's office and not taking no for an answer when a customer needed something done" (Nocera, 1995, p. 120).

In the summer of 1990, the conflict came to a head: "All the top executives except Morgridge went to Valentine and said in essence, either she goes or we go. For Valentine, of course, that was an easy call. . . . [Len] Bosack walked out with his wife" (Nocera, 1995, p. 120).

In the contract that the Bosacks had signed with Valentine, there were no projections on their employment with the company they had founded. They had agreed to a provision that gave the right to Valentine to purchase the Bosacks' shares.

" 'It was not my intention to get rich. My intention was to not be poor,' said Sandy Learner. . . . 'We worked 20 hours per day, saying the check is in the mail over and over to our vendors. In 1987 we finally got money from our seventieth or eightieth venture capitalist. . . . Then I was fired by the venture capitalists in August 1990, and Len walked out in support of me. After financing the business on our credit cards for three years, we had a four-year vesting agreement! With some hassling, we were finally allowed to vest, got the stock and sold it' " (Cringley, 1997, p. 306).

In December 1990, they sold their two-thirds shares of Cisco for about $170 million. Soon after leaving Cisco, Sandy and Leonard Bosack divorced.

LIFE STAGES OF COMPANIES

The case of Cisco illustrates an important concept in financing high-tech companies—that there are distinct stages with different financial requirements in the life of a successful high-tech company: (1) new venture startup stage, (2) first-mover stage, (3) dominant-player stage, and (4) mature market stage.

We examine here the first two stages, new venture startup and first-mover.

NEW VENTURE STAGE

As we see illustrated in the Cisco case, there are several critical points, financial milestone goals, that any new business venture must pass through along the way to commercial success, and all of these must be properly financed. The financial milestones for a successful new high-tech venture are: (1) acquisition of start-up capital, (2) development of new product and/or service, (3) establishment of production/delivery capabilities, and (4) initial sales and sales growth.

Acquisition of Startup Capital

Capital is the resource necessary to begin and operate a productive organization. Startup capital is required for establishing a new organization and hiring initial staff, developing and designing the product/service, funding production capability and early production inventory, funding initial sales efforts and early operations. Startup capital can be in the form of (1) the founder's personal wealth, borrowing, and/or sweat equity (e.g. the Bosacks' home mortgage, credit cards, and Sandy Learner's job at Schullumberger), or (2) venture capital investments from individuals (called investment "angels") or from venture capital firms (e.g., Sequoia Capital).

Startup capital is seldom sufficient for rapid growth, and therefore further capital requirements are usually necessary for commercial success. This is why the Bosacks approached Sequoia Capital. Often, as is the Bosacks' case, this is how founders can lose control of a new business to venture capitalists.

Product/Service Development

As we earlier saw, a critical milestone in a new ventures is the design and development of a product or service. This will be a major drain on the startup capital. Ordinarily, design and development should be far along before startup capital can be attracted. However, development problems or design bugs that delay the introduction of a new high-tech product or service make serious problems in starting a new firm, because such delay also eats into initial capital. Moreover, if the delay is so long that competitors enter the market with a similar new high-tech product or service, then the first entry into the market advantage is lost. In the case of Cisco, the Bosacks' major development work was done at Stanford University. When they went into business, they already had a product prototype developed and ready for final engineering design to produce and sell the new routers. This is why Cisco was a quick financial success, founded in 1985 and profitable by 1986.

Production/Delivery Capabilities

We recall that a third milestone for new ventures is to establish the capability to produce the new product or service, and this also requires investment in prodution facilities. In the case of a physical product, parts or materials may be purchased or produced and the product assembled. The decision to purchase parts or materials or to produce them depends on whether others can produce them and whether or not there is a competitive advantage to in-house production. Establishing an in-house production capability for parts or materials will require more initial capital than purchase would require, but is necessary when the part or material is the innovative technology in the product. However, the establishment of any new production capability will also create production problems: problems of quality, scheduling, and on-time delivery. Capital will also be required to debug any new production process. In the case of Cisco, early design and production of the routers were financed by the Bosacks' personal savings and by advance payments from customers for their unique and urgently needed product. Later, the perceived need for more capital for expansion during rapid growth motivated the Bosacks to find and have Valentine invest in their company.

Initial Sales and Sales Growth

Initial sales and growth are the next critical milestone; and this also requires investment in establishing a sales force. The larger the initial sales and the faster the sales growth, the less room there is for competitors to enter. An important factor influencing initial market size and growth is the application of the new product or service and its pricing. Another marketing problem is establish a distribution system to reach customers. Distributions systems vary by type, accessibility, and cost to enter. Planning the appropriate distribution system for a new product or service, the investments to utilize the system, and its cost influence on product or service pricing is important for the success of new ventures. Generally, it costs less to reach industrial customers than to reach general businesses or consumers. This is one reason that a large fraction of successful new high-tech ventures are those in which industrial customers provide the initial market. These are usually industrial equipment suppliers or original equipment manufacturers selling to large manufacturing firms. This allows a new firm to get off to a fast start but eventually limits the size of the firm and places the firm in a vulnerable position. A part supplier to a commercial customer may find the customer deciding to integrate vertically downward by producing its own parts. Moreover, a small firm with only a few industrial customers is very sensitive to cancellation of orders from any one of them.

CASE STUDY CONTINUED:

Cisco Systems

As new CEO of Cisco in 1988, Morgridge's strategy was to instill a culture of tight control capable of building Cisco in rapid growth. His earlier experience at Honeywell had taught him that grand schemes were seldom attained; it was the tight control in yearly planning that made things attainable. He introduced planning: "At Cisco, we build a one-year plan with 80 to 90 percent assurance we'll meet or exceed our goals, so it's not a stretch. Then we modify the plan, because we're conservative" (Bunnell, 2000, p. 32).

Morgridge also continued to build on Cisco's early culture of working closely with customers, primarily large businesses. His salespersons were technically competent and fixed any customer problem. Morgridge kept expenditures in control, providing modest salaries to employees with stock options. In 1991, Morgridge hired John Chambers as senior vice-president of worldwide operations. In 1995, Morgridge moved up to chairman of the board of directors and promoted John Chambers to CEO of Cisco.

Chambers had a law degree from West Virginia University and an MBA from Indiana University. After graduation in 1977, he took his first job in sales at IBM. He saw IBM's decade of strategic errors in the 1980s. IBM failed to fully exploit the rise of the personal computer and computer networking to reinvent IBM's computer businesses successfully. (IBM did not begin serious reorientation until the 1990s, under new leadership.)

From his IBM days, Chambers had learned lessons of both what to do and what not to do. A good to-do lesson was IBM's efforts to satisfy customers. Another good lesson was a salesperson's need to sell information technology at all the multiple levels of the customer organization. A third good lesson was how important software was to IBM's successful mainframe hardware business. But Chambers also learned some lessons of what not to do, some bad practices at IBM. One bad practice was IBM's relative neglect of small businesses. From this, Chambers saw the importance of selling not only to big businesses but also to small businesses. Chambers also learned to avoid the overly restrict command-and-control structure of IBM, which made it difficult for IBM to make timely and appropriate decisions.

Chambers decided to move to another company, but unfortunately, he joined Wang Laboratories in 1983, just as it was beginning to fail. We recall that Wang had pioneered word-processing workstations but soon lost its market to word-processing software on the new personal computers. In 1986, An Wang retired as the company was trying to find new markets using minicomputer technologies. But in 1990, An Wang had to came back from retirement to try to save the failing company, as minicomputers were also being replaced by personal computers. He asked Chambers to become the senior vice-president of U.S. sales and field service operations. Unfortunately, soon after, An Wang died. Then Chambers had to try to control the company's continuing decline. He presided over five layoffs of 4000 people as sales fell from $3 billion a year to $42 million. Chambers's stock options in Wang became worthless. From both the IBM and Wang experiences, the important lesson that Chambers learned was to adapt to the flow of technology advance.

In 1991, Chambers quit Wang, accepting the offer to join Cisco. Later when he became CEO of Cisco, he attended to the challenge of keeping Cisco advancing on progress in information technology: "When Cisco's technology started to become dated in the early 1990s, the company saw it coming and adapted. . . . Routers were still a hot ticket, but there were at least two new networking technologies. . . . One was called switching—primarily a box which some small companies were already starting to manufacture. The other was something called asynchronous transfer mode or ATM" (Nocera, 1995, p. 120).

Chambers formalized the acquisition of new companies as a strategy to continue to get new applied knowledge capabilities and new product lines into Cisco. The information technology challenge was in tying local area networks (LANs) into wide area networks (WANs). The fast Ethernet technology was still the preferred LAN technology, but for WAN networking, asynchronous transfer mode (ATM) switches was becoming preferred by customers. An ATM switch was a hardware-based switch that transmitted data faster than routers and could be used to connect a finite number of LANs together, with resulting high-speed communication between LANs. Moreover, ATM switches allowed digital emulation of traditional switch-based phone networks and could bridge data communications to telephone communications. Then Ethernet technology was hooking up computers into LANs, while ATM switch technology was hooking LANs together into WANs, and routers were hooking all into the Internet.

This began happening even before Chambers became CEO. He had discovered the need for Cisco to move rapidly when he had visited one of Cisco's largest customers, Boeing: "One day Chambers . . . was visiting a long-time Cisco customer and discovered to his horror that Cisco was about to lose a $10 million order to a competitor that was manufacturing switches. 'What do I have to do to get that order?' Chambers remembers asking the man. 'Start making switches,' the man replied. So Cisco did. It bought a startup called Crescendo Communications" (Nocera, 1995, p. 120).

Afterward, as CEO, Chambers launched an aggressive Cisco policy in continuing business acquisitions to gain new technologies. With a rising stock market in the 1990s and a high value on Cisco stock, Chambers could use Cisco's stock to acquire other companies for new technology. This kept Cisco on the cutting edge of technology through that decade.

FIRST-MOVER STAGE

We recall that after a new venture is started successfully, its next challenge is to become a large company in its market, a stage that requires significant new capital. We also recall that for a new high-tech business in a new industry, a first-mover strategy is necessary for dominance and long-term survival. In running Cisco, Morgridge and Chambers paid attention to the aspects required to make a new venture into a first mover:

1. Continuing to acquire new technology
2. Developing large-scale production capacity
3. Developing a national distribution capability
4. Developing the management talent to grow the new firm

In the transition from a successful startup to a first mover in a new industry, there are additional financial requirements for investments that must be made in the first-mover stage. These are:

5. Production and distribution expansion
6. Meeting competitive challenges with new product models
7. Production improvement
8. Organizational expansion and management development
9. Attaining capital liquidity

Production Expansion

As the new market grows and sales are successful, production expansion must be planned and implemented in a timely manner or sales will be lost to competitors because of delivery delays. Production expansion will usually require a second round of capital raising, for the initial capital seldom provides enough for expansion.

The exception is when production can be outsourced. The rapid growth of the market and high margin of Cisco's unique products allowed Cisco to finance production expansion from cash flow. Cisco's hardware products were standard commodity-type minicomputers, all of whose parts could be outsourced. Cisco used outside vendors to produce their physical product. All of Cisco's proprietary advantage lay in its software, not in its hardware. This meant that Cisco did not need much capital for hard-goods production facilities.

New Product Models to Meet Competition

In a very few areas and rare cases, a patent on a new product or process is basic and inclusive enough to lock out all competitors for the duration of the patent. This is true in the drug industry and occasionally elsewhere. However, most new high-tech ventures are launched with only partial protection from competition by patents, and competitors soon enter with me-too products. These are likely to be introduced with improved performance or features and/or at a lower price. The entrance of competitors into the new market is a critical time for new ventures. They must at that time meet the competitive challenges with new product models. Cisco met competitive challenges by continuing to acquire new technology products.

Product Improvement and Diversification

A new firm must upgrade its first-generation products with new products to keep ahead of competitors in product performance and features. It must also continually lower its cost of production to meet price challenges by competitors, and it must diversify its product into lines to decrease the risk that a single product problem will kill the firm. The round of capital raised for production expansion also needs to provide for product and production improvement. Cisco's major product diversifications occurred through a strategic and aggressive policy of acquisition of potential competitors.

Organizational Development

As an organization grows in size to handle the growth in sales and production, it is important for the firm to expand organizational structures and train new management. This is an important transition, as the early entrepreneurial style of organization and openness and novelty of culture needs to mature toward a stable but aggressive large organization. In a small firm, coordination is informal and planning casual. In a large firm, both coordination and planning needs formalization, and this increases overhead costs, requiring more working capital In the Cisco case, the transition from the management and organizational styles of the founders to the traditional control of experienced strategic managers occurred abruptly (and rather violently) when the venture capitalist took control of Cisco and installed seasoned professional managers.

Capital Liquidity

The final step for success in a new firm is to know when and how to create liquidity of capital assets and equity. One means is to go public, and another means is to sell the firm to a larger company. Liquidity of capital enables the founders of the firm and early employees to transform equity into wealth. In the case of Cisco, the initial public offering of Cisco in 1990 was successful and provided the founders with a personal fortune, even though they lost control of the company. The venture capital firm, Sequoia Capital, leveraged its modest $2.5 million investment into billions of dollars by the late 1990s.

FINANCIAL FUNDAMENTALS: CAPITAL AND PROFITS

As we saw in Cisco, control in a company is always through finance. We recall that the concept of economic value adding (EVA) had two subconcepts: value to customers and value to shareholders. Finance in a company is always eventually about value to the shareholders. Because of the importance of finance to business (and to technological innovation), we should review briefly the fundamental concepts of managing the financial aspects of business. Over the long run, all businesses must add significant economic value to investors, or they will fail. Profitability is essential to continuing survival of any business.

Capital

As an input into an enterprise, capital is an investment in the business. Investments can be in the form of debt or equity. *Debt* is a loan from a lender that needs to be paid back in principal and interest according to a schedule. Debt can be in the form of fixed loans or revolving lines of credit. *Equity* is in the form of stock, either preferred or common, with or without voting rights. In accounting practice, the capital of a firm can be measured in several ways:

- By *share price,* the price at which a share on the stock market in which the share is listed can temporarily be brought or sold
- By *market capitalization,* the total number of shares issued times the current market price of a share
- By *equity,* the value of corporate assets minus liabilities
- By *company sales price,* the current market price of a buyer's offer per share times the total number of shares issued

Share price is the current monetary value of a piece of ownership (stock) of a company that can be traded in a public stock market. Share price measures the current market value of pieces of the company. But this cannot simply be added up to obtain the entire value of the company, because whenever large amounts of stock are sold in a market, the share price at that time usually declines sharply. Accordingly, the *market capitalization* of a company (share price times number of outstanding stock) is not really a very useful figure, but does make an exciting accounting number because it can be very big—perhaps billions of dollars.

Similarly, *equity* is not a very useful absolute number but is useful as a relative number: Did the equity of the company increase or decrease from last year? If increasing, the company is prospering; if decreasing, the company is trouble. The reason that equity is not a useful absolute measure of company value is that the stock price listed in the equity portion of company financial statements is measured at the offering price of the shares (according to standard accounting principles).

Company sales price is a useful number, but only when another company makes an offer to buy the business. Then a sales price can be compared meaningfully to the market capitalization at the time of the offer.

Profits

Profits are a measure of the value to the company of the productive operations of the company. Profits derive from revenues received from the sales of products and/or services of the business less all costs and taxes. Revenues received are the payments that customers actually make on the purchase of the products or services. Profits are calculated from revenues received by subtracting out expenses, such as:

- Costs of supplies to produce the goods or services
- Costs of the direct transformative operations of the business
- Costs of the marketing operations of the business
- Costs of the overhead operations of the business
- Costs of research and development for new or improved products or services
- Depreciation costs to replace equipment and software in the operations of the business
- Debt service
- Payment of taxes
- Costs of executive stock options

What is left after subtracting all expenses from revenue received is the profit, or *earnings,* and should be reported as such on financial statements.

CASE STUDY:

Enron

Finance is the repository of future wealth in a merchant society. The bright side of finance in high-tech firms is captial growth. But as in any financial activity, there can be a dark side—financial fraud. As we earlier saw in the e-commerce stock bubble of the last decade of the twentieth century, the U.S. stock market was enamored of anything calling itself high-tech. Then one of the darlings of the U.S. business world was Enron, which advertized itself to be a new high-tech business introducing trading into the commodity business of energy. Enron became the classic case of financial fraud masked as high tech.

In 2001, Enron's stock peaked at $80 a share on January 8 and fell to zero on December 2. At the time it was the biggest bankruptcy in U.S. business history and a financial scandal. What was revealed by the scandal was that Enron had "questionable business practices" covered up by "aggressive accounting practices."

These accounting practices deliberately obscured the fact that Enron was never really profitable. Meanwhile, Enron's executives had made a lot of money: "In the three years before Enron's collapse, 29 senior executives and directors sold more than $1 billion in stock, much of it after exercising options" (Oppel, 2002b, p. 2). Meanwhile, Enron employees and the shareholders of Enron lost money, a lot of it. Enron's story unfolded in bits and pieces as news events:

January 20, 2001

> Earlier, Enron and its chairman and CEO, Kenneth Lay, had been publically regarded as successful: "As a chilly rain soaked the streets of Washington . . . the motorcade of [the new president] moved past the reviewing stand. . . . In the exclusive Pioneers box on the parade route, Ken Lay watched the festivities. He was there amidst an elite group of about 200 men and women who had each raised $100,000 [for the president's campaign]. . . . Also Enron and its top two executives had kicked in $300,000 for the inauguration" (Eichenwald and Henriques, 2002).

February 2, 2001

> But Enron had never really been a sound business: "Late on Feb. 2, a special committee formed by the Enron board to investigate the company's labyrinthine finances released a tough critique of its executives. Particularly disturbing were off-the-books partnerships constructed by Mr. Fastow that were backed by Enron stocks. The report found an across-the-board failure in ethics and oversight. The off-the-book partnerships were designed to deceive by moving debt off the balance sheet. They also improperly enriched employees involved in the partnerships.

February 5, 2001

> Arthur Andersen was Enron's auditor and met to discuss Enron as one of their big clients but saw nothing wrong at Enron: "Arthur Andersen officials discuss Enron's 'aggressive' accounting practices and potential conflicts of interest at a meeting to decide to retain the energy-trading company as a client" (Brown and Sender, 2002, p. C1).

March 2001

> Andersen's auditors approved of Enron's aggressive accounting practices in off-the-book partnerships and did not insist that they be revealed as Enron liabilities in Enron's audited public accounting: "What made Enron's stock price so important was the fact that some of the company's most important deals with the partnerships run by Mr. Fastow . . . were financed, in effect, with Enron stock. Those transactions would fall apart if the stock price fell too far. When Enron's stock share price [dropped to] around $70 a share in early March, [there] was a risk of triggering the provisions [in the partnerships requiring Enron to reimburse partnerships with substantial cash]. Enron's deals [called Raptors] . . . were keeping roughly $504 million in red ink off Enron's books (Eichenwald and Henriques, 2002, p. A26).

March 26, 2001

Andersen's audits had approved Enron's off-book partnerships backed with Enron stock. But Enron's partnerships had invested in technology stocks during the prior decade's Nasdaq stock market boom, and these stocks had dropped. Therefore, the value of the partnerships had dropped precipitously, and Enron stock had been pledged to partners. When Enron stock fell below $70 a share, Enron would have to pay the partners cash. "One reason the Raptors (partnerships) were so shaky may have been the fact that . . . they had all ready paid out more than $160 million to Mr. Fastow's [previous] LJM partnerships" (Eichenwald and Henriques, 2002). The later Raptor partnerships were created by Fastow to pay off his earlier Enron LJM partnerships—a pyramid scheme.

April 17, 2001

Enron's then CEO (succeeding Lay), Jeffrey Skilling, presented the results of Enron's first quarter of 2001 to investors, but he did not reveal the partnerships. Richard Grubman, with Highfields Capital Management, asked Skilling: "why was Enron only reporting its profits and not releasing its balance sheet that would list assets and liabilities." Skilling replied that Enron did not do that. Grubman said: 'You're the only financial institution that can't produce a balance sheet or a cash flow statement with their earnings.' Mr. Skilling paused [then answered]: 'Well, thank you very much,' he said. 'We appreciate it.' Then Mr. Skilling turned to his colleagues in Houston and muttered a vulgarity. The group in Houston laughed" (Eichenwald and Henriques, 2002, p. A27). Soon afterward, Skilling resigned as CEO and was replaced by the earlier CEO, Kenneth Lay.

August 20, 2001

"Enron Vice President Sherron Watkins [also a former Andersen employee] writes anonymously to Kenneth Lay [who is again CEO of Enron] with concerns about potential conflicts of interest and accounting practices" (Brown and Sender, 2002, p. C1). Later Ms. Watkins will testify about Enron to a congressional committee: "In more than four hours of testimony, she described a culture of intimidation at Enron where there was widespread knowledge of the company's shaky finances but no one felt confident enough to confront Mr. Skilling, who resigned in August, or Mr. Fastow, who was ousted in October" (Oppel, 2002a, p. A1).

Fall 2001

After the Watkins memo, Lay asked Enron's law firm, Vinson and Elkins, to look into the charges. But Vinson and Elkins lawyers saw no evil: "The facts disclosed through our preliminary investigation do not, in our judgment, warrant a further widespread investigation by independent counsel and auditors" (Brown and Sender, 2002, p. C1). But Enron was already in critical financial trouble.

October 12, 2001

Andersen began to anticipate trouble for itself. Andersen lawyer Nancy Temple sends an e-mail to Andersen auditors: "E-mail by in-house lawyer reminds risk-management partner Michael Odom of the document-and-retention policy" (Brown and Sender, 2002, p. C1). Andersen employees then destroy audit records about Enron, and this destruction will result in a federal indictment of Andersen.

October 22, 2001

Enron's stock price drops to $20 a share: "As one thread pulled away, the whole garment unraveled. . . . During . . . the price collapse, Enron employees with huge percentages of their retirement funds in Enron stock could not sell their shares because . . . employees' accounts were temporarily frozen" (Eichenwald and Henriques, 2002, p. A26).

November 19, 2001

Enron files a latest quarter report which revealed a cash drain of $2 billion: "The bottom [for Enron] dropped out" (Eichenwald and Henriques, 2002, p. A26).

December 1, 2001

Enron files for bankruptcy-court protection: "At 7 p.m. Saturday, the board of Enron met. . . . Finally, the motion to declare bankruptcy was put before the directors. They unanimously supported it" (Eichenwald and Henriques, 2002, p. A27).

December 22, 2001

Government investigators began looking at the Enron case: "As government investigators look for the roots of Enron's demise, [former] Enron chief executive, Jeffrey Skilling . . . broke his silence today to deny any responsibility or wrongdoing" (Oppel, 2002a, p. C1).

January 10, 2002

Andersen's document destruction comes to light: "Andersen discloses to investigators that a 'significant but undetermined number' of documents relating to the Enron audit had been destroyed in recent months" (Brown and Sender, 2002, p. C1).

January 15, 2002

Andersen fires the Andersen partner in charge of auditing Enron, David Duncan, "saying that he led 'an expedited effort to destroy documents'" (Brown and Sender, 2002, p. C1).

January 16, 2002

Investigators from the U.S. House Energy and Commerce Committee and the Justice Department questioned Duncan: "Duncan tells the investigators he called the February meeting because he was aware the Enron account posed 'significant risk' . . ." (Brown and Sender, 2002, p. C1).

January 24, 2002

A U.S. congressional committee subpoenas Duncan to testify, and Duncan takes the "Fifth": "The right not to be forced to incriminate oneself is a fundamental one in American justice, and one that is normally respected. A witness who will not testify does not provide information, and so there is little reason to force him to show up and invoke his rights under the Fifth Amendment. But Congress has been willing to subject some people—normally those deemed to be beneath contempt—to televised humiliation: . . . suspected Communists . . . Mafia bosses . . . Now it is auditors. 'Enron robbed the bank; Arthur Andersen provided the getaway car, and they say you [Duncan] were at the wheel,' [said] Representative Jim Greenwood" (Norris, 2002b, p. C1).

January 25, 2002

Enron former vice-chairman, J. Clifford Baxter, commits suicide in his locked car in the driveway of his Dallas home. Baxter was one of the Enron executives who personally benefitted from the Enron off-balance-sheet partnerships. "Before he was found dead, Baxter was 'obsessed' with worries that the scandal surrounding the company's collapse would forever tarnish his reputation. . . . Baxter was one of 29 past or current Enron officials named as defendants in lawsuits by shareholders who lost billions of dollars when the company went bankrupt" (Duggan and Romano, 2002, p. A1).

January 28, 2002

More of Enron's questionable business practices become public, including their use of derivatives to avoid taxes: "As the significant role that derivatives played in Enron Corp.'s downfall comes into focus, lawmakers and regulators are lining up in favor of more oversight of these risky investments. . . . [Enron used derivatives] 'to create false profit and loss entries for the derivatives Enron traded' " (Schroeder, 2002, p. C1).

February 7, 2002

House Energy and Commerce Committee holds hearings on Enron's collapse: "The room was packed with angry former employees in search of answers, battalions of newly retained criminal defense lawyers with their white-collar clients, scores of reporters, and an aggressive panel of Republican and Democratic lawmakers" (Labaton and Oppel, 2002, p. A1).

February 12, 2002

Next, Kenneth Lay appeared before the U.S. Senate Commerce Committee: "After being subjected to a bipartisan oral barrage from 21 senators, Kenneth L. Lay, the former chairman of Enron, asserted his Fifth Amendment right against self-incrimination. . . . [He was] likened to Charles Ponzi and compared unfavorably to a carnival barker . . . [and] accused of running a conspiracy and gouging the energy consumers of California." And Senator Byron Dorgan asked: "How is it that 29 Enron executives at the top were able to earn $1 billion in stock sales in 2001 while people at the bottom lost everything?" (Oppel and Kahn, 2002, p. A1; Schmidt, 2002, p. A1).

February 12, 2002

The issue of the role of the accounting firm of Andersen in the Enron debacle centered on the liability of an accountant. Does the accounting firm merely have to follow acceptable principles of accounting, or should it ensure that the accounts fairly and completely communicate the real financial condition and performance of a publically traded company? "Signaling a tough stance on accounting irregularities, a top securities regulator warned that corporations strictly following accounting rules still could be accused of securities fraud if the filings don't accurately reflect a company's underlying economic condition. . . . Auditors who issue clean bills of health are required to certify that a company's financial statements fairly represent the client company's financial performance" (Liesman, 2002, p. C1).

January 24, 2002

Exploiting accounting "loopholes" to boost apparent earnings and sales growth had not been limited to Enron but became the financial "game" of the 1990s. In a congressional hearing, Arthur Levitt, a former chairman of the Securities and Exchange Commission (SEC), summarized: "Enron's collapse did not occur in a vacuum. Its backdrop is an obsessive zeal by too many American companies to project greater earnings from year to year. When I was at the S.E.C., I referred to this as a 'culture of gamesmanship.' A gamesmanship that says its O.K. to bend the rules, tweak the numbers and let the obvious and important discrepancies slide. . . . First, we must better expose Wall Street analysts' conflicts of interest. . . . Second, company boards often fail to confront management with tough questions. . . . Third, many accounting rules need to be undated . . . to give investors a better understanding of the underlying health of companies" (*New York Times,* 2002, p. C9).

March 14, 2002

The U. S. Justice Department indicted Arthur Andersen for obstruction of justice.

June 16, 2002

After a one-month trial of Andersen, "a federal jury convicted Arthur Andersen today of obstruction of justice for impeding an investigation by securities regulators into the financial debacle at Enron. Soon afterward, Andersen informed the government that it would cease auditing public companies . . . effectively ending the life of the 89-year-old firm" (Eichenwald, 2002a, p. A1).

August 21, 2002

The first of the Enron executives was indited: "Nearly nine months after Enron filed for bankruptcy, federal prosecutors last week took their first step in building a criminal case against top officials . . . by striking a plea bargain with one of the former chief financial officer's Andrew S. Fastow's most trusted lieutenants. Mickael J. Kooper pleaded guilty to charges of conspiracy to commit wire fraud and money laundering. . . . It was clear the prosecutors are moving methodically up the hierarchy, much as they do in organized-crime investigations. . . . Fastow is viewed as the crucial link to any top-level prosecutions" (Grajek, 2002, p. H2).

Finally, it had all emerged. Enron's reported earnings during the 1990s were false. When this began to came to light in 2001, Enron's credit ratings fell. Enron could no longer finance its enormous debt and failed. Enron had reported itself as a high-tech firm that was both profitable and growing. But it had only engaged in a kind of pyramid scheme, constantly borrowing money to pay previous debts accumulating in basically unprofitable businesses and trades and even in fraudulent trades. Curiously, at the time it was debated whether Enron's business and accounting practices were merely unethical or actually illegal. The ambiguity in the debate derived from the widespread practice in the 1990s of inflating reported earnings. Since 1990, U.S. corporate executives were being rewarded with large stock options as an incentive to increase the price of corporation stock. As an incentive, it worked. But with it came the practices that bent the rules of good accounting practice (i.e., aggressive accounting). Enron went bankrupt because they never had real earnings and positive cash flow but were only reporting such through a misuse of accounting. As Norris (2002c) commented: "You can't fake cash flow. Well, actually, you can. . . . Accountants rose to the challenge. If investors wanted to see operating cash flow, well, by jiggery, they would see it. Cash flow mirages are crucial parts of what went on at Enron. . . . Enron found ways to borrow money and report the cash as if it were real operating cash flow" (p. C1).

The pressure on U.S. corporate executives such as Lay and Skilling to project greater earnings for growth had come from a confluence of three U.S. economic practices:

1. The U.S. stock market valued growing companies with share prices at very high price/earnings ratios.
2. The U.S. personal income tax code had a tax rate structure that taxed corporate dividends at roughly twice the rate of short-term stock appreciation.
3. CEOs chose directors who in turn rewarded the CEO with enormously large stock options.

In the second half of the twentieth century, U.S. income tax law was changed to lower personal income tax on stock gains to 18% while keeping top rates of dividends received from corporations taxed at 36%. This meant that to the wealthy U.S. corporate shareholder, gains in share price were more valuable than dividends on shares. A taxpayer needed nearly twice the dividend income than share gain in the stock for the same after-tax take-home income. Accordingly, most U.S.

corporations in the last quarter of the twentieth century simply stopped paying dividends, retained all earnings, and even borrowed heavily to grow by acquiring other businesses. Growth in sales and revenues was the game.

The way the game was played by some CEOs, such as at Enron, was to create a corporate record of continuing growth in sales and revenues by any means possible. Without rapid growth in the share price, executive stock options were worthless. CEOs did not have to provide annual dividends to shareholders but could use all the corporate earnings to buy growth through acquisitions. Then CEOs could reward themselves by cashing in options to become multimillionaires. The U.S. financial structure had encouraged what Levitt called a *culture of gamesmanship:* "The profits were an illusion. The multi-million-dollar rewards for executives were real. Over the last few years, executives at some companies released inaccurate earnings statements and, before correcting them, sold large amounts of stock at inflated prices" (Leonhardt, 2002, p. 1).

In June 2002, several large U.S. companies were under criminal and/or civil investigation as a result of false practices: Adelphia Communications, Computer Associates, Dyergy, Enron, Global Crossing, Quest, Rite Aid, Tyco International, WorldCom, and Xerox (Berenson, 2002). In January 2003, President George Bush proposed abolishing income tax on stock dividends received by individuals.

FINANCIAL FUNDAMENTALS: EARNINGS, CASH FLOW, EQUITY, WORKING CAPITAL, CORPORATE SHARES, AND RETURN ON INVESTMENT

Financial numbers are critical to running a business, particularly a high-tech business. But the numbers must be real, not a result of accounting fraud. Basic financial numbers are earnings, cash flow, equity, working capital, corporate shares, and return on investment.

Earnings

By itself, the accounting category of *earnings* is not a very critical financial number. The reason for this is that earnings always have subtracted from income a category called *depreciation* (which is always made as large as possible to reduce taxes). In accounting theory, depreciation is the potential future cost of replacing present production equipment. Depreciation is a theoretically a pot of money in the business formally set aside for replacing equipment, but it need not be used for this. Accordingly, adding back depreciation to earnings gives a measure of the *cash flow.*

Cash Flow

Cash flow is important, for this is the money that can be used (or played with) for various purposes: investments, dividends, paying off debt, acquisitions, and so on.

Cash flow is the critical measure of useful operating capital in an organization. It is the money on hand to continue to operate the business.

Equity

After accounts are settled, a balance sheet is computed to show the financial health of the organization at the time summarized in the basic accounting formula *equity = assets − liabilities*. Equity is the ownership value (capital) of the business. Assets are the positive elements of capital, such as buildings, equipment, cash on hand, and retained earnings. Liabilities are debts to be paid, such as short- and long-term loans. An important measure of the vulnerability of a company to changes in earnings is the *ratio of debt to equity*. A high ratio indicates that the company is vulnerable in business downturns, as debt service is high.

Working Capital

In addition to cash flow, a business may have a *line of credit* with a lender, such as a bank. A line of credit allows the business to draw up to the maximum of the line any time (as long as payments to the credit line have been keep up to date). The sum of the cash flow and the balance in a line of credit is the total of money available to management in a fiscal quarter of the year and is called the working capital of the business.

> **Businesses go bankrupt when working capital is insufficient to pay expenses and debt service and no further debt can be obtained.**

This is what had happened to Enron as in the fall of 20001, Enron had no real earnings or positive cash flow. Enron was borrowing money, increasing debt. Without more loans, Enron's working capital was insufficient for expenses and debt service. Exposure of Enron's real financial condition prevented Enron executives from borrowing more money and forced bankruptcy.

Corporate Shares

The capital value of a privately held company is made liquid by changing the company into a publically held company by selling stock in the company through an initial public offering (IPO). For investors and executives in a new high-tech company, the IPO is a "cash-in" time for all the initial investments and hard work in creating a new company. (For example, in the decade of the 1990s, the excitement about the Internet and new e-commerce created hundreds of new companies that went public before January 2000 and hundreds of new millionaires who cashed in on the Internet frenzy of the time.)

But for an already public company, the only way that an executive can cash in on the increasing capital value of a company is through the board of the company authorizing stock options as payments to executives. The stock options are new stocks and so dilute the value of existing shares. Existing shareholders therefore pay for the stock option bonuses of executives. But in the 1990s, U.S. stock options were not accounted as expenses and so were relatively hidden from investors.

This is how Enron executives made $1 billion from selling Enron stock options just before bankruptcy. Stock options for U.S. executives were large and widespread. For example, in the year 2001, Microsoft issued $3.3 billion stock options for its top executives; Cisco Systems, $2.6 billion; Nortel Networks, $2.5 billion; IBM, $1.9 billion; Intel, $1.6 billion; and Lucent Technologies, $1.5 billion (Morgenson, 2002).

Return on Investment

After investments of capital in a company, a return-on-investment (ROI) is made when cash is received from the company to the investor. This return can be in the form of a sale to the public of stock (issued for the investment or for services as the executive stock-option). A return on investment can also be in the form of a dividend paid to a shareholder from earnings of the corporation. The ROI is calculated as the ratio of return (R) divided by investment (I): $ROI = R/I$. For example, if a stock is purchased for $10 a share (I) and is sold for $20 a share (S), the ROI = $(S - I)/I = (20 - 10)/10 = 10/10 = 1$ or as a percentage, a 100% ROI. [A return of 100% means double your money, as you received your original investment ($10) and also gained an equal amount ($10 + $10 = $20) for a gain of 100% over the original investment.]

Earnings from a prior year's operations in a business increases its capital value. Earnings can be returned (in whole or in part) to shareholders as dividends or retained within the corporation for investment (retained earnings). Earnings are published as a per share earning (E), which is the total earnings (TE) that year divided by the number of outstanding stock shares (s): $E = TE/s$. The increase in capital value of a business can thus be calculated as a kind of per-share $ROI = E/S$, where E is the earnings per share and S is the share price on the stock market. For example, if the stock is selling for $20 a share (S) and the company publishes per-share earnings of $2 (E), the per-share $ROI = E/S = \$2/\$20 = 0.1$. As a percentage, the per-share ROI is 10%. Thus, purchasing a share of stock at $20 when the company has just earned $2 per share that year means that you have made a stock investment with an expected per-share return of 10% per annum if the company does the same level of business in the next year.

This per-share ROI is a way to estimate the value of a stock share in a stable business and a stable stock market. It tells the investor what his or her expected return could be as a percentage of their investment. This is why the financial pages of newspapers in listing share prices on exchanges often publish a ratio of price/earnings (P/E) ratio. The inverse of the P/E ratio is the expected annual per-share ROI. For example, the traditional level of P/E ratio for stable businesses in the very early part of the twentieth century was a P/E of about 10. This meant that the business could return a 10% annual dividend to each shareholder. However, a growing business (say, doubling in sales and earnings annually) could be priced at a P/E of 20. This meant that the company was expected to double next year, so that your stock purchase this year would yield 10% next year (since you paid twice what it was worth last year, expecting the value of the company to double in one year).

SUMMARY OF FINANCIAL MEASURES

For managing technological innovation in high-tech companies, the following financial measures are important:

- Volume of sales
- Revenue
- Gross margins of products
- Costs
  - —Production
  - —Marketing
  - —Overhead
  - —Debt service
  - —Depreciation
  - —Taxes
- Earnings
- Cash flow
- Debt/assets ratio
- Share price
- P/E of shares
- ROI
  - —Dividends
  - —Share price gain
  - —Stock dilution

SUMMARY AND PRACTICAL IDEAS

In case studies on

- Cisco Systems
- Enron

we have examined the following key ideas:

- Life stages of companies
- New venture stage
- First-mover stage
- Financial fundamentals

These theoretical ideas can be used practically for seeing the financial challenges in high-tech businesses and in launching new high-tech ventures.

FOR REFLECTION

Examine the long history of a once very successful high-tech firm such as Xerox or DuPont. How did the financial challenges of the firm change as the technologies within the firm changed from innovative to mature?

19

TECHNICAL PROJECT MANAGEMENT

INTRODUCTION

Now we look at the management style of how new technology is implemented: project management. Innovative change in a business always occurs through a technical project. It is in discrete technical projects that (1) research is performed, (2) new technology is invented, (3) high-tech products are designed, and (4) production processes are invented or improved. Managing a discrete innovation project is called *technical project management*. In modern management theory, there are three distinct styles of management:

1. Bureaucratic management for running large organizations
2. Entrepreneurial management for starting and building organizations
3. Project management for introducing finite change in an existing organization

Project management is the style of running technical projects because they are focused on *creating a specific incident of change*. Kerzner (1984) emphasized that a project should have (1) a specific objective, (2) definite start–stop dates, (3) a funding limit, (4) a specific resource requirement, and (5) a specific customer.

In addition to project management, there is also *program management*, which is the management of not one but of several projects simultaneously—a program of projects. Project and program management are complex topics, but we can briefly review their essential features in terms of: (1) *stages and phases of projects*, (2) *project risks, future value*, and *cost*, (3) *sets of projects*, (4) *implementation*, and (5) *intellectual property*.

CASE STUDY:

Pilkington's Float Glass R&D Project

The case study of Pilkington nicely illustrates a successful technical project (Layton, 1972). The historical setting was in the 1950s when an invention to radically change the manufacture of plate glass was developed and substituted as a new production process throughout the glass industry. In 1952, Pilkington was the major glass manufacture in the United Kingdom, controlling 90% of that market with annual sales of 113 million pounds sterling. At that time, the world's glass industry was oligopolistic—dominated by a few large producers, with major shares of their respective national markets. Other major glass firms in the world included Libby–Owens–Ford in the Unitd States, with annual sales of £187 million; PPG Industries in Canada, with annual sales of of £477 million; Asahi in Japan, with annual sales of £265 million; and Saint Gobian in France, with annual sales of £436 million.

Up until 1952, flat glass was manufactured in one of two ways. The first way was to draw glass upward as a ribbon from a bath of molten glass; this produced a low-cost, fire-polished glass. Yet it had considerable optical distortion, since the glass surface wavered as it was drawn upward. The second method was to roll cooling glass horizontally between rollers, which pressed the glass to the right thickness but left marks from the rollers. Then expensive polishing and grinding had to be done, which produced better glass optically but at a much higher cost.

Such was the state of the technology when a young Alastair Pilkington, a cousin of the owning family, went to work for Pilkington. Alastair began by working in the sheet glass division and became familiar with the inexpensive method of glass production, drawing glass upward. Next he became a production manager in the plate glass division, where he saw the very expensive process of grinding and polishing plate glass. We recall that the process of radical innovation has a series of logical steps, proceeding from invention, to functional prototype, to engineering prototype, to pilot plant (for a process), to marketing and sales. This project went through these steps.

1. Invention

Alastair's inventive mind began thinking about alternative ways of producing plate glass. He saw glass pulled vertically. Why not float it horizontally? This was his inventive idea! Molten glass could be floated out on a bed of molten metal. He chose to try a hot bath of tin. Molten tin had a low enough melting point to remain molten as the glass solidified on it. Moreover, the tin could be kept free of oxide if the atmosphere were controlled. Alastair presented his idea to his superior, the production director of the flat glass division, who took the idea to the company's board. The board recognized the potentially vast economic return if the idea worked and approved a budget for an R&D project.

2. Applied Research and Functional Prototype

A small project group was created in 1952, consisting of Richard Barradell-Smith and two graduates, reporting to Alastair Pilkington. They built a small pilot plant, costing £5000, that produced 10-inch-wide plates. They found that when glass cooled on molten tin, it was precisely 7 millimeters thick (because of the physical balance between the surface tensions and densities of the two immiscible liquids, molten glass and molten tin). This was lucky because 60% of the flat glass trade was in 6-millimeter-thick glass. By stretching the glass ribbon a little as it cooled, they could reduce the natural 7-millimeter thickness to the desired commercial 6 millimeters.

3. Engineering Prototype and Testing

Next, they built a second pilot plant to produce 48-inch-wide plates (this was commercial width). It cost £100,000 (five times more than the first 10-inch machine). These two machines and their research took five years, until 1957, to get working right. By then the research team thought that they had perfected the production technique enough to try a production-scale plant.

4. Production Prototype and Pilot Plant Production

In 1957, the R&D team was expanded to eight people, with three graduates from different disciplines. The group was moved into the plate glass factory and built a plant costing £1.4 million (14 times the cost of the engineering prototype).

The major purpose of research is to reduce technical risk before production-scale investment is committed, because technical problems always occur. At this plant scale of production, the control of the atmosphere had to be improved to prevent oxidation of the tin. Next was a problem that had to be solved regarding how to flow large quantities of glass out rapidly onto the tin. There was also a problem at this scale of stretching the glass to the right thickness and pulling the cooled glass plate off the molten tin. At each scale-up, previously solved technical problems had to be re-solved. (This is typical of engineering problems in production.) This production-scale problem-solving phase of the development took 14 months, from 1957 through 1958, and produced 70,000 tons of unsalable glass.

Then, finally, success! All the problems were solved and salable glass was produced. Again a problem occurred, but now they understood the process and solved the last technical problem. Production was ready.

5. Initial Production and Sales

In 1959, the plant went into commercial production. Another plant was built in 1962 and another converted in 1963. The total expenditure and time on the development and commercialization of the new process had cost £1.9 million over 10 years. Pilkington had innovated a commercial process to produce plate glass at the inexpensive cost of sheet glass. They obtained worldwide patents on the valuable process. What business strategy should they choose for commercial exploitation over the world? All their competitors were vulnerable. The process was so radical in improved quality and lower cost that they could take away any competitor's market.

Pilkington decided to license the new process to competitors in other countries rather than trying to expand internationally. Several reasons went into the decision: the great amount of capital required for worldwide expansions, and the location needs for plants—near supplies of iron-free sand and sources of soda ash, quarries for dolomite and limestone, and cheap fuel—all resources their competitors had lined up in their own companies. From 1959 to 1969, all producers purchased licenses from Pilkington. For a time, the license revenue provided Pilkington with a third of the company's profits; and after a period of time (generally, 17 years) Pilkington's patents expired.

STAGES OF TECHNICAL PROJECTS

There are three types of technical projects: research, design, or operations. Each has different technical stages for their type of project. Research-based projects are usually called *R&D projects;* design-based projects are usually called *engineering projects;* operations-based projects are usually called *systems projects.* R&D projects innovate new technologies, engineering projects reapply existing technologies in product model design and sometimes add in new technologies and systems projects develop new or improved operations, often as software. Each of these types goes through different kinds of stages. The three types of projects are all important in R&D management because technological invention may begin as an R&D project, then move into an engineering project for development and design of a new product, or into a systems project for the improvement of operations or development of a new service.

R&D Projects

The technical stages of R&D projects are:

1. Basic research and invention
2. Applied research and functional prototype
3. Engineering design and testing
4. Product testing and modification
5. Production design and pilot production
6. Initial production and sales

Since stages 1 to 3 are usually called *research,* while stages 4 to 6 are called *development;* and thus arose the term *research and development* (R&D). Each stage of innovating a new product is expensive, with the expense increasing by an order of magnitude at each stage. Management decisions to continue from research to development are therefore very important.

Engineering Projects

The technical stages of engineering projects are:

1. Customer and needs identification
2. Product concept
3. Engineering specifications
4. Conceptual design

5. Detailed design
6. Prototype engineering design
7. Production design.
8. Testing
9. Production

Engineering design uses working technologies, applying such to customer needs. Accordingly, engineering design begins with identifying customers and their needs. These functional needs must be formulated as engineering specifications, which the creative activity of conceptual and detailed design addresses. What emerges is a prototype-engineered design of a product or process. This design needs to be modified for production and then tested.

Systems Projects

The technical stages of systems projects are:

1. Customer and function identification
2. System definition
3. System requirements and specification
4. System architecture design
5. System component design
6. System control design
7. System prototyping and programming
8. System testing
9. System implementation

Systems projects tend to emphasize operations as opposed to the artifact orientation in traditional engineering projects. In fact, a device such as an airplane, where operations requirements are complex, may be called a *system project*. Software projects that produce coding for operations are also usually called *system projects*. Projects designing operational systems, such as transportation systems or communication systems or information systems, are also usually called *systems projects*.

In most system projects, known technologies are used and applied, and therefore system projects usually begin with customer and functional identification. From this a system must be identified with boundaries and specifications. System design activities then include design of architectures, components, and controls. Completed system designs must be prototyped and programmed and then tested and implemented.

PROJECT MANAGEMENT PHASES

In addition to the technical stages of an innovation project, several management phases occur in any project. Technical projects need to be planned, implemented, and completed, requiring the following management phases:

1. Planning phase
2. Staffing phase
3. Performance phase
4. Implementation phase

Project Planning Phase

The planning phase should include participation by all potential stakeholders in the project: research and development personnel, marketing and production personnel, finance personnel. Buy into the plan by relevant parties is important for the eventual success of the project. What results formally from this buy-in is a project plan and an allocation of resources to perform the project.

The project plan should state succinctly the goal of the project, the means and timing of the project, the resources required, and customer for the project:

- What's to be done
- Who will use it
- How and when it is to be done
- Who's to do it
- How progress will be monitored
- How the customer will get it
- What it will cost

Accordingly, a project plan should include the following sections:

1. Project objectives
2. Project customers
3. Project schedule
4. Project staffing requirements
5. Project review procedures
6. Project implementation plan
7. Project budget

Project Staffing Phase

Once a project plan is approved by higher management and a budget allocated, the next project management phase is staffing and beginning the project. Project staffing requires:

1. Recruiting
2. Developing the subculture of the project team

3. Organizing the activities
4. Coordinating and liaison and developing shared supervision of team members between the project and the functional homes of the team members

The quality of the team will determine the quality of the project outcome. A technical project team needs to be technically multidisciplinary and business multifunctional to cover both the technical and business dimensions of the project.

Project Performance Phase

Project monitoring requires:

1. Developing measures of technical progress in the project
2. Developing a modular task structure for organizing the work
3. Developing techniques to track technical progress (e.g., PERT charts)
4. Developing modes for problem solving and debugging during the project
5. Developing a testing program for modular testing of the system
6. Developing proper documentation for the project as the project progresses

Project Implementation Phase

Project completion requires:

1. Prototyping the project results
2. Demonstrating the prototype's performance
3. Transferring the completed project to the project's customer
4. Assisting the project's customer with implementation of the next phase of the project product or into use
5. Rewarding project team members for contributions to the project

Project success requires:

1. Achievement of technical performance
2. Results achieved within the window of opportunity
3. Achievement of results within a budget that can produce economic benefit
4. Results that satisfy the project customer
5. Effective rewards to project personnel for successful performance

PROJECT MANAGER

The project manager is responsible for formulating and planning the project, for staffing and running the project, and for transferring project results to the customer. In doing this, the technical project manager needs to coordinate and integrate activities across multiple disciplinary and functional lines. In addition, all this must be performed within a finite budget and according to a firm schedule. One of the problems is that the project manager has responsibility but with only limited authority. The project manager does not control personnel positions, has a limited budget, and must 'sell' the project to upper management and to project personnel and also satisfy customers. As Kerzner (1984) summarized the difficulties of the role: "The project manager's job is not an easy one. [Project] managers may have increasing responsibility, but with very little authority" (p. 10).

Gemmill and Wilemon (1994) studied kinds of frustration felt by technical project leaders and found that the most frequent kinds were (1) trying to deal with the apathy of team members, (2) wasting project time when the team goes back again over the same issues, (3) getting the team to confront the difficult problems, and (4) getting appropriate support for the project from the larger organization.

Beltramini (1996) examined a set of product development projects and concluded that many factors were contributory to the success of a project team, among which were (1) how the team was organized, (2) how information was handled, (3) the timing of project and its performance, (4) training of the participants, and (5) resource availability.

There are many software tools for project managment, particularly for the time and budget and resource aspects of project management, such as project scheduling software. Also embodied in some software are techniques for determining the priority of tasks and risks of completion, such as the *program evaluation and review technique* (PERT). Other forms of software, such as *groupware,* facilitate the communication and recording of project documentation.

TECHNICAL RISKS IN PROJECTS

In innovating new technology, project management techniques should be adapted to technical risks. For example, Shenhar (1988) has emphasized identifying the degree of technical risk in a project: "The technology used in projects is one of the main aspects which should receive special attention, since there are great differences among projects. Some projects use well-established technologies, while others employ new and sometimes even immature ones, and involve enormous uncertainties and risks" (p. 2).

Shenhar suggested classifying a development project by the degree of risk in the technology being implemented, from low-tech projects, to medium-tech projects, to high-tech projects, to superhigh-tech projects. A *low-tech project* is a project in which no new technology is used. A *medium-tech project* is a project in which some new technology is used within an existing system but is a relatively minor change in the whole technology of the system. A *high-tech project* is a project that uses only key technologies as components but in which the integration of these technologies is a first-time achievement. A *superhigh-tech project* is a project in which new key technologies must be developed and proved, along with their integration into a new first-time system.

For low-tech projects, project planning should use the experience of prior projects of the same type for estimating scheduling, budgeting, and staffing and for arranging for completion and transfer. Here a PERT chart is very useful, wherein past experience can be used to estimate probabilities of risk within the chart.

For a medium-tech project, it is important to determine how much competitive advantages the new technology provides in the project: nonessential or essential. If nonessential, a substituting technology should be within the scope of the design and ready if substantial problems develop with the new technology. If essential, it is important to plan and schedule the development and testing of the new technology before system integration is far advanced. The scheduling of system development should not be begun until confidence is demonstrated that the new technology will perform properly and on time for the entire project schedule. Here probabilities in a PERT chart should be used only for established technologies, not for the new technology.

For a high-tech project, it is important to plan and schedule the system integration in a modular, subsystem fashion, so that subsystems can be tested and debugged thoroughly before final system integration and assembly.

For a superhigh-tech project, the project planning should be in the form of a series of go/no-go decisions. Each stage of project development should consist of developing and testing a key technology before the full project system integration program is launched. Whenever a key technology fails, a no-go decision should be made to keep the full integration program on hold until either the key technology can be made to work or an adequate substituting technology can be implemented.

In summary, for low-, medium-, and high-tech projects, development can be scheduled for fixed dates (with appropriate regard to project design for testing in a modular fashion a new-technology component or subassembly); but for a superhigh-tech project, full system development should not be scheduled until all key technologies have been proven successfully or substitutions have been put in place.

SOFTWARE PROJECT TASKS

Jones (2002) studied the distribution of project time in the various tasks in software creation and noted that the actual coding in writing software was a quarter or less of the time required for software development projects. Moreover, the up-front conceptualization of the software project (design tasks) also took about a quarter of the time. Thus, design and coding of software together took about half the time spent on developing software. The testing task took another significant portion of the time. Time of different tasks is important to commerical success. For example, in a survey of software projects in 1994, 8000 projects were reviewed in 350 companies, and only 16 of these were delivered on time and on budget (Leishman and Cook, 2002). Evans and co-workers (2002) listed seven common reasons for failure to complete a software project on time and on budget (p. 16):

1. Failure to apply essential project management practices
2. Unwarrented optimisim and unrealistic management expectations (not recognizing) potential problem areas

3. Failure to implement effective software processes
4. Premature victory declarations due to pressure to deliver timely software products
5. Lack of program management leadership willing to confont today's challenges to avoid tomorrow's catastrophes
6. Untimely decision-making, avoiding making time-critical decisions until too late
7. Lack of proactive risk management to contain failures

RETURN ON PROJECTS

Budget and time spent on projects affect the potential profitability of project results. Technical projects cost money, and one needs to estimate the potential value of a project. This is relatively direct for technical projects that result in new products; for these, a *technical project financial chart* is useful (Figure 19.1). This chart consists of a horizontal time axis and a vertical financial axis. On the vertical financial axis are plotted (1) the cumulative profits from commercialization of a technical project and (2) the dollar volume of sales of products developed from the technical project.

Such a financial chart graphs both the projected costs and the project returns as sales volume over time. When the project begins at a time $t = 0$ on the horizontal time axis, the costs of the project increase down the negative vertical financial axis of the graph until the point when the product resulting from the project is first

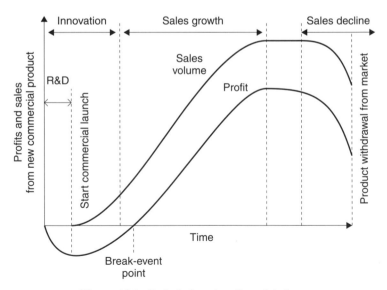

Figure 19.1. Technical project financial chart.

commercialized. This point is the commercial launch of the project, and sales of the product begin. As sales volume grows, the project costs are covered by the profits from sales until a *break even point* occurs. At this time, the cumulative profit from sales of the product just offset the project investment costs. Further on, cumulative profits from the project continue to climb positively on the vertical financial axis as sales volume grows. If, finally, the sales of the product decline from product obsolescence, the cumulative profit will level off and even end when the product is withdrawn from the market.

> **Financial projections of technical projects for new products should estimate when and how a return on investment will occur.**

One management technique often used to estimate the long-term value of a project compared to the short-term cost is called a *discounted-future-value analysis*. For example, if one has a choice of two projects, either of which can create a new product line, and there are limits to investment such that both cannot be developed simultaneously, there are several criteria upon which to choose between them, some regarding the business and some regarding the timing.

A technique to compare the timing differences is to calculate what is called the *discounted future return rate* of the projects. This calculates not only the cumulate financial return from commercializing the project but also the rate at which the return occurs. Returns occurring earlier than later returns can be seen as financially more valuable because earlier returns can be increased by reinvestment of earnings. Thus, when comparing financial returns, a later financial return should be discounted by its later time of return.

For example, a simple way to compare discounted future returns is to assume that an investment could be alternatively placed into a project or into an interest-bearing bank account. If the return on the project is x years in the future, this return can be compared to the alternative deposit in the account, with its return compounded by a fixed interest over the same x years. What this comparison does is to show if the project investment can return more income than an alternative simple interest-bearing account.

This technique should be used cautiously, however, because some R&D investments are necessary simply to stay in business, and alternative forms of investment are not feasible. For example, in 1982, Hayes and Garvin cautioned about misusing discounting techniques: "Highly sophisticated analytic techniques now dominate the capital budgeting process at most companies. . . . As these techniques have gained ever wider use in investment decision making, the growth of capital investment and R&D spending in this country has slowed. We believe this to be more than a simple coincidence. We submit that the discounting approach has contributed to a decreased willingness to invest. . . . Bluntly stated, the willingness of managers to view the future through the reversed telescope of discounted cash flow analysis is seriously shortchanging the futures of their companies" (p. 71).

PROGRAM MANAGEMENT

An engineering department may sometimes have several design projects running at the same time, or a research and development lab will have many research projects running simultaneously. Such simultaneous projects may be related to one another. A group of related projects is often called a *program*. The management of the program provides oversight for the management of the group of individual but related projects.

For example, in a R&D laboratory, there are many projects, and a formal process for formulating, selecting, monitoring, and evaluating projects is usually established. Figure 19.2 illustrates the levels of management involved in R&D projects. Senior management sets research priorities and business directions. Middle managers then define project requirements around which they and technical staff generate research ideas. Technical staff then draft R&D proposals that are sent to middle and senior management, who select which projects to fund. Funded projects are performed by technical staff and monitored by middle management. Periodic review of project portfolios by senior and middle managers aids in selecting research projects to terminate or to continue into development and implementation, based on technical progress and commercial importance. Program management is concerned with the strategy, selection and coordination, and oversight of a group of related projects.

Program Strategy

Program strategy consists of the direction and rationale for a group of projects and how they relate to one another. For example, at the U.S. National Science Foundation, research awards were organized by programs that were disciplinary focused, such as chemistry awards, physics awards, and mathematics awards. Projects funds in each program were related to each other as disciplinary projects. In industrial

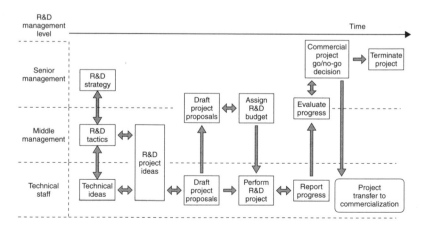

Figure 19.2. Formal R&D progress in a research lab. (Modified from Costello, 1983.)

research, project portfolio selection is ordered in programs around morphological areas of the core technologies of the firm. Program strategy provides the criteria for project selection.

Project Selection

Within programs of disciplines or morphological areas, there is still a problem of choosing an appropriate portfolio of R&D projects in a program. Projects within a program are compared on their technical merit and commercial potential according to the selection criteria of the program. There has been a long history of attempts to provide quantitative models for doing this, called *formal models for project selection,* and periodically there are reviews of this topic. For example, in 1994, Schmidt and Freeland, reviewing progress in formal R&D project selection models, grouped models into two approaches: decision event or decision process. The *decision-event model* focuses on outcomes: Given a set of projects, the model determines the subset that maximizes an objective. But the history of actual use of these models is poor, as Schmidt and Freeland state: "Classical R&D project selection models have been virtually ignored by industry. . . . This fact became apparent in the early 1970s" (p. 190).

The *decision-process approach* focuses on facilitating the process of making project selection decisions rather than attempting to determine the decision. Schmidt and Freeland suggest that formal project selection models should assist in coordinating decisions about selecting and monitoring a project portfolio. Other reviewers of the formal project selection literature agree. For example, Cabral-Cardoso and Payne wrote (1996): "Although a large body of literature exists about the use of formal selection techniques in the selection of R&D projects . . . much skepticism exists about their value."

They, too, argued that a judicious use of formal techniques can improve communication and discussion about projects in their selection. What constitutes judicious use is wherein lies the problem. All formal decision methods of R&D project selection techniques require estimation of probabilities as to project characteristics such as technical risk and likelihood of commercial success. These are often impossible to quantify because of the uniqueness of innovation projects. The risk of quantitative decision errors occurs when valid quantitative data cannot be input, and so the old adage: "garbage in–garbage out." The danger is that formal selection methods may as often result in the wrong project selection as in the right one. Also generally ignored in the project selection literature has been the problem of validating a project selection decision model. All models, decision or otherwise, need to be validated by experience or experiment before they should be used for serious business.

Program Oversight

Program managers need to monitor the progress of technical projects within a program without supervising the individual projects. The task of the project manager is project supervision, not that of the program manager. Monitoring, or oversight without supervision, requires that the program manager have milestones for progress of the project that are reported by the project managers. Oversight of projects within a program are by program goals and milestones to be achieved on the way to the goal.

INTELLECTUAL PROPERTY

An important aspect of technical projects is intellectual property. When significant inventions occur in a technical project, it is the project manager's responsibility to see that proprietary rights to the invention are pursued properly. We briefly review patent law.

Intellectual property plays an important role in industry, especially in the early period of a new industry. Patents on innovative ideas in the new industry provide a powerful competitive edge to the patent-holding company, at least for a time, which is the duration of the patent. There are four classes of intellectual property recognized in law: (1) patents, (2) copyrights, (3) trademarks, and (4) trade secrets. *Patents* protect inventions of new and useful *ideas. Copyrights* protect the *expression* of ideas but not the ideas expressed. *Trademarks* are *registered identifications* of products or corporate identities. *Trade secrets* are commercially important information that is gained of one company by another *without permission and wrongful means,* such as spying.

For a patent, a government provides the legal right of exclusivity to use a patent to a patent holder for a finite term, in return for the inventor disclosing full details of the invention. The patent concept evolved in England in the eighteenth century in order that the details of an invention would not be lost to society due to excessive secrecy by the inventor. The rights to use an invention are protected and limited only to the holder of the patent to the invention or to those to whom the patent holder may license use.

In the United States, only the inventor may file for a patent. In many other countries, anyone can file for an invention, whether the inventor or not. In Europe and Japan, the first to file is awarded the patent. In the United States the need to establishing priority to invention has often resulted in patent conflicts, because technological inventions have often involved the cross-fertilization of ideas by several people. Often, there have been cases of independent simultaneous invention of the same device by different people. The reasons behind multiple invention lie both in the patterns in scientific and technical progress and in societal problems and market needs. The progress in science or technology is usually widely shared and can often stimulate new ideas for inventions by several persons who are familiar with the details. Thus, ideas in technology push are often shared by researchers who can imagine the potential applications of the research. In addition, the problems and market demands of a situation can simultaneously motivate different people to invent solutions. Thus ideas in market pull are also often shared by several inventors.

In the United States the most common form of patent is called a *utility patent.* It may be granted by the U.S. government to inventors, domestic or foreign, individuals or corporations. Utility patents provide the right to exclude anyone other than the inventor (or those whom the inventor licenses) from making selling, or using the patented invention for a specific term. This term is normally 17 years, with additional time for certain patents that require Food and Drug Administration approval. The steps in acquiring a patent consist of first filing a patent disclosure and patent application with the U.S. patent office. The patent office performs a review and grants or denies the patent. In the case of denial, the applicant may appeal. In most countries, public disclosure of information about the patent before filing the patent disclosure and application invalidates the patent application. In the application, the claim to invention must include the conception of an idea and reducing the idea to workable form. Workable form may be an actual device, built and working, or merely a clear description of a workable form in the patent application (Bell, 1984).

In the U.S. system an invention is patentable only if it satisfies three criteria: utility, novelty, and nonobviousness. To be useful, the courts have generally followed a common notion of someone using it: "The issue of simple practical usefulness was addressed, and largely settled, in *Lowell* v. *Lewis,* heard by Justice Story in 1817. . . . Judge Story held [that it] didn't have to be extremely useful or the most useful. . . . It just had to have utility. Clearly the [Perkins pump] was useful; people used it" (Lubar, 1990, p. 11).

Judge Story added that the patent should not only be useful but should not hurt society. Also, to be novel, the invention must not simply be a rearrangement of prior art. A revision of the patent law in 1952 added a new standard for patentability of nonobviousness: "It declared that an invention would be unpatentable 'if the differences between the subject matter as a whole would have been obvious at the time the invention was made to a person having ordinary skill in the art to which said subject matter pertains' " (Lubar, 1990, p. 15).

Since this definition was itself nonobvious, a further change in the patent process occurred: "In 1982, to help settle this confusion, a whole new court was created, the Court of Appeals for the Federal Circuit, to handle appeals of patent cases. . . . One of the ways it overcame the problem of defining invention was by putting a greater emphasis on what are called secondary criteria, especially commercial success, in determining the patentability of an invention" (Lubar, 1990, p. 16).

Natural phenomena and scientific laws cannot be patented. However, new material forms can be patented as products, and the processes to produce them may also be patented. New life forms can also be patented, as can the biotechnology process needed to produce them. However, the new biotechnology has created new concerns about patent law; for example: "The confusion over product and process patents has affected the biotechnology industry so much that a few companies [in 1991] have been asking Congress to intervene. Genentech, for one, is cheer leading a reform effort, sponsored by Representative Rick Boucher (D-Va) and Senator Dennis DeConcini (D-Az). Their bills would change the law to state that 'process' and 'product' patents can cover the same thing, if the product is truly original. The aim is to make it easier to get patents on genetically engineered proteins, even when the natural analogs may have been well-characterized in the literature. Some think it would be a mistake for Congress to intervene on such a fine point, and the big companies in genetic engineering quietly oppose the bill" (Marshall, 1991, p. 22).

Computer software is also a new area of technology that has been creating changes in patent practices. Initially, software was held not to be patentable, only copyrightable. Then an exception was that when software was part of a process, it may be patented. This exception has been creating whole new categories of patents: "Last August [1989], Refac International Ltd. sued six major spreadsheet publishers, including Lotus, Microsoft and Ashton-Tate, claiming they had infringed on U.S. Patent No. 4,398,249. The patent deals with a technique called 'natural order recalc,' a common feature of spreadsheet calculations that allows a change in one calculation to reverberate throughout a document. . . . Within [a] few years, software developers have been surprised to learn that hundreds . . . of patents have been awarded for programming processes ranging from sequences of machine instructions to features of the user interface" (Kahn, 1990, p. 53).

The patent office has distinguished between mathematical algorithms and computer algorithms: "While patents are not awarded for algorithms, which are considered 'laws of nature,' the Patent Office draws a fine distinction between 'computer algorithms' (which are patentable) and 'mathematical algorithms' (which are not)" (Hamilton, 1991, p. 23).

This has been one of the major effects of the new Court of Claims and Patent Appeals established in 1982. Another area of patents that this court added to U.S. patent law was for business processes. In 1998 the U.S. Court of Appeals for the Federal Circuit ruled in favor of a patent on a method of managing mutual funds filed by the Signature Financial Group. In 1998, over 1000 patents were filed for business methods, and over 2000 filed in 1999. For example: "Mr. Walker is one of the patent office's best customers. Walker Digital earns money from licensing its inventions—including most notably, the patents for Priceline.com's name-your-own-price Web site. . . . Walker Digital [says that] its portfolio of 66 patents and 400 pending patents has earned it a $1 billion valuation with private investors" (Angwin, 2000, p. B4).

Patents protect the commercial exploitation of novel inventions. In the U.S. system an invention is patentable only if it satisfies the three criteria of utility, novelty, and nonobviousness. Although natural phenomena and scientific laws cannot be patented, new material forms can be patented as well as new life forms. Currently in the United States, software algorithms may not be patented except as part of a transforming process.

SUMMARY AND PRACTICAL IDEAS

In a case study on

- Pilkington's float glass R&D project

we have examined the following key ideas:

- Technical stages of projects
- Management phases of projects
- Project manager

- Technical risks
- Discounted future value
- Software cost
- Program management
- Matrix organization
- Transferring technology
- Patent law

These theoretical ideas can be used practically in understanding how to manage projects for innovation. All innovation is discrete, and project management is the proper management style for discrete activities of change.

FOR REFLECTION

Define a project within an existing organization for a new product, process, or service. Create a project plan and project costs and benefits.

20

FORMULATING
TECHNOLOGY STRATEGY

INTRODUCTION

We have reviewed (1) the *imperative* of technological change, the *'why'* and (2) the required *capabilities* to innovate, the *'what'* of technological innovation; and now (3) we turn to the *'how'* of technological innovation, *technology strategy*. We look at how technology strategies can be formulated as select areas for research, in the technology side of scientific technology strategy. We shall see that technological progress can be planned to look for improvements in the technology's physical phenomena or schematic logic—new physical morphology or new logic. This progress provides the technical basis for next-generation technology products.

CASE STUDY:

Inventing the Silicon Transistor

The transistor was invented in AT&T's Bell Laboratories as a germanium transistor—physical morphology on the semiconducting material of germanium. But its performance was sensitive to temperature. The technology issue was: What good would transistorized electronics be if their circuit performance changed dramatically from the heat of the day to the coolness of the night, from the heat of summer to the cold of winter? The historical setting of this case was 10 years after the invention of the germanium transistor (which we will describe in the next chapter). Many electronic engineers appreciated the new technology but yearned for a more reliable, less-temperature-sensitive version of the transistor. Since the transistor used a semiconducting material, an obvious strategy was to look for a different physical morphology as an alternative semiconducting material. Many researchers were thus trying to make a transistor not from germanium but from its sister element, silicon—new physical morphology.

  One of the groups looking for a silicon version of the transistor was at a small

U.S. company, Texas Instruments (TI). Patrick Haggerty was president of TI in 1952, and the company was a maker of seismographic detection instruments sold to oil companies for use in oil exploration. Their instruments used electron vacuum tubes but needed to be portable and rugged and use little power. As a technology strategy for TI, it was obvious to Haggerty that transistors would be a desirable replacement for tubes in TI's products.

Haggerty assigned to Mark Shepard a research project to develop a germanium transistor that could be sold for $2.50. Shepard developed it, and TI produced a pocket radio using it in 1954. But TI did not follow through with this product with a major marketing effort, and sold few units. (However, soon afterward, SONY introduced its own independently developed germanium transistorized pocket radio and proceeded to exploit this new product commercially to a big-time product.)

Meanwhile, Haggerty knew that the germanium transistor needed to be replaced. He hired a physicist from Bell Labs, Gordon Teal, who had been researching silicon. Silicon was brittle and difficult to purify as a material for making transistors. Haggerty told Teal and another researcher, Willis Adcock (a physical chemist), to develop a silicon transistor. Many other research groups were seeking the silicon transistor; and it was not an easy artifice to make. But Teal and Adcock succeeded.

In May 1954, Teal took the new silicon transistor to a professional conference, where he listened to several speakers tell of their difficulties in trying to make a silicon transistor. What he heard made Teal happy, for as yet no one else had succeeded. When Teal's time to speak came, he stood before the group and announced: "Our company now has two types of silicon transistors in production. . . . I just happen to have some here in my coat pocket" (Reid, 1985, p. 37). Teal's assistant came onto the speaker's stage carrying a record player that used a germanium transistor in the electrical circuit for the amplifier. The tiny germanium transistor had been wired visibly outside the record player with long leads. Teal plugged in the player, put on a record, and started it playing music. Next, the assistant brought onto the stage a pot of hot oil and set on the table beside the record player.

Teal picked on the connected germanium transistor and dunked it dramatically into the hot oil. Immediately, the music stopped as the germanium transistor failed in the oil's hot temperature. Then Teal picked up one of the silicon transistors that he had earlier taken from his pocket and placed it on the demonstration table. Teal took a soldering iron (which the assistant had also supplied) and replaced the germanium transistor with the new silicon transistor. The music from the record player sounded again. Then Teal picked up the silicon transistor and dunked it into the pot of hot oil, as he had previously with the germanium transistor. The music did not stop! It continued! The silicon transistor could stand the heat! The meeting exploded in excitement! Texas Instruments had done it! Finally, a useful transistor—the silicon transistor!

We recall that physical phenomena are useful to technology when arranged as useful for manipulation. Performance of a technology depends on both the phenomenon used and its arrangement. In the case of the transistor, both the electron vacuum tube and the transistor provided a means of amplifying and controlling electrical signals. However, the performance differed between the two electronic devices. The efficiency of the germanium transistor was much higher than that of the electron vacuum tube and it was much smaller. Yet the germanium transistor was very sensitive to temperature and therefore unreliable. The invention of a silicon transistor then improved the reliability of the device.

TECHNOLOGY PERFORMANCE PARAMETERS

We recall that a quantitative measure of the usefulness of a technology is called a *technology performance parameter*. There are seven ways of considering a technology's performance: (1) application, (2) capability, (3) performance, (4) quality, (5) safety, (6) resources, and (7) potential.

The *application* of a technology identifies the activity to be performed by a customer in a context. For example, the electronic control devices of the electron vacuum tube and transistor were initially applied to communications, telephone and radio. An application will require certain capabilities of the technology.

The *capability* of the technology identifies the functional transformation of the technology that transforms inputs to the technology system into outputs. For example, application of the tube and transistor to radio communications required the capability of these devices to provide signal amplification, signal detection, and signal oscillation generation.

The *performance* of the technology expresses how well the technology performs its capability of functional transformation. For example, one important performance measure of the electron tubes and transistors was their frequency response, the range of frequencies over which capabilities such as amplification could be used.

The *quality* of a technology expresses how dependable the technology is in use. Electron tubes were notoriously undependable, as the metal grids corroded over time, making the tube fail. The germanium transistor was undependable because its performance was temperature sensitive.

The *safety* of a technology expresses how much danger arises from the use of the technology. For example, both electron tubes and transistors are relatively safe to use, since neither burst into flames during use.

The *resources* used by a technology expresses what kind and the quantity of supplies consumed by the performance of the technology. In the example of electron tubes, they required a heater current to stimulate electrons to leave the cathode, which consumed electrical power. Transistors did not require such a heater element and so consumed much less electrical power in use.

The *potential* of a technology expresses the possibility of future improvement of the technology. For example, the size of the electron tube was reducible from the size of an apple to the size of a peanut but little further. In contrast, the size of the transistor was reducible from the size of a raisin down to that of a bacterium (microscopic) and even smaller.

In the functional description of a technology, one can describe the different ways of performance, and measuring one of these provides a performance parameter for the technology. A change in physical phenomena (e.g., a change from electron conduction in a vacuum to electron conduction in a solid, going from the tube to the transistor) changes the functional parameters. The structure of the physical morphology (e.g., the *npn* or *pnp* structure of a transistor) also affects the performance parameters.

Physical technology performance parameters express the functionality of the technology system based on a physical morphology of the technology.

CASE STUDY:

Molding Technologies

Any technology system is a configured system of a physical morphology. But for future progress in performance, to which part of a system configuration should a technology strategy attend? The technique for aiming technology strategy is in examining alternative configurations of a technology system for deciding where focus scientific research. We next look at a planning technique for searching alternative configurations of a technology system. As an example, Foray and Grubler (1990) provided such an analysis of the various physical processes in molding technologies—different configurations of a molding technology system: "A morphological analysis starts with building a morphological space for a particular set of technologies or products, in order to understand and thus not to 'miss' a technological route of possible future development . . ." (p. 537).

This first step is to define the technology system carefully in terms of its *functional capability* (system transformation): "First, the problem to be solved [for the functional capability desired] must be stated with great precision: in our case the problem consists of realizing ferrous metal products by a casting process" [molding technology] (Foray and Grubler, 1990, p. 537).

The second step is the abstraction of the key parameters of the functional capability (system transformation). Foray and Grubler considered four parameters: P_1, the nature of the pattern; P_2, the nature of the mold cavity; P_3, the stabilization force; and P_4, the bonding method. Next, they listed the principal alternatives for each parameter (p. 537):

- P_1 *(pattern):* permanent (P_{11}) or lost (P_{12})
  "The molding methods can be classified according to the nature of the pattern. . . . The pattern is used for a large number of castings. . . . The pattern is used once only."
- P_2 *(mold cavity):* hollow (P_{21}) or full (P_{22})
  "The molding methods can be classified according to the fact that the mold cavity is hollow (the pattern is extracted before the casting) or full (foam polystyrene as expandable pattern is gasified by the molten metal during the casting), and by taking the place of the expandable pattern, the molten metal fills in the full cavity as it would fill in a hollow cavity)."

- P_3 *(stabilization force):* chemical (P_{31}) or physical (P_{32})
  ". . . the kind or kinds of bonding systems used for stabilization of individual granules of molding material."
- P_4 *(bonding method):* simple (P_{41}) or complex (P_{42})
  "Both chemical and physical bonding methods can be described as simple (mechanical and inorganic chemical binder) or complex (magnetic field, vacuum and organic chemical binder)."

Although all combinations of configurations could be considered for possible progress, Foray and Grubler (1990) emphasized that one gets to more interesting technical alternatives faster by utilizing logical relations between system parameters: "Each parameter corresponds to a given level of aggregation (or integration). The four levels can be ordered hierarchically, according to the relation between the parameters. For example, the stabilization force (P_3) influences the bonding method (P_2), or the nature of the pattern (P_1) influences the nature of the model cavity (P_4)" (p. 537).

They next created combinations of alternatives ordered according to a logical hierarchy beginning with P_1 (pattern), progressing to P_2 (mold cavity), then P_3 (stabilization force), and finally, P_4 (bonding method). Figure 20.1 summarizes their ordering of combinations.

What this physical morphology logic tree shows is the different combinations of technical parameters that can define different technical processes (a process system), 16 in all for the four parameters. After eliminating impossible solutions from these 16, they had all possible processes in the actual technological state of the industry. An example of one of these combinations was: "The GP process involves investing an injection molded foamed-polystyrene pattern in a free flowing magnetizable modeling material. Immediately prior to pouring, the molding material is rigidized by a powerful magnetic field. During casting, the polystyrene pattern volatizes in the face of incoming metal stream which occupies the void left by the falsified pattern. Shortly after the casting has solidified, the magnetic flux is switched off and the flask containing the casting is taken to the knockout station" (Foray and Grubler, 1990, p. 540).

PHYSICAL MORPHOLOGICAL ANALYSIS OF A TECHNOLOGY SYSTEM

Technology strategy can be formulated by examining alternative physical configurations of a technology system. This case is an example of a formal technique in technology-push planning called a *morphological analysis*. We recall that the term *morphology* literally means the "science of form," as *morphe* is the Greek term for "form." However, we have been using this term of physical morphologies to indicate the *mechanical structures and processes of the physical phenomenon* of the configuration of a technology system. We have been so using the term for historical

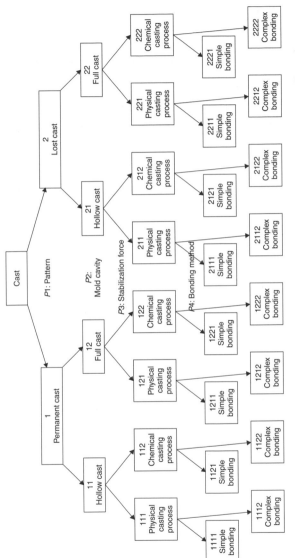

Figure 20.1. Morphological ordering of ferrous casting technologies (four parameters allow six possible technologies).

reasons, dating back to 1948, when F. Zwicky proposed his technique which he called *morphological analysis* (Zwicky, 1948). Zwicky wrote that one can systematically explore technology-push sources for technical advance in a system by logically constructing all possible combinations of physical alternatives of the system.

He said that first, a clear statement of the function of the technology must be made. A general logic scheme of the functional transformation must be designed. Then a list can be made of all the physical processes mappable into the logical steps. Morphological alternatives of possible alternative configurations of a technology system can then be searched by all conceivable variations on the functional architecture of the system and of all possible physical substitutions in the functional transformational steps. This requires an abstraction of the physical configuration of the technology system in terms of components, connections, materials, control, and power. One then makes a list of all conceivable alternatives of each of these as different physical processes. One then makes a list of all possible combinations for alternative constructs.

Zwicky used the example of alternative configurations for jet engines, identifying each major system element (Jantsch, 1967, p. 176):

1. Intrinsic or extrinsic chemically active mass
2. Internal or external thrust generation
3. Intrinsic, extrinsic, or zero-thrust augmentation
4. Positive or negative jets
5. Nature of the conversion of the chemical energy into mechanical energy
6. Vacuum, air, water, earth
7. Translatory, rotatory, oscillatory, or no motion
8. Gaseous, liquid, or solid state of propellant
9. Continuous or intermittent operation
10. Self-igniting or non-self-igniting propellants

Zwicky computed that all possible combinations of these features would be 36,864 alternative combinations. Many of these alternatives would not be technically interesting, so Zwicky performed this evaluation in 1943 with fewer parameters and reduced his alternatives to be examined to only 576 combinations. (Zwicky's procedure in its full combinatorial form is cumbersome.)

> **Technology-push research opportunities can be planned through a morphological analysis of possible alternative physical configurations of a technology system.**

To summarize the procedure for conducting a morphological analysis of the physical aspects of a technology system. First, the functional context analysis of a technology system needs to be examined:

1. Identify the major generic applications of product/production/service systems that embody the technology system.

2. From the applications perspective, identify desirable improvements in performance, features, and safety that would be recognized by the users (final customers) of the product/production/service systems.
3. Determine the prices that would expand applications or open up new applications.
4. Determine the factors that facilitate brand loyalty when product/production/service systems are replaced in an application.

Next, abstract the principal parameters of the desired functional transformation to meet the functional requirements for use of the technology. Different logical combinations of these parameters can be examined (that are mutually physically consistent):

1. One should begin with an existing boundary and architecture that provides the functional transformation defining the technology.
2. Next, one uses the applications system forecast to classify the types of applications and markets currently being performed by the technology system and to envision some new applications and markets that could be performed with improved performance, added features, and/or lowered cost of the technology system.
3. One can then identify the important performance parameters and their desirable ranges, desirable features, and target cost to improve existing applications and to create new applications.
4. These criteria then provide constraints and desired ranges under which alternative morphologies of a technology system can be bound.
5. Examine critical elements of components and connections whose performance would limit the attainment of the desirable performances, features, and cost.
6. The performance of critical elements that do not permit desired system performance can then be identified as technical bottlenecks.
7. One can examine alternative base materials and base power sources for alternative morphologies, again selecting only those that would fulfill desirable system performance.
8. One can also examine alternative architectural configurations and alternative types of control that might attain the desired higher performances.
9. Combining and recombining various permutations of critical elements, base materials and power, and alternative architectures and control systems provide a systematic way of enlarging the realm of possible technology developments.

CASE STUDY:

Inventing the Integrated-Circuit Chip

In addition to examining alternative physical configurations of a technology system, technologies sometimes may be improved by examining alternative logic schemes of the system. An example of this is the invention of the integrated-circuit (IC) chip. The historical context was the middle of the twentieth century after the invention of the transistor was enabling more complex circuits to be designed because of the smallness and reduced power consumption of the transistor compared to the older electronic tube. In the late 1950s, electrical engineers were using the new silicon transistor for many advanced electronic circuits. The transistors were so useful that a new problem arose in electronics. The new complex circuits that engineers could then dream up might require so many transistors that no one could physically wire them together (with the technology of soldering transistors into printed-circuit boards). So a new natural limit was on the horizon for transistorized electronics—how many could be wired together. In 1958, this limit was on the mind of two researchers: Jack Kilby at Texas Instruments and Robert Noyce at Fairchild.

Jack St. Clair Kilby had grown up in Kansas. As a kid he loved technical things and built a ham radio set using electron tubes (Reid, 1985). In September 1941 he went to college at the University of Illinois to become an electrical engineer. But three months later, the United States was at war with Japan, and Kilby joined the U.S. Army. He ended the war as a sergeant, assigned to an Army radio repair station on a tea plantation in India. He returned to college, graduated as an electrical engineer, and went to work at a firm named Centralab, located in Milwaukee.

One evening he attended a lecture at nearby Marquette University given by John Bardeen, one of the inventors of the transistor. The new technology astonished Kilby, and he began reading everything he could find on the solid-state artifice. In 1951, Bell Laboratories announced the licenses, and Centralab purchased one for the $25,000 fee. They sent Kilby to Bell Labs for a five-day course in semiconductors. From 1951 to 1958, Kilby designed transistorized circuits for Centralab. As these circuits increased in complexity and numbers of transistors, Kilby saw the need for technical breakthrough beyond hand-wiring transistors together.

But Centralab did not have the research capability that Kilby would need to tackle the problem. In 1958, he sent his résumé to several larger firms, one of which went to Willis Adcock of Texas Instruments. Adcock hired Kilby, since TI was also worried about the integration problem of transistors. Adcock had a micro-module research project going at TI to approach that problem, to which he assigned Kilby. But Kilby didn't like their approach. One thing was clear to Kilby, since TI led in silicon technology, a good solution to transistor integration should use silicon technology—if it could. So he had an idea! If transistors were made of silicon, could not the other components of a circuit, such as resistors and capacitors, also not be fabricated on silicon! Why not make the entire circuit on a silicon chip!

It was an inventive idea, for then the technology for fabricating resistors used carbon as the phenomenal material and plastics and metal foil for capacitors. On July 24, 1958, Kilby wrote down in his lab notebook what he called a monolithic idea: "The following circuit elements could be made on a single slice [of silicon]: resistors, capacitors, distributed capacitors, transistor" (Reid, 1985, p. 38). He entered rough sketches of how to make each component by arranging silicon material properly.

Kilby then showed his notebook to Adcock, who said that if he could demonstrate that the idea worked, an R&D project to develop it would be authorized.

Kilby carved out a resistor on one silicon chip and then carved a capacitor on another and wired the two chips together. They worked, showing resistance and capacitance in the circuit. Adcock gave his approval to build an entire circuit on a chip, and Kilby chose to build an oscillator circuit, one that could generate a sinusoidal signal.

On September 12, 1958, Kilby was ready. A group of TI executives assembled in the lab for a demonstration. The circuit on the chip was hooked to an oscilloscope that would display the form of the signal from the circuit. Before the circuit was turned on, the oscilloscope displayed a flat line, indicating no signal output. Kilby switched on the new integrated circuit on the silicon chip and it worked—a wiggling line appeared on the screen, the sine-wave form of an oscillator circuit: "Then everybody broke into broad smiles. A new era in electronics had been born" (Reid, 1985, p. 39).

Sometimes, a basic invention is made independently about the same time by another inventive mind aware of the technological needs and opportunities. This happened with the invention of the semiconductor integrated-circuit (IC) chip. That same summer in 1958, Robert Noyce, who was a physicist and then president of Fairchild Semiconductor, was also worrying about the problem of transistor integration. Fairchild was a new company to manufacture transistors (begun by one of the inventors of the transistor, Schockley, after he left AT&T). That following winter, on January 23, 1959, Noyce conceived a solution and wrote down in his lab notebook his monolithic idea: "It would be desirable to make multiple devices on a single piece of silicon, in order to be able to make interconnections between devices as part of the manufacturing process . . ." (Reid, 1985, p. 41).

Both Texas Instruments and Fairchild Semiconductors filed for the patent on the IC chip. Years of legal argument followed about who had the right to the patent. In 1966, several semiconductor chip manufacturing firms met together; and TI and Fairchild agreed to grant licenses to each other and share in inventing the device. Later, Kilby and Noyce both received (in addition to many other prizes) the National Medal of Science for inventing the semiconductor IC chip.

SCHEMATIC LOGIC IN PHYSICAL TECHNOLOGIES

We see in this case that it was not a change in the physical morphology (transistor structure and process) that created this advance in transistor technology progress, but it was a change in schematic logic of the technology as expressed in the physical morphology. It was the integration of all circuit elements on a single silicon chip that transformed modern electronics from wired transistor circuits to IC chip circuits. Circuits were constructed on a chip, and chips were then wired together on a board to create electronic devices.

Technology strategy can seek improvements in technological systems by:
- **Substituting physical phenomena**
- **Altering configurations of the physical morphology**
- **Improving the schematic logic of a system**

CASE STUDY:

Designs of Early Microprocessors

Now we look at the challenges of designing morphological forms of products as successive technological generations. The historical case of early microprocessors [central processing unit (CPU) chips for personal computers] provides a nice illustration of generations of a high-tech product. Early CPU chips were designed in the 1970s and 1980s, and Wilson (1988) compared the designs of the following CPU chips: Mostek 6502 microprocessor, introduced in 1978; Intel 8086 microprocessor, introduced in 1983; and Motorola 68000 microprocessor, introduced in 1985.

The basic architecture of all early computers was the von Neumann architecture. However, in the evolution of microprocessor designs, compromises were made due to technology and cost constraints. When personal computers were innovated, superior computing technology existed in higher product lines, such as the supercomputer, the mainframe computer, and the minicomputer. The question for the designer of microprocessors was how much computing performance could be designed into a product at a targeted cost.

MOSTEK 6502 Microprocessor

We recall that the first microprocessor was designed by Frederick Gaggin in the early 1970s at Intel, but Intel management did not market the chip aggressively. Historically, it was two later chips that powered the first generation of personal computers, one of which was the Mostek 6502 used in the first Apple personal computers. The design of the Mostek 6502 was technically limited by the available transistor density in chip fabrication in 1977. Then large-scale integration (LSI) production technology for chips allowed only up to 10,000 transistors on a chip. For this reason, the microprocessor used 8-bit word lengths for instruction and data.

In designing the chip, Mostek engineers first considered the necessary features for the arithmetic and logic unit (ALU). As a minimum, the ALU should perform at least addition and substraction of binary numbers. To perform the steps of addition or substraction for binary numbers, a designer can list all the steps and find that there are fewer than 256 steps and more than 16 steps. This suggested to the designer of the Mostek 6502 to consider using one byte of information to name a particular step in an addition operation, for an 8-bit word length (a byte) could distinguish between 256 names. Thus the 6502 chip designer choose an 8-bit operation code (Wilson, 1988, p. 215).

The next design consideration was the size of the registers in the memory that will hold the binary numbers to be added or subtracted from one another. Memory size then was limited by the numbers of memory chips that could be used economically with a microprocessor. In the middle 1970s, 16K chips were the largest random access memory (RAM) chips available. Four of these chips could provide a 64K memory, and this was judged economically feasible by the microprocessor designers of the time. To address this size memory, the address registers in the microprocess would require at least 16 bits for naming all the memory cells. Accordingly, the designer of the 6502 chip thought of instruction lengths in the computer of 24-bit size (8 bits of the operand code plus 16 bits of the memory address code).

But this was a problem. The 24-bit instruction size would use a lot of the precious 64K-bit memory for programs, leaving little memory for data. So the designer of the Mostek 6502 thought of a clever way to get around this technical limitation: "The number 24 looks nasty; not only is it not very binary, it's also very big—our programs will use up lots of expensive memory. And besides, we want the machine to do more than loads, adds, and stores—we'd like, for example, to be able to look at each member of a collection of values in memory. We can do this by inventing another operation—an *indirect load*" (Wilson, 1988, p. 215).

Component technologies (in this case, chip density) require designers to optimize a design under both technological and economic constraints. Yet such inventions to optimize designs under economic constraints also create new problems: "But such an indirect load has two problems. First, it involves two memory accesses, and the first rule of microprocessor design says that it's always easy to build a processor that can do something much faster than affordable memory can cycle" (Wilson, 1988, p. 215).

Today's optimization solution in product design for a technical problem will become tomorrow's performance bottleneck in the product.

Thus in implementing indirect address, a new performance bottleneck was designed into the processor. Also, playing with the address location in memory meant destroying the values in the microprocessor's accumulator, which requires more loading and storing of information to and from the memory (and this is a major source of slowness in computing). Again, the designer of the 6502 chip thought of a clever solution by giving the processor another register to be used as an address register.

But again problems! There weren't enough transistors in the large-scale integration chip technology in 1975 (when the 6052 chip was being designed) to *afford* a 16-bit register, so the designer managed with only 8 bits.

The present state of technology limits the design performance of high-tech products.

In the design of the 6052 microprocessor, another design optimization was made by the designer. Instructions were made 16 bits long, instead of 24 bits, by restricting the address portion of the instruction to 8 bits. That, however, restricted the microprocessor to using the bottom 256 locations in memory as indirect values. But the advantage was that programs could then be shrunk by this by 33%, and the computer went faster (with fewer memory cycles to read instructions from memory as the instructions were smaller).

The designer of the 6052 also added a program register (of 16-bit size) to indicate the next instruction to execute. Some instructions were also added to manipulate this register (so that a program could jump from one instruction to another rather than simply sequencing instructions). The designer also added the capability of performing subroutines by adding another register (plus some instructions) that provided for a "call": "A *call* saves the current program counter in memory whose address is specified by the new register, and then increments that register, while a *return* decrements the register and reloads the program counter from the indicated memory location." (Wilson, 1985, p. 217).

Thus both the limitation of computer memory and the cost of memory required the designer of the Mostek 6502 to optimize his design, with restrictions on instruction word length and register size. But these restrictions then created new bottlenecks: (1) limited addressable memory and (2) slowness of performance (due to ALU accessing memory through indirect addressing). Still given the economics and technical limitations of LSI chip technology in 1975, the Mostek 6502 chip was a very good design: "The resulting machine [described above] is very close to the 6502, originally designed by Mostek and used in the Apple II and the Commodore PET. . . . It used fewer than 10,000 transistors" (Wilson, 1985, p. 217).

NEXT-GENERATION TECHNOLOGY PRODUCTS

We have emphasized that technology discontinuities provide the opportunity for competitive discontinuities—beating the competition through next-generation technology products. To optimize product performance under cost constraints, a designer using a new technology may invent new techniques that solve a particular problem at the time of the design. Yet every clever solution for optimizing a product design, to get around some economic or technical limitation, usually will also create a new bottleneck of technical performance.

> **The performance/cost optimization in one generation of products becomes the technical bottleneck to be solved by the next-generation product.**

This is what makes product evolution possible from one generation of technology to another. Product generations will continue as long as the embedded technology systems continue to make large, discontinuous leaps in performance. The activity of designing a product requires (1) understanding the technologies that will be embedded in the product but also (2) requires decisions on the trade-off between desirable aspects of a product design and the constraints of economics and limitations of current technology.

We recall that product design specifications are determined by the class of customers and their applications. In technology innovation, the most easily approached first market is an established market in which the new technology can substitute, since some knowledge of customer applications and product requirements already exist for that market. Products/services with embedded new functionality create brand new markets. Improved performance products/services of existing functionality substitute in existing markets.

> **For radically new products, the first critical product decision is identifying the customer and application of the product.**

Still it is difficult for a new-technology product or service to establish the customer requirements for the new product or service. What does the customer want? *Seldom is this question effectively answered by simply asking the customer.* It helps to ask, yet the customer may not fully know until he or she tries the new product.

> **The more radical the technical innovation in a product/service, the less likely the customer can know product/service requirements before actual experience in using the new product/service.**

Moreover, the more radical the functional capability of a new-technology product/service, the more unlikely the initial substitution market will prove to be eventually the most important and largest market.

> **A critical product decision is identifying the necessary performance criteria and essential features of the product/service for the application.**

Usually, the cost of an innovative product will be higher than the price of existing products for which it is to substitute.

> **A second critical product decision is estimating the acceptable sales price to a customer of the product and a required production cost for profitability at such a sales price.**

As technology progresses, new products embedding the technology substitute for existing products and also enable new product applications. Product specifications should be determined by the applications. As the application systems develop, product requirements change. This is a chicken-and-egg problem. Which comes first in technical progress, product specifications or application system requirements? The answer is both, since they are interactive in new technologies.

When radically-new-technology products are innovated, the immediate applications that are found for them soon indicate the performance limits of the technology. This provides incentive to improve the performance of the technology guided by the applications. (In the case of the early microprocessors for personal computers, it was primarily memory address limitations that applications highlighted.) As performance in the product improves and costs decline, new applications open up for the product. In a new technology, products often go through several generations of next-generation-technology products. Yet in each generation, there will be trade-offs between desired performance and cost.

Designs of Early Microprocessors

We return to our case study, examining some of the later design trade-offs in generations of CPU designs.

INTEL 8086 Microprocessor

The next example in the technology generations of microprocessors was the Intel 8086. This was designed in 1980 and used to power the IBM personal computer of 1985. By 1980 more transistors were available on a chip to improve the microprocessor design, and one of the first design choices made by the designer of the 8086 was to increase all register sizes to 16 bits. By then, denser memory chips had been produced and memory was cheaper. However, the 16-bit registers could still only address 640K bytes of memory. The ideal then would have been 32-bit address registers, but again the economics and current technology prohibited that choice to the designer of the 8086. So another clever design trick was used. Some bits were added to the 16-bit address to create a larger bit address. Then the processor could use a small program to write into this additional register. This allowed one to skip around in a big memory (of say 1 million bytes). But it still meant that one could not "see" more than a 640K-byte size of that larger memory at any one time: "So we'll provide one magic register for addressing code and one for data. And sometimes the code will want to play with two different collections of data, so we'll add a third. That's not very binary, so we'll give the machine four such base registers. . . . That machine, of course, is the 8086 design. . . . The 8086 uses about 30,000 transistors (Wilson, 1988, pp. 218–219).

In the 8086 design, these four additional base registers were each of 16-bit size, and a megabyte of memory was addressable—but only in 640K segments. A base address was shifted four places left and then added to the machine's 16-bit address. Yet even with this clever design, new performance bottlenecks were introduced: "What are the bottlenecks in [the] design? First, the machine is using that memory too often. . . . Those loads and stores are crippling performance. . . . Second, that megabyte of memory has people writing more ambitious programs, so they write in a high-level language to finish in a reasonable time. . . . Programs are getting bigger . . . using procedures all over the place" (Wilson, 1988, pp. 219–220).

MOTOROLA 68000 Microprocessor

Continuing progress in transistor chip density provided the designer of the next-generation Motorola 68000 chip with new opportunities. First, the designer of the

68000 attacked the memory-access bottleneck by giving the new processor several data registers on the same chip so that they could be accessed quickly. But each of these new registers required instructions to use them: "Rather than our one-address instructions (one-address because the instructions literally specified just one operand, the other being the accumulator), we'll have two-address instructions that say things like 'take 13 and add it to register 8.' . . . We've gained performance because we no longer need to go messing around putting things into an accumulator, adding stuff into that and copying it out again" (Wilson, 1988, p. 220).

These new registers could also be expanded by having more transistors available for making the new registers of 32-bit length. Moreover, these registers could be divided into address registers and data registers. Now doing these neat new things for increased performance also created now complexities in the design: "Trouble is, we can now call for some pretty complex operations in one instruction. Things that used to be done by a whole sequence of instructions now fit into one. So let's implement the thing as an interpreter. We'll build a much simpler machine whose instructions are unbelievably crude but very quick to execute, and have a tiny program in a ROM on-chip. . . . The little program is called *microcode*" (Wilson, 1988, p. 222).

The designer of the Motorola 68000 innovated a new feature in microprocessors, microcode. This was an innovation in logic in the form function of the microprocessor. The chip processing technology of very large scale integration (VLSI) in 1985 provided more transistor capability, making possible more functional capability in the design.

The microcode capability with the new registers allowed complex sequences of operations performed swiftly as a single but complex instruction: "We can have several different instruction sizes, with the longer ones specifying a complex sequence of operations involving multiple steps (e.g., add this constant to an address register, look into memory, take the value, use that as an address, read from memory, and multiply register 7 by that value). Real power at last! This machine . . . is 68000-flavored. A 68000 needs around 70,000 transistors (Wilson, 1988, p. 222).

HIGH-TECH STRATEGY AND A NEXT-GENERATION INNOVATION PROCESS

We recall that the radical process provides dramatic, discontinuous impacts on the economy, enabling the creation of new markets or the entry of existing markets. In contrast, the cyclic product development process with incremental innovation process provides quiet, steady, continuous impacts on the economy, eventually enabling the capture and domination of markets. But since technology is implemented in either innovation process, how can both processes be managed within the firm? One way to do this is to use the concept of 'generations of technology', for this requires both radical and incremental innovation. Figure 20.2 sketches how a linear

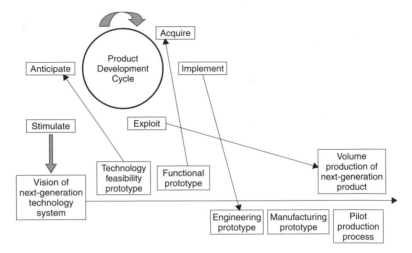

Figure 20.2. Next-generation technology innovation logic.

innovation process for next-generation technology can provide technical information for a cyclic innovation process:

- In the cyclic innovation process the *stimulation of need for new technology* fosters the *vision of a next-generation technology* in the linear innovation process.
- The *technical feasibility prototype* in the linear innovation process provides the grounds for *anticipation* of new technology in the cyclic innovation process.
- The *functional prototype* in the linear innovation process provides the information for *acquiring* new technology in the cyclic innovation process.
- The *implementation* phase in the cyclic innovation process designs the *engineering prototype* in the linear innovation process (which with concurrent engineering practice also fosters the *manufacturing prototypes and pilot production.*
- *Volume production* in the linear innovation process of the next-generation product provides the opportunity for *exploitation* of the new technology in the cyclic innovation process.

By establishing 'technical feasibility', the linear innovation process can provide *anticipation* to the cyclic innovation process that a new generation of a technology is possible. Once such demonstrations can be pointed out, the next logical problem is estimating the timing of the introduction of new technology. And this is a critical problem, since the timing of technological innovation is one of its most important competitive assets. The step of functional prototyping in the linear innovation process establishes the timing for introduction of a next generation of a technology to be *acquired* by the product development cycle. The engineering prototype of a new-generation-technology product next allows the product development cycle of a strategic business unit to implement the new technology.

In the product development cyclic, the critical step is exploiting a new technology. When one produces and markets the new product, the design, manufacture, timing of the introduction, pricing of the new product, and establishing distribution are all critical events. The more radical the innovation in the new products, the more likely a correction to these events will be necessary. Accordingly, the accumulating market experience in the new product will probably require a fast recycle though the product development cycle. The more the technology discontinuity in a next-generation-technology, the more necessary is a redesign of a new product.

For competitive discontinuities, after the product development cycle has produced a new product dependent on the new technology:

1. The cycle of technology anticipation, acquisition, implementation, and exploitation must start all over again soon after a new product is introduced.
2. Competitors will see the new product and its impact on the market and will quickly introduce me-too products either with improved features or at a lower price.

Next-generation-technology products can be created when the scientific understanding of the phenomenon manipulated in the technology is better understood. This why science is essential to modern scientific technology.

> **Science allows improved understanding of the phenomena underlying a technology to enable radical technological progress.**

SUMMARY AND PRACTICAL IDEAS

In case studies on

- Inventing the silicon transistor
- Molding technologies
- Inventing the integrated-circuit chip
- Designs of early microprocessors

we have examined the following key ideas:

- Technology performance parameters
- Physical morphological analysis
- Schematic logic of physical technologies
- Next-generation products
- Next-generation product innovation processes

These theoretical ideas can be used practically in looking systematically at a technology system to formulate technology strategy for its improvement.

FOR REFLECTION

Select a major technology system (such as an automobile, airplane, computer technology, etc.). Discuss the history of progress in the technology, identifying where in the technology system progress was made, by whom, and how.

21

PHYSICAL TECHNOLOGY PARADIGM

INTRODUCTION

Now we will go deep into science. We have seen how high-tech strategy depends on new technologies for next-generation products, and these are dependent on scientific progress in improved understanding of the natural phenomena underlying technologies—scientific technology. We have also seen that in between radical innovations, incremental innovations in products and processes and services are necessary to defend against competition. We look at the science side of scientific technology—how it is possible to plan scientific progress by examining the foundations of *scientific technology*—its *scientific paradigms*.

We recall that there arc four types of technologies: material, power, biological, and information. We will see that each type uses principally a different scientific foundation, a different paradigm, and there are four scientific paradigms: those of *mechanism, logic, function,* and *language*. A paradigm in science provides the representational basis for the perspective of a scientific discipline. The principal paradigm in physics and chemistry is *mechanism*. In biology it is both *mechanism* and *function*. In computer science, the principal paradigms of software are *logic* and *language*.

First we review the physical paradigm of *mechanism* upon which modern science began. The paradigm of mechanism was constructed by the great scientists of the seventeenth century, such as Descartes, Galileo, Leibnitz, and Newton. Then in the eighteenth century, the modern scientific disciplines of physics and chemistry began, based on the new scientific paradigm of mechanism, which includes *kinematics, dynamics, prediction,* and *theory*.

> **The secret of how technology strategy can be formulated for *improving the physical morphologies* of technologies lies in exploration of the *mechanisms* of physical aspects of technologies.**

Inventing the Transistor

A case that nicely shows how the physical morphology of a technology can be improved by the understanding of its mechanisms is the case of the invention of the germanium transistor in 1948. The invention occurred in an industrial research laboratory. It was planned as an industrial research program to *search deliberately* for a technical substitute for the electron vacuum tube. The technology strategy used a new scientific theory, quantum mechanics, to enable a new technological search. Earlier, we looked at the cases of (1) the improvement of the transistor in the invention of the silicon transistor and (2) at the invention of the transistor and integrated circuit. Now we look at the invention of the transistor itself.

At the beginning of the twentieth century, electronics was a new technology made possible by a key technological component, the electron vacuum tube. The electron tube was a kind voltage-control valve; the tube controlled electronic signals sent through the tube by modifying (manipulating) the voltage of the signal (Figure 21.1). An electron can be forced from one point to another if there is a difference of voltage between the two points. Voltage is the potential of a force

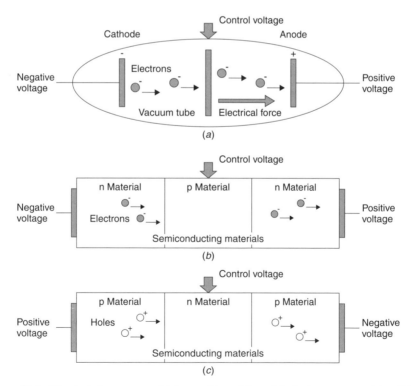

Figure 21.1. Physical phenomena of tubes and transistors: (*a*) electron vacuum tube; (*b*) *npn* transistor; (*c*) *pnp* transistor.

difference within a space to force motion of an electron (negatively electrically charged particle) over the distances of the space. The first application of the tube as a voltage-control device was to amplify electrical signals. This provided the capability of long-distance telephone communications. Later the tube would also provide the capability for radio communications, which was developed soon after the invention of the electron vacuum tube.

Early in the twentieth century, American Telephone & Telegraph (AT&T) had become the centralized U.S. telephone company and needed to provide long-distance telephone service. Engineers and scientists at the Bell Telephone Laboratories leapt on the electronic tube and innovated the first mass production of electron vacuum tubes for the amplification of phone conversations. Soon it was apparent that there would be natural limits to the new artifact. Electron tubes consumed a lot of energy to heat the negative plate to boil off electrons; they were large and not very rugged and did not last long. Some Bell Lab scientists were thinking of replacing the electronic vacuum tube with a better device. But what? In 1936, Mervin Kelly was appointed director of research at Bell Labs and established a research group to search for a way to replace the electron vacuum tube (Wolff, 1983).

Kelly had gone to work for the Western Electric engineering department after he had received a Ph.D. in physics from the University of Chicago in 1918. He was born in 1894 in Princeton, Missouri, where his father was principal of the high school in that small town. When Kelly graduated from high school, he enrolled in the Missouri School of Mines and Metallurgy. There he decided that physics and mathematics were more interesting to him than mining. In 1914, he graduated and enrolled in graduate work at the University of Kentucky. He also taught for two years to support himself, leaving with a master's degree in physics and a new bride. In 1916, they went to the University of Chicago, where Kelly worked as an assistant to Robert Millikan. Millikan was the first person to accurately measure the electrical change of a single electron, using his now famous *oil-drop experiment*. Kelly assisted in this experiment and received his doctorate. Later, Millikan received a Nobel Prize in Physics for the experiment.

When Kelly graduated, the director of the engineering department at Western Electric was Frank Jewett (who himself had earlier been a student of Millikan). Jewett hired Kelly. Later, in 1925, when Bell Labs was established, Jewett became its first director and Kelly was also transferred to the new lab. In 1936 Kelly succeeded Jewett as director.

As Kelly was trained as a physicist, he had been following the exciting developments in physics of that decade: the new quantum mechanics (Bardeen, 1984). Kelly saw a technological opportunity in quantum mechanics, as it was providing theory to explain the behavior of electrons in materials. Kelly envisioned that the new theory applied to certain semiconducting materials, such as germanium and silicon, might provide a way to invent something to replace the electron vacuum tube.

Earlier, when radio was being developed using the electron vacuum tube, one invention was the discovery that germanium could be used to detect radio signals.

Another discovery had been that silicon crystals could convert light into electricity and change alternating current to direct current. Both these interesting materials had been called semiconducting materials, because their electronic conductivity was between the high conductivity of metals and the nonconductivity of dialectic materials. Kelly created a research program in solid-state physics to explore the further use of germanium and silicon, from the viewpoint of the new theory of quantum mechanics. One of the new hires that Kelly made for his new program was William Schockley, who in 1936 received his Ph.D. in sold-state physics from MIT: " 'Kelly gave me an eloquent pep talk. . . . He pointed out that relays in telephone exchanges caused problems. . . . He felt electronics should contribute. . . . Dr. Kelly's discussion left me continually alert for possible applications of solid-state effects in telephone switching problems' " (Wolff, 1984, pp. 72–73).

Kelly's support and Schockley's' alertness paid off, but not until after World War II. During the war, Kelly was responsible for 1200 military research programs performed at Bell Labs. After the war, Kelly reestablished the solid-state physics research program and invited Schockley back to Bell. Schockley had spent the war at the Pentagon, working on military problems. Kelly also appointed James Fisk assistant director of physical research at Bell and told him to make a major and broad research effort in all aspects of solid-state physics and the underlying quantum mechanical theory. Fisk set up three research groups: physical electronics, electron dynamics, and solid-state physics. William Schockley and Stanley Morgen led the solid-state physics group. They hired John Bardeen and Walter Brattain as members of the group.

Schockley's group learned how to alter the conductivity of germanium by adding other atoms (doping) to the germanium crystal during its growth. Next Schockley proposed an artifact (transistor) to control the current in doped germanium crystals by applying an external electric field perpendicular to the direction of current in the crystal, much as a voltage applied to the grid of a electron vacuum tube controls the flow of electrons in the tube. The reason that Schockley thought this would be practical for germanium was that he had used the new quantum mechanics theory to calculate the expected effect of the controlling field on the electron flow. His calculations showed that it would be large enough to provide amplification. Amplification meant that the electron flow through germanium would closely follow the increases and decreases of the controlling field, thereby mimicking its form, but larger.

Schockley made a sample, but it didn't work—it didn't amplify. Schockley asked Bardeen, a recent graduate in physics from Columbia, to check his calculation. Bardeen did and suggested that perhaps electrons were being trapped at the surface of the crystal, which would prevent them from seeing the calculated effect of amplification. Next, Britain performed experiments on the artifact which confirmed Bardeen's explanation. They were on the right track. They needed to play with the structure of the artifact, trying other configurations; and one finally worked, providing a small amplification. A new kind of electronic valve was almost invented, but the amplification was too small. The group continued to invent alternative configurations. One configuration placed two closely spaced gold strips on the surface of a doped germanium crystal and an electrical connection on its base. Relative to that base, one gold strip had a positive voltage (called the *emitter*) and the other a negative voltage (called the *collector*). They applied a signal between emitter and base. Voilà! There was amplification in the voltage between the collector and the base. They called this a *transistor*. It could do everything the old electron tube could do and would eventually replace it.

AT&T acquired a basic patent for the transistor and made transistors for its own use. AT&T licensed others to make them for other uses. At that time, AT&T was a legal telephone monopoly and prevented from entering other businesses. The electron vacuum tube began the first age of electronics, and the transistor began the second age of electronics. This change from tubes to transistors is an illustration of *technology discontinuity,* providing a dramatic improvement in technical performance. This change was made by the invention of a new device to perform a technical function similar to that of an older device but based on a new physical phenomenon (electron transport in solid-state materials as opposed to electron transport in a vacuum).

Later, Kelly became chairperson of the board of Bell Telephone Laboratories, retiring in 1959. In that year, Kelly was awarded the John Fritz Medal for "his achievements in electronics, leadership of a great industrial research laboratory and contributions to the defense of the country through science and technology." The medal was an award from the American professional engineering societies (other recipients include Edison, Marconi, Wright, and Westinghouse). Schockley left Bell Labs and began the first company to manufacture transistors, the Fairchild Company. Schockley, Brattain, and Bardeen were awarded the Nobel Prize in Physics in 1956.

PHYSICAL PHENOMENON OF A TECHNOLOGY

We recall that underlying any physical technology is a physical phenomenon. In the dictionary, *phenomenon* is defined as an object or an observable fact, susceptible to scientific description and explanation. Kelly had decided to use a new scientific theory (quantum mechanics) to improve the explanation of electron conduction in germanium—as a possible new phenomenal basis for inventing a technical substitute for the electron vacuum tube. The electronic control device of the electron vacuum tube was based on the physical phenomenon of electron transport in a vacuum. Kelly had wondered if the new physical phenomenon of electron transport in a semiconducting solid-state material might provide a basis for an alternative electronic control device. It did and Kelly's group invented a new device.

Earlier, the electron tube had made possible the entire engineering field of electronics. Figure 21.1 illustrates the physical morphological structure of electron vacuum tubes and transistors.

In the 1850s, the physical phenomenum of electron conduction in a vacuum (then called "cathode rays") was observed by a German scientist, Julius Plucker, and a Swiss scientist, Auguste de la Rive. They had placed a metal plate, the cathode, at one end of an elongated glass tube and another metal plate, the anode, at the other end of the tube. Next a battery was connected across the two plates with negative voltage on the cathode and positive voltage on the anode and put an ammeter in the line between the battery and tube. The ammeter measured current flow of electrons through the wire and it showed zero current flow. This meant that the electrons that could flow from the positive terminal of the battery through the wire to the negative battery terminal were not flowing. The electrons could not jump the physical gap within the glass tube between the tube's cathode and anode. There was no flow of electrons within the tube across the airgap.

Then a vacuum pump was connected to an outlet of the glass tube and began to pump the air out of the tube. When the air was mostly evacuated from the tube, current flow suddenly registered in the ammeter. This meant the electrons were flowing through the wire from the negative battery terminal into the metal cathode plate and then across the vacuum gap to the metal anode plate and back through the wire to the positive terminal of the battery. The scientific discovery was that electrons could flow in a vacuum; which was finally established by J.J. Thomson in 1897 (e.g. that the cathode rays were corpuscular electrons).

About 30 years later, a technologist, Lee de Forest, figured out how to use this phenomenon—how to manipulate these natural states of electron flow in a vacuum for human use. De Forest added a small grid of wires perpendicularly midway between the negative and positive plates (Figure 21.1) and found that a small voltage between the grid and the negative plate had a huge influence in the size of the current between the negative and positive plates. This small grid voltage controlled the larger plate current. It made the electron vacuum tube into a type of control valve for controlling the flow of electrical current and the attendant voltage drop across the plates. A positive voltage on the grid would accelerate more electrons from the cathode to the anode, and a negative voltage on the grid would decelerate or impede the flow of electrons from cathode to anode. Positive voltages attract negatively charged electrons, and negative voltages repel negatively charged electrons. The utility of de Forrest's invention was that his electron vacuum tubes could be used to amplify signals. A small signal applied to the grid of the tube would be reflected accurately in the voltage fluctuation of the anode and thus amplified when the voltage of the anode was much greater than the voltage on the grid.

This electron tube provided a new functional capability for technology, amplification, and control of electrical signals. However, to get good amplification, de Forest also found he needed to add an electrical heater near the cathode, as the hotter the cathode, the more electrons boiled off. The cathode heater improved the efficiency of the electron tube. This description of how a vaccuum tube works is a descripion of the *mechanism* of the phenomenon.

Bell Labs' transistor was much, much smaller, as the first commercial transistors were about an eighth of an inch in diameter and a quarter of an inch tall. Later, transistors would become invisible, measured not in inches but in millionths of an inch. Also, transistors did not require a cathode heater to get the electrons out of the metal cathode and flying out into the vacuum of space in the tube. Thus transistors consumed much, much less energy as heat in their processes. They eventually could be put onto the same chip in densities of millions of transistors. A quantum theory description of the transistor would be a description of the mechanism of the phenomenon.

PARADIGM OF MECHANISM

In modern science, all physical phenomena are depicted as mechanisms. The paradigmatic approach of *mechanism* is a description of the world as mechanical processes. The approach of mechanistic reductionism provides a general strategy for improving the physical basis of any technology. This strategy is to go smaller in understanding the phenomena involved in physical morphologies of a technology. As one goes down the spatial scale from macro to micro levels of physical nature, one usually finds ways to control macro phenomena by manipulating micro phenomena. Going smaller in understanding technology at a smaller level of scientific phenomenon has usually resulted in improvements to technology.

> **The generic strategy for all physical technologies is to go *smaller* in researching the mechansims of a technology system.**

This is the reason for the linear portion of the technology S-curve for all physical technologies.

> **The linear portion of the technology S-curve for a physical technology expresses the time needed to pursue scientific research in going smaller and understanding nature at micro levels.**

Let us review the concept of the *scientific paradigm of mechanism*. The term *paradigm* was popularized by a philosopher of science, Thomas Kuhn, to emphasize that any scientific discipline has an underlying representation of the world, the paradigm of the discipline. For the physical scientific disciplines of physics and chemistry, their paradigm is to represent the world as natural mechanisms. This paradigmatic concept of mechanism is complex, as it contains within it several ideas: (1) a spatial framework for description, (2) a kinetic framework for motion, (3) a dynamic framework for explanation, (4), scales of explanation, and (5) causal prediction.

Spatial Framework of Physical Mechanism

Space and time are the two basic ideas in scientific representation of physical things. All physical things in the universe are individuated from each other in a spatial and temporal framework. Two physical objects in nature are said to be different because they can exist at different points in space at the same time.

> **Space is the concept of how physical objects can coexist in nature at the same time.**

The scientific construction of a spatial description uses geometry and algebra together in the mathematical topic of analytical geometry. This is illustrated in

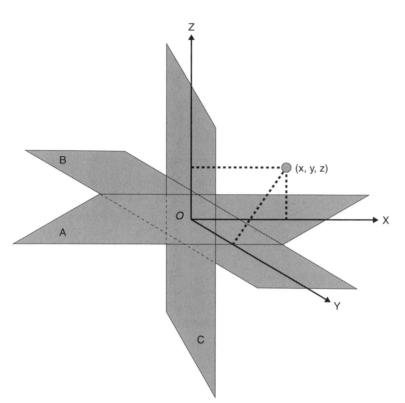

Figure 21.2. Representation of a three-dimensional space.

Figure 21.2, wherein three geometrical planes (A,B,C) are mutually perpendicular. This provides a geometrical description of a three-dimensional space. Next, three mutually perpendicular arrows (X,Y,Z) are positioned on each plane so that their intersection (O) describes the midpoint of intersecting planes. The three arrows (X,Y,Z) and origin (O) become a reference frame in which an algebraic measure is overlaid on each arrow. Each now becomes what is called a mathematical *vector*. A space described by vectors is called a *vector space*. In Newtonian mechanics, we live in a three-dimensional vector space (length, width, and depth). In relativistic physics, we live in a four-dimensional vector space (the threesdimensions of space and a fourth dimension, time).

> **The position of any object phenomenally observed in a given space can be described mathematically by a set of spatial coordinate numbers (x,y,z) on a reference frame (X,Y,Z) of the space.**

Description of position is the first step in any mechanistic representation of physical things in nature.

Kinematics of Mechanism

In addition to things co-existing phenomenally in space, things change their positions in space over time. A mechanistic description of phenomena needs to add to its spatial description a temporal description, a *kinematics,* which is the motion of objects through space. As illustrated in Figure 21.3, the motion of an object is described as a spatial path occurring as successive pictures of the space frame. This is like old motion pictures on celluloid film. Each frame of the film (spatial representation) shows the configuration of all objects at a given time. The picture frames of space are run in sequence past the eye of an observer. The observer perceives the motion of objects from their rearrangements in space.

Kinematics is the concept of the temporal sequencing of spatial reconfiguration.

For example, in Figure 21.4, we can see successive time frames (T_1, T_2, T_3, T_4) showing the motion of an electron from negative cathode to positive anode in the electronic vacuum tube. As a second example, Figure 21.5a illustrates the kinematics of the solar system and of the atom. In both cases, a smaller body orbits around a larger body. In the solar system, planets such as Earth orbit around the sun. The kinematics of the solar system is the continual circling of the planets like Earth around and around the sun in a path of the form of an ellipsoid. This planetary

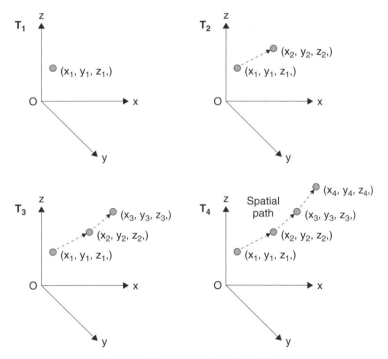

Figure 21.3. Space–time description.

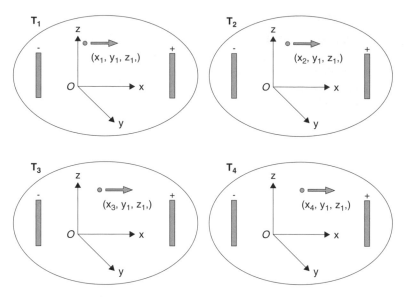

Figure 21.4. Space–time description of electron transport.

ellipsoid summarizes the kinematic path of a planet around the sun and determines the length of the solar year of the planet (with Earth completing an orbit every $365\frac{1}{4}$ days).

In another example, that of an atom (Figure 21.5b), electrons circle the nucleus of the atom. In a classical mechanics (Newtonian mechanics) model of the atom, an electron would be described as a particle circling the nucleus, similar to the passage

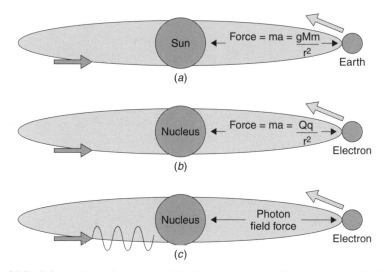

Figure 21.5. Solar and atomic systems: (a) planetary system; (b) atom–classical; (c) atom–quantum mechanical.

of planets around a sun. However, this classical model of an atom did not fit the real observations of the orbits of electrons in an atom, because it predicts that electrons could be found in an orbit of any size (as planets can be found in an orbit of any size around suns), and in fact, electrons can be found only in discrete orbits. To describe an atom having only discrete orbits, Niels Bohr postulated a model of discrete electron states, a quantum model of the atom. Later, Bohr's model was explained by having the electrons travel in a kinematics of motion not as a particle but as a wave. The energy of an electron determines a wavelength, and an electron of a given energy can occur only in an orbit wherein the wavelength nicely fits the length of the orbit (so that the wave path reinforces itself as the electron goes around and around). This wave motion of the electron was another feature of the new quantum mechanics needed to describe physical morphologies at the tiny sizes of atoms and electrons.

Force and Energy Dynamics of Mechanism

The kinematics of mechanisms describes motion of objects in space, but what causes changes in motion of objects in space? Change in motion is caused by a force. The explanation of change in motion is the mechanistic topic called *dynamics,* whose three basic laws were summarized by Newton:

1. Every moving object in space tends to remain in the same motion unless acted upon by a force.
2. The effect of a force upon an object is to accelerate the object in direct proportion to the object's mass: $F = ma$.
3. For every action, there is an equal and opposite reaction.

As the space–time framework is fundamental to mechanistic *description,* the concept of force is fundamental to mechanistic *explanation.* All explanations of the universe operating as a mechanism is through depiction of the forces between objects in the universe.

For example, in Figure 21.1, the force accelerating the electrons from the cathode to the anode is depicted as an arrow, a force vector. This electrical force results from an external energy source such as a battery connected to a cathode and an anode. The electrical force between the plates accelerates the electron from the negative to the positive plate.

As a second example, in Figure 21.5a, the dynamics of the solar system is shown as the gravitational force between the sun and Earth, as any two physical objects with mass attract each other. Because the sun is so massive compared to Earth, the gravitational force between the two ($F = gMm/r^2$) swings Earth in an elliptical orbit around the sun. Because of the second law of dynamics, the earth tends to travel inertially at a tangent to its orbit; and because the sun's gravitational attraction is radial in direction, the result of the gravitational force is to turn Earth around the sun.

Dynamics explains by means of forces how changes occur in the kinematics of objects moving in space.

Also in Figure 21.5 we see how if the atom had dynamics similar to those of the sun, an electron would orbit the atomic nucleus under a similar electrical force ($F = Qq/r^2$). But since dynamics at the microscopic level follows quantum mechanical laws, it is a force field (of wave/particle photons exchanged between the nucleus and its electrons) that supplies the attraction to maintain the wavelike moving electrons in their discrete atomic orbits.

> **The dynamics of a physical mechanisms depend on the spatial scale of the phenomenon: Newtonian mechanics for macro systems and quantum mechanics for micro systems.**

There are known to be different kinds of forces that operate at the different scales of space:

- Gravitational force
  —binding matter together
- Electromagnetic force
  —binding atoms and molecules
- Weak nuclear force
  —decaying neutrons
- Strong nuclear force
  —binding nucleons into atomic nuclei and binding subparticle quarks into fundamental particles

In the quantized field theories of modern physics, these forces are all different kinds of fields composed of virtual particles. For example, the electromagnetic force is composed of virtual photons, boiling off and returning to a charged particle (such as an electron or proton) in such a short time that the law of the conservation of energy is not violated. The time a virtual field particle (e.g., photon) can boil off and return to its charged source particle (e.g., electron) must be less that the inverse of its energy ($dt \leq h/dE$). All forces in modern physics are explained as virtual field-quantized-particles boiling off a charged source particle and then quickly returning.

The particles of light, photons, are the quantized field particles of the electron-magnetic field between electrically charged particles (e.g., electrons, protons). The quantized field particles of the weak nuclear force are the Z and W bosons. The quantized field particles of the strong field are gluons that bind quarks into elementary particles.

We should note that modern physics also describes the dynamics of physical systems as energy states (as well as force interactions). In modern physics, an idea complementary to that of force is that of energy. As force alters the rates of motion of physical things, so energy alters the rates of events of physical things. The relationship of force to energy is thus: Energy is the result of a force times the length of path over which the force is exerted. This is also called *work done,* so that energy = work done = force × distance. The dynamics of a physical thing can be formulated either as interactions of forces or as states of energy. In modern physics, the change in physical system states can be formulated as the total energy (Hamiltonian energy) operating on a system as the rate of change of states of the system: $Hs = ds/dt$ (where s is the probable state of a system, H is the total energy of the system as the sum of the kinetic energy and potential energy, and ds/dt is the partial derivative of the state with respect to time).

> **Spatial description, temporal kinematics, force and energy dynamics, and scales of explanation—are three key ideas of the paradigm of physical mechanism.**

Scale

Physical processes are depicted in this mechanistic picture from very, very small spaces up toward very, very large spaces. This is the microscopic–macroscopic explanatory strategy of science. In the very smallest space we have to date, subparticle space, the fundamental particles are made up of smaller particles, quarks and gluons. The very, very, very tiny quarks circle about each other and interact with each other through the gluon field to constitute the fundamental particles of protons and neutrons. Quarks and gluons have not been observed outside the internal spaces of the fundamental particles, but their effects have been seen in very high energy collisions between fundamental particles and in the spectrum of fundamental particles. This is the *quark-level scale* of space. In the next size up, the fundamental particles of the universe exist in space: electrons, protons, neutrons. This is the *particle-level scale* of space. In the next spatial size up, protons and neutrons bind into stable states of atomic nuclei. The nucleus of atoms is composed of the neutrons and protons bound together by the strong nuclear force. The weak nuclear force causes the decay of neutrons to protons, thereby creating the phenomena of nuclear radioactivity. This is the *nuclear-level scale* of space.

In the next spatial size up, the atomic nuclei and bound orbiting electrons form atoms. The atom is constructed of negatively charged electrons orbiting the positively charged nucleus. This is the *atomic-level scale* of space. With the invention of the tunneling electron microscope, we can observe this scale of individual atoms directly.

In the next spatial size up, molecules are formed from combinations of atoms that bond together by the exchanging outer electrons (valent bonding) or sharing outer electrons (covalent bonding). In addition, molecules also stick together through the distant effects of the electromagnetic fields of the atoms (van der Waals bonding). Gases are molecules in a contained space, flying too fast to stick together. Liquids are molecules stuck together but vibrating strongly enough to prevent rigid structural configurations. Solids are molecules shivering, but stuck together in rigid structural configurations in crystalline forms (regular order), glassy forms (irregular order), or polymeric forms (long molecular chains). This is the *molecular-level scale* of space. After the invention and development of nuclear magnetic resonance imaging technologies, we can observe this scale of molecular structure directly.

In the next spatial size, atoms or molecules stabilize in liquid or solid configurations as domains or polymeric structures. This is the *domain-level scale* of space. We can observe domain-sized structures using the electron microscope and scanning ion probe.

In the next spatial size we find the microscopic level of the organization of matter as aggregates or organisms. This is the *cellular-level scale* of space. We can observe this cellular scale directly using optical microscopes. We humans exist on a macro scale of space of organism systems. We experience nature through the physics of our senses. For example, we experience gases as air, vapors, or winds. We experience liquids as fluids, water, rain, rivers, lakes, and oceans. We experience solids as rocks, dirt, and materials.

Finally, there are two more scales of space above this macro level in the mechanistic picture of nature, the planetary and cosmic levels. At the *planetary-level scale* of space, nature appears as environmental systems of physical processes and ecological systems. At the *cosmic-level scale* of space, the physical universe is formed of stellar systems, galaxies, and clusters of galaxies.

All of the foregoing scales, from quark and particle through atom and molecule to domains, cells to organisms, macro scale to planetary and cosmic—together, these provide the scales in the mechanistic representation of nature.

Prediction

The importance of the kinematics, dynamics, and scale of *mechanism* is that a mechanistic model of the forces and/or energy states of a physical system can be constructed. This can allow one to calculate subsequent physical states from prior states and predict the temporal evolution of a physical system. The dynamics of force and energy connect states of the physical system in a causal sequence. A cause-and-effect relationship between two events A and B is said to exist if the occurrence of a prior state A is both necessary and sufficient for the occurrence of a subsequent B. Then A is said to cause B as an effect.

> **Prediction of natural events is possible when physical theory provides an explanation of physical events as causally related.**

Prediction of future states of a physical system is possible when all the mechanistic states of a system can be calculated from a model of the forces in the system. Prediction is very powerful for both science and technology. If a technologist can model the physical phenomenon of the physical morphology of the technology, the technologist can predict how and when future states of the physical morphology occur under technological manipulation. For example, exact models have been constructed of the electron transport in transistor structures (physical morphology of the transistor). Electrical engineers use these models when they design transistorized electrical circuits to predict precisely what will happen to an electrical signal as it is processed by the circuit. The U.S. National Aeronautics and Space Agency planned the trips to the moon using precise predictions as calculations of flight paths to get astronauts there and back.

Summary of the Paradigm of Mechanism

> **The five key ideas that together make up the paradigm of physical mechanism are:** *spatial description, temporal kinematics, force and energy dynamics, scales of explanation,* **and** *causal prediction.*

PHYSICAL THEORY

The paradigm of mechanism makes modern physcial theory possible. Physical theory allows all physical morphologies of any technology to be represented as mechanisms and enables manipulations of nature by the technology to be predictable. In the paradigm of mechanism, a generic technology strategy for the physical aspects of all technologies can be devised as a scaling strategy—improve technology by better understanding nature at a smaller or greater scale. Physical phenomenon at one spatial scale can be explained by physical mechanisms at a smaller spatial scale.

> **A generic technology strategy for improving any physical technology is to understand nature mechanically at a smaller scale.**

We recall that science can be defined as the activity of discovering and understanding nature. Now we can see that the scientific paradigm of mechanism provides a perspective for observing physical nature and understanding nature in theoretical representations of physical mechanisms. A theoretical representation of a mechanism has (1) a *description* of nature as special and temporal kinematics and (2) an *explanation* of nature as energy dynamics, which in mathematical form allows (3) *prediction* of nature.

> **Physical theory provides a scientific representation of nature as a mechanism consisting of description, explanation, and prediction of nature.**

Now let us return to an earlier depiction of science and engineering (Figure 2.2), shown again as Figure 21.6, where the difference is now the linguistic focus of the interactions. There we indicated that scientific *theory* provided a way of representing the things (T_1 and T_2) in nature and their interactions. In creating a mechanistic theory, physical scientists developed a specialized kind of language of physics to abstract and generalize things in nature as physical objects. This specialized theoretical language of physics constructed precise meanings of terms of physical experience, such as space, time, matter, velocity, acceleration, momentum, force, energy, and so on. This set of terms and relations between the terms (e.g., force = mass $\times$ acceleration) all together provide the scientific discipline of physics as a

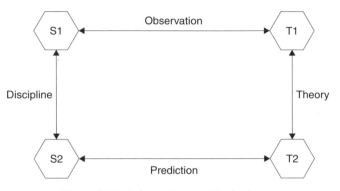

Figure 21.6. Information model of science.

kind of specialized language for representing physical nature as mechanisms. Physical scientists call the specialized language of physics such topics as Newtonian mechanics, special relativity, or quantum mechanics.

These linguistic topics differ by differing relationships in the respective linguistic topics. For example, in Newtonian mechanics, the conserved relationship that is seen by all observers as the same is the spatial length of an object and also the time interval ($x^2 + y^2 + z^2$ and t). But in special relativity mechanics, the conserved relationship of an object seen by different observers is the square of the spatial length minus the square of the time interval times the speed of light [$x^2 + y^2 + z^2 - (ct)^2$]. Physicists reconcile the Newtonian and special relativity theories by asserting that Newtonian mechanics is an accurate representation when relative speeds of things in nature are much less than the speed of light, and that special relativity mechanics is accurate when speeds are near the speed of light. Therefore, Newtonian mechanics is a special case of special relativity mechanics (an approximation at slow speeds).

In general, one can view any scientific theory as a kind of language, a specialized formal language that provides precise representations of nature. Scientific theories are semantically specialized languages developed by disciplinary communities to represent the natural content of disciplinary fields of observable nature.

> **Scientific theory consists of semantically specialized languages used to precisely describe semantic fields of natural things.**

It useful to see scientific theories as semantically specialized languages so that one can see how mathematics is essential to science. Mathematics provides another kind of specialized language for scientists to use to observe nature and construct theory, but in quantitative relations it is a syntactical specialized language for science to use to express nature quantitatively.

For example, consider the basic Newtonian equation of motion, $\underline{F} = m\underline{a}$. The physical term *mass* (which is the property of matter to respond inertially to force) is expressed as a scalar number in the equation. This means that in measuring and expressing the quantity of mass, physical scientists express mass in the mathematical topic of an algebra. Physical masses can be measured as a scalar number and then added and subtracted or multiplied and divided as an algebraic set of numbers. In contrast, both force and acceleration must be measured and expressed as vector numbers, $\underline{F}$ and $\underline{a}$. Vector numbers and vector spaces are a different mathematical topic than scalar algebras. Vectors must be expressed not by a single number (as can a scalar such as mass) but by a set of numbers. In three-dimensional physical space, a position vector of a massive particle must be described by a set of three numbers (x,y,z) as the projections of the particle position onto the base vectors of the space. Moreover, the vector expression of acceleration, $\underline{a}$, needs to be expressed further as a differential number of velocity (the rate of change of velocity), so that the mathematical topic of calculus must also be used (in addition to algebra and vector spaces) in expressing the Newtonian equation of motion, $\underline{F} = m\underline{a}$.

> **Science uses mathematics as a syntactical specialized language for expressing *quantity* of nature.**

This is an important way to view scientific theory—as the construction of semantically specialized languages of natural content expressed in quantitative syntactical languages. It relates scientific theory construction to language development by scientists.

> **Scientific mechanism is a paradigm for representing physical nature.**

SUMMARY AND PRACTICAL IDEAS

In a case study on

- Inventing the transistor

we have examined the following key ideas:

- Mechanism paradigm
- Spatial framework for description
- Kinetic framework for motion
- Dynamic framework for explanation
- Spatial scales of explanation
- Causal prediction
- Physical theory as a semantically specialized language
- Understanding nature at a smaller scale
- Mathematics as a specialized syntactic language

These theoretical ideas can be used practically in understanding how the paradigm mechanism in the physical sciences provides a language for representing nature, such that the manipulation of nature by a physical technology can be predicted by physical theory. This is the basis of scientific technology.

FOR REFLECTION

Read a history of a scientific discipline to see how the paradigm of the discipline emerges through the construction of theory upon scientific experimentation with nature.

22

BIOLOGICAL TECHNOLOGY PARADIGM

INTRODUCTION

We have seen that the generic scientific technology strategy for improving physical technologies is to elucidate mechanisms of physical phenomenon at smaller scales. Is there a generic technology strategy for improving biological technologies? The answer is yes, but it is more complicated. One must not only elucidate the mechanisms in biology but also interpret the *functional benefit* of the mechanisms to an organism. The paradigm ideas of *mechanism and function* are both needed to fully explain the phenomenon of life or of *animate matter*.

Earlier we saw in the case of the discovery of DNA how both the *mechanism* and *function* were evident from the *structure* and *operation* of the DNA molecule. DNA's structure consisted of identical double strands in a helix which in operation uncoil and split for the replication of a cell, mitosis. To examine a generic technology strategy for biotechnologies, we need to add to the paradigm of *mechanism* the paradigm of *function*. And we will find, the idea of b*iological function* is a complex idea, containing subideas such as *behavior, action, will, mind,* and *reason*.

Reductionist Methodology in Biology

In Chapter 21 we described how the use of the spatial scaling provides explanations of the processes of mechanisms of a larger phenomena in terms of the mechanisms of smaller physical phenomena. This explanatory approach in science has been called a *reductionist methodology*. Reductionist methods underlie the scientific disciplines of physics and chemistry and biology. For example, Kell and Welch (1991) noted the reductionist approach in molecular biology: "Molecular biology is widely reputed to have uncovered the secret of life, with DNA being heralded as the quintessential component of biological systems. . . . The biologist's

391

problem is the problem of complexity. According to conventional wisdom, the way to study complex systems is by breaking them down into their parts" (p. 15).

Reductionism assumes that living systems are arranged hierarchically in space, and larger organizations of life can be explained in smaller levels of organization—as an example: "Hydrolysis of energy-rich molecules causes protein filaments to slide, sliding filaments cause muscles to contract, muscle contraction causes the acquisition of food and the escape from predation, and group behavior centered around common defence and protection of food supply causes the formation of social systems. This analytical approach may be termed 'hierarchical reductionism' " (Kell and Welch, 1991, p. 15).

Kell and Welch (1991) asserted that although this approach provides much insight, it in itself cannot provide a complete explanation in biology: "Methodologically speaking, reductionism is fruitful; it yields much utilitarian detail about observable phenomena. However, reductionism can never be complete; there is always a 'residue' [in biology] . . ." (p. 15).

What they meant by *residue* is that biological *explanation* requires the idea of function in additon to that of mechanism. As a biological explanation, the idea of *function* is not a mere intellectual residue from the idea of *mechanism*.

In science, the explanatory ideas of *function* and *mechanism* are independent, coequal ideas.

SCALE AND SYSTEMS OF BIOLOGICAL MECHANISMS

Just as spatial scale is used as a explanatory principle in the physical sciences, so is scale used in the life sciences. At the microscopic scale, the smaller unit of life is the cell, and below the cell at the molecular level are structures and processes of organic chemistry. Above the cell are multicellular organisms and organs of larger, more complex beings. Individual organisms belong to a species, and several species form populations of an ecology. How the populations of an ecological niche interact forms an ecological system. Scale also enables biological complexity. A multicelled system can develop specialization between cells that creates complicated organisms.

Structures and processes in biology are described as biological systems. In an earlier chapter we reviewed the concept of system as a totality with dynamic dynamics states in an environment. The concept of *system* is also important to biological explanation. The mechanisms of a biological organism are partitioned into functional systems of the organism for sustaining and reproducing life. For example, at the macroscopic level of the human organism, mechanisms are partitioned into the various functional systems of the human body: skeletal system, muscular system, digestive system, circulatory system, immune systems, neural system, reproductive system, and so on.

Each mechanistic system of an organism provides a different functional capability in the organism.

First Evidence of Mechanisms of Cancer

An illustration of how mechanism and function together provide an explanation in biological science is the case of the first understanding of the mechanism of the connection between environment and cancer. In the period from 1950 through 1970, many statistical studies had suggested a correlation between environmental factors and cancers, such as cigarette smoking and lung cancer. Yet the exact mechanism of such interactions was not known. How did irritants in the smoke cause lung cells to turn cancerous? For many years, cigarette producers maintained that it was not proven that cigarettes caused cancer. Still the connecting mechanism had not been observed between cigarettes and cancer. But in 1991, there was a discovery of a connecting mechanism between a cancer and an environmental factor: "Scientists have detected a molecular 'hot spot' that is strongly linked to liver cancer, one of the commonest and most lethal malignancies in the world. The spot is a tiny region of a single gene where toxins that infiltrate the liver seem to home in, sabotaging the gene and touching off cancerous growth" (Angier, 1991, p. A1).

Then liver cancer was one of the top five cancer killers worldwide, although relatively rare in the United States and Europe. The scientists suspected that either the hepatitis B virus or a toxin from a fungal poison called aflatoxin could cause liver cancer. Contaminated grain and other food improperly stored in warm, moist places grow molds that produce aflatoxin.

The mechanism of this cancer is as follows. An earlier infection of a victim with hepatitis B virus had left the virus's DNA implanted in the DNA of the victim's liver cells. There the viral DNA remained inactive in the liver cells until the ingestion of the liver cell by a chemical that could turn on the viral DNA. This happened when the victim ate rice contaminated with the fungal toxin. Toxin entered a liver cell infected with the viral DNA and activated the viral DNA chemically. That viral DNA then created a cancerous condition in this cell by turning off the cancer defenses of the cell and dividing into rapidly growing cancer cells.

The mechanism of the toxin chemical *to the virus* signaled a *functional response in the virus's DNA* to both turn off the cancer defenses of the host cell and to replicate.

This model of a cancer mechanism consists of a retrovirus resident (in a victim's cell from an earlier viral infection) that is later triggered by an environmental factor. The ingestion of a carcinogen chemical from the environment, such as the fungal toxin, chemically turns on the resident retrovirus to create a cancerous cell. We recall from Chapter 21 that prediction was possible when mechanisms are explained in causal models. In this case there were two causal factors (virus infection and carcinogen chemical) that needed to be present together to cause cancer. This is the reason that the many statistical studies of the strong correlation between cancer and cigarettes could never prove the relationship, since the studies looked only for a single causal factor. Only complete mechanisms prove causal relationships in biological phenomena.

BIOLOGICAL FUNCTION

Bacterial cells and viruses are the smallest units of life. They are constructed from information encoded in molecules that replicate themselves mechanically, DNA and RNA. As noted in Chapter 2, the DNA molecule is structurally in the form of a double helix of identical chains of polymers constructed of different sequences of four amino acids. These four amino acids provide information for the construction of proteins by the RNA in groups of three—the *genetic code*. Here is the basic difference in the *scientific explanations* between inanimate and animate matter. The mechanisms of inanimate matter do not convey information to other inanimate matter. Inanimate matter only undergoes physical processes, but the mechanisms of animate matter can convey information about biological function. To explain animate physical processes, we need to describe both the *physical* processes and the *relevance* of these processes (function) to the animate being.

> Biological *function* is a description of the *relevance* of physical processes to an *animate being*.

This is why the information aspect of the triplet coding of the amino acids in DNA is important. It conveys information about the function of the DNA mechanism—its value to a living being in constructing useful proteins.

> In biology, *information* conveys *functional relevance* as *value* about a physical mechanism to a living organism.

We recall that technology requires understanding about the value of the physical morphology, a logical scheme that orders natural states for a functional transformation. Thus in biology, the *physical mechanisms* of living organisms are *a kind of technology* to the organisms.

CASE STUDY:

Genome Duplications

The issue of how organisms can become *increasingly complex* on larger spatial scales is a major challenge in explaining the *evolution* of life. The case of studying genome duplications provides an illustration of this challenge, as Pennisi (2002) reported: "One of biology's greatest mysteries is how an organism as simple as a one-celled bacterium could give rise to something as complicated as a human. Thirty years ago (1970), prominent geneticist Susumu Oho of the City of Hope Hospital in Los Angeles (California) put forth what many of his colleagues then considered an outrageous proposal: That great leaps in evolution—such as the transition from invertebrates to vertebrates—could occur only if whole genomes were duplicated" (p. 2458).

Earlier, geneticists had realized that for organisms to evolve, they needed to keep old functions operating while gaining new functions. The mechanism of duplications of genes could provide this way of continuing old functions while gaining new ones. It was in 1970 that Oho took the bigger step of proposing that entire sets of gene, genomes, could be duplicated as well as single genes. His hypothesis has been difficult to verify, but in the 1990s, researchers studying comparative genomics (entire genomes of a species) began investigating Oho's hypothesis.

For example, at the University of Reading in the United Kingdom, Peter Holland and Rebecca Furlong were studying a fishlike invertebrate species, the *Amphioxus* (Pennisi, 2002). This species might be one of the nearest ancestors to vertebrate fish. They found that in comparison, fish vertebrates have several copies of some genes of which the *Amphioxus* has only one copy.

FUNCTION AND BEHAVIOR

Charles Darwin's theory of the species is a *functional explanation* of the complexity of mechanisms in the various species of life. Mutations in the genes of organisms create different organic mechanisms. Organisms with these mechanisms propagate as surviving species when the mutated mechanism provides a species with a functional advantage for survival in an environment.

> **In evolutionary theory, function is the value to survival of an organism in its environment of a biological mechanism.**

Thus the *behavior* of individuals in a species is an adaptation to its environment. We next examine the idea of *behavior* as an idea in the *paradigm of biological function*. For this, we turn to the writings of John Dewey, one of the philosophers of the American school of pragmatism around the turn of the twentieth century. Dewey emphasized how both biological and cultural functions are connected to mechanisms of life in a basic relationship of *energy between life and environment:* "Whatever else organic life is or is not, it is a process of activity that involves an environment. . . . An organism does not live in an environment; it lives by means of an environment. . . . For life involves expenditure of energy and energy expended can be replenished only as the activities performed succeed in making return drafts upon the environment—the only source of restoration of energy" (Dewey, 1938, p. 25).

What Dewey was noting was that the explanatory idea of *function* in biology derives from a description of life not merely as mechanism but also as biological *action,* action by living organisms. Living matter alters its environment through behavior—behavior as willful and purposeful activity of life. Inanimate matter makes up (composes) an environment in a purely physical and unintentioned manner. Yet animate matter acts within and upon an environment in an intentioned manner, through active behavior. *Action* is an animate organism's willful engagement of its environment.

Intention and action are properties of animate matter.

The phenomenon of animate matter cannot be fully described without a description of the intention (purpose, function) of the activity of an organism.

Concepts such as *function*, *will*, and *intention* are concepts that science needs to use (in addition to the concept of mechanism) to *describe* life *fully*.

For example, the behavioral action of an amoeba is to move its pseudopods out to encompass organic matter to obtain its food. The action of a plant is to grow leaves and orient them toward the sunlight to absorb the energy of light, while also growing roots into the soil to absorb the materials of water and nitrogen for growth. The action of a herbivorous gazelle is to graze on the green grasses of the plains; and the action of a carnivorous lion pack is to seize and eat a gazelle. The actions of the laborer going to work and punching in the computerized time clock, or that of the white-collar worker signing off biweekly on the computerized time sheet, or that of the executive exercising stock options are all like—the noble lion, the graceful gazelle, the peaceful plant, the lowly amoeba, the struggling worker, the wealthy executive—all alike in seeking food and energy. Action is the basic description of matter as animate—action necessary for survival, prosperity, and reproduction. In action, the concept of *function* requires an *actor* (one who acts). Any action by an actor occurring in the present is intended to attain a future value from the action.

All action is performed by an actor in the present for future anticipated value to the actor.

Life is organized into different levels of complexity, from the levels of DNA to that of the cell to that of the individual to that of group and society. But at each level, the essential phenomenon of action is to act in the present for future value. For example, even in bacteria there are elaborate operations in these organisms to gather matter and energy from their environments and to construct proteins and lipids and sugars for their own material being and biological processes. As organisms evolved in complexity, they developed capabilities of sensing and motion in their environments to actively seek out materials and energy.

The seeking out of materials and energy for living is the biological ground of animate "purpose."

CASE STUDY:

Fire Ant Society

The first discovery of a gene that actually regulates biological behavior directly occurred in a research project attempting to understand the social behavior of ants:

"Genes regulating behavior are very hard to pinpoint. . . . But fire ant researchers at the University of Georgia say they've characterized a gene that may single-handedly determine a complex social behavior: whether a colony will have one or many queens" (Holdern, 2002, p. 1434).

The discovery of how a gene can control behavior was in how ants sense the pheromones that identify a queen ant: "Fire ants have two basic kinds of social organization. A so-called monognyne queen establishes an independent colony after going off on her mating flight, nourishing her eggs with her own fat reserves without worker help until they hatch and become workers themselves. Polygyne queens, in contrast are not as robust and fat. They need worker aid to set up new colonies. They spread by budding from one primary nest into a high-density network of interacting colonies. Monogyne communities permit only a single queen, and those with a resident royal kill off any intruding would-be queen. Polygyne colonies can contain anywhere from 2 to 200 queens and accept new queens from nearby nests" (Holdern, 2002, p. 1434).

The differences between the two kinds of queens, monogyn and polygyne, was explained as a difference in a single gene that ants possess, their Gp-9 gene. This Gp-9 gene encodes information to know how to build a pheromone-binding protein (which is crucial for recognizing fellow fire ants). In a monogyne colony, all the ants have two copies of one version of the gene, called a *B allele*. But in polygyne colonies, at least 10% of the ants have a different version of the Gp-9 gene called a *b allele*. So there can be two kinds of queens in the ants, those with the gene *BB* and those with the gene *Bb*.

The behavioral result of these different versions of the gene is that ants in a polygyne community will kill off any potential *BB* queens. But ants in the *Bb* portion of the colony will accept *Bb* queens. Then the *Bb* queens can escape attack from the *Bb* workers, because *Bb* workers aren't good at recognizing the *Bb* queens as competition queens to the *BB* queen. The *b* allele encodes for a faulty pheromone-binding protein that does not bind the queen's pheromone.

The importance of this research was that it did connect a specific mechanism with a specific behavior: "This is the first time scientists have nailed down the identity of a single gene of major effect in complex social behavior. . . . Andrew Boruke of the Zoological Society of London said: 'The research opens for the first time the study of genes influencing social behavior across the whole span of the biological hierarchy' " (Holdern, 2002, p. 1434).

BEHAVIOR AND WILL

Essential to the concept of behavior in organisms with neural systems is the concept of the *will* of the organism. The term *will* is an old-fashioned term for the functional concept that traditionally describes the intention of an action, a deliberate action by an organism. Functional concepts such as *will* are needed in biology to explain complicated behavioral patterns of *sentient beings,* beings conscious of their state of being. Sentient consciousness can be kind of 'hardwired' in the genes of behavior, as in the case of the fire ant. Or consciousness can be a mixture of hardware and software, as in the learned behavior of more intellectually complex organisms, such as (compared to the fire ants) the fiery-tempered humans.

Living behavior needs to be described not only by the mechanism of the living processes (pheromone-binding protein in the ant case) but also by the function of the process (identification of the queen ant). From the perspective of the observer of the organism (the fire ant scientists), this behavior is described as functional (function of the number of queen ants and how the fire ant colonies grow). But from the perspective of a *BB* fire ant queen, this same functional behavior could be called will, the will of the *BB* queen to have no competitors. In sensing out other competing *BB* queens and killing them, the *BB* workers can be said (from the queen's perspective) to be carrying out the queen's will.

Parallel tales of this type of exclusive leadership behavior can also be seen in the human species. For example, in the history of England's monarchs in the sixteenth century, the Protestant Queen Elizabeth had her contentious and Catholic cousin, Mary Queen of Scots, imprisoned and then beheaded. Pushing the metaphor, one might say that the fiery colony of England would have only one monarch, and the English workers bearing the banner of the Protestant allele carried out the Queen's will to have no other queen but Protestant Elizabeth. Previously, a sister, Mary (called by the Protestants, Bloody Mary), had been on the English throne and had killed many Protestants. Protestant workers then feared to let another Catholic queen rule the English hive.

Now this is only a crude metaphor because no single gene codes for religious and political affiliations in human colonies. But there is a parallel. Humans have been genetically hardwired for family, kinship, and tribal affiliation. The social phenomenon of this kind of group behavior (e.g., competitive national games such as football, territorial wars, genocide) probably is all based in some kind of genetic human adaptation for intense tribal affiliation. Human will is a functional description of how humans decide on a plan of purposeful action. For organisms less complex than humans, much of their will (as in the will of the fire ant) has been programmed in their DNA—is hardwired behavior, so to speak. This kind of hardwired willful behavior in animal species has been traditionally called *instinct*.

Instincts provide the primitive hardwired biological instructions for attaining the basic purposes of animate beings.

Yet in complex organizations of life, instincts are not sufficient to explain behavior, as in humans. In the human species, the activities of learning and planning add to and even go beyond their basic instinctive instructions for action. In the complex cognitive and social capabilities of humans, their purposes are not all purely instinctual but are also learned and deliberate. Instinct combines with learning and reason in the human, so that human will is a mixture of instinct and reason. The concept of *reason* needs to be added to any theory of action to describe how actions can be deliberately chosen by some sentient being as a result of a prior analysis of an action as a *means* and *end*.

The *means* of an action is the *way* the action is carried out by an actor and the *ends* of an action is the actor's *intended outcome* of the action.

This is where planning as a kind of behavior affects human action. Humans can reason about possible futures as means and ends: (1) to satisfy both instinctive and learned needs as *ends,* and (2) to plan action as *means* to bring about such futures.

The reasoning of humans about future action is called *planning;* and the means–ends description for future action from such reasoning is called a *plan.*

CASE STUDY:

Mechanisms of the Human Brain

The brain is a central mechanism of the body in the neural systems of higher biological organisms for both instinctive and reasoned behavior. The mechanism of the brain is marvelous and complex. The brain consists of a high density of neural cells connected in synaptic structures. These structures are organized into complex sets of folded planes that are interconnected three-dimensionally. There are 10^{12} (a million billion) neurons that communicate with each other. The total weight of the brain is about 1300 grams.

Thought processes occur as the effect of electrical and chemical changes among these communicating neurons. Moreover, different regions of the brain process different kinds of sensory information and operate different mental activities. The brain is but one component of a larger neural system of the body, with its connection to the rest of the neural system. The entire neural system operates with both chemical and electrical signals.

The functions of the mind in the brain are spatially localized. The lower involuntary functions of the brain center in the medulla, which controls breathing, swallowing, digestion, and cardiovascular coordination. The hypothalamus area is involved in the expression of many basic instincts. The cerebellum plays an important role in coordinating motor and voluntary movement. The cerebral cortex (or neocortex) is the most recently evolved portion of the brain in higher mammals and is involved in the sensory and cognitive processing function of the brain. The processing of visual information takes up a large part of the brain. Many of the higher functions of the brain, such as moral judgment, are processed in the frontal lobes of the brain. Locomotion is a coordination between balance located in the brain and motor synapses located in the spinal chord. Control of the heart itself is located primarily in the structure of the heart. Control of sleeping and waking is coordinated in hormonal activity in the brain. Two areas of the brain, Broca's and Wernicke's, are involved in speech processing. Attention in vision uses a large part of the neocortex. Tasks that involve more than one sense require coordination across different regions of the brain. For example, a task of reading visual text will stimulate activity in both the speech and visual areas of the brain.

The neocortex of the brain is divided into two hemispheres, with each hemisphere coordinating the kinesthetics of different sides of the body. The right hemisphere coordinates the kinesthetics of the left side of the body; and the left hemisphere coordinates the kinesthetics of the right side of the body. In visual processing, both halves of the brain receive information from both eyes. But at the primary step of visual processing in the cortex, the left brain begins to integrate information from the right side of the visual field, from views by both the left and right eyes. Complementarily, the left side of the visual field is first integrated by the visual region of the cortex of the right hemisphere with views from both left and right eyes. Coordination between the two lobes occurs through communication in the corpus callosum, which connects the two hemispheres.

In addition, the two hemispheres of the brain in humans have evolved different reasoning specializations. The left side of the human brain has specialized in verbal reasoning and the right side in spatial reasoning. For example, people who (either by accident or by surgery) have had the connection between the left and right brain severed show mental dysfunction. When they view things only with their right eye, they lose the ability to perceive the connectedness of spatial forms; yet they can perceive spatial connectedness through their left eye. This implies that the right brain is more specialized than the left brain in spatial reasoning. Similarly, the verbal reasoning ability of the left brain appears to far surpass the verbal reasoning ability of the right brain. Normal brains process the meaning of written and spoken language primarily in the left half of the brain and process spatial reasoning and motion primarily in the right half of the brain. Connected to each other, the two half brains coordinate linguistic and spatial cognitive functions. For example, left–right brain studies of normal people have further identified that some people have cognitive preferences or talents for either spatial reasoning or verbal reasoning. For example, some students will learn and practice spatial and motor skills faster and better than other students, who in turn may learn and practice linguistic and analytical skills with talent.

BRAIN AND MIND

While the mechanisms of the brain are the physical basis of sentient behavior, an additional functional idea about the brain is necessary to describe how brain functioning—*mind*. This requires a description of brain as a kind of information processing system, a mind. The brain is a mechanism, and the mind is the function of the brain mechanism.

In the last half of the twentieth century, scientific research made significant progress in understanding the mechanisms of the brain and also in understanding mind as a cognitive activity. Psychologists study this problem, as do biologists and sociologists. Economists even have a model of human minds as being economically rational. Moreover, the new scientific discipline of computer science even has new techniques about the mind under the term *artificial intelligence*. They have also developed pattern recognition techniques, called *neural networks,* proposed to model

biological processes. Some electrical engineers are even creating a new area of ro-
botic neural circuitry called *neuromorphic engineering*. From all these efforts, one
can expect that eventually there will be mentally sophisticated robots, emulating
much of the activity of the human mind, as presently computers emulate much of
the computational and graphical representational capabilities of the mind.

Yet central to all these disciplinary perspectives on the mind is a shared assumption that the function of mind is to process information. Thus the concept of
information has become central to any model of the mind. *Information input* to the
mind begins with the sensual mechanisms of the neural system of the body, of which
the brain is a central processing unit. *Information output* is in the form of decisions
to action. *Information feedback* is in the form of monitoring and evaluating the consequences of action. In between sensation and action is the process of *mental reasoning*. What modern research is doing now is detailing the nature of the sensual
inputs to the mind, the mechanisms of mental reasoning, and feedback control of
the mind over bodily action.

An information input–output model of a mind that interacts with an external world
can be depicted as *sensing* (receiving experiential inputs through the senses) and
reasoning (processing sensory inputs into mental outputs that describe and/or initiate action). For example, suppose that an automobile were in the field of vision of
a person and reflected light to the eye of the person, then the automobile could become an external object in the awareness of the person's mind. This would begin
by the person's eye receiving portions of the reflected light, with the iris of the eye
controlling the intensity of the light and the lens of the eye focusing an image of
the auto onto the retina. The working of the eye constitutes the sensory mechanism
of the sensing process. Next, as the light triggers electrical discharge in patterns of
retinal cells, the optical nerve encodes these patterns as information that is transmitted to the brain by the optical nerve. This encoding and transmission of information by the optical nerve provides the sensory processing of the visual sensing
portion of the mind. To perceive an external world, the mind needs sensory mechanisms that physically interact with the external world and neural mechanisms that
encode information and transmit it to the brain.

Next, in the visual region of the brain, information processing begins to detect
and assemble the features of objects. Visual processing begins detecting edges and
textures of shapes of objects and assembling these into a scene of objects, and this
provides a visual object pattern that next requires interpretation as to its meaning.
From prior experience with automobiles, the person's mind can then recognize the
visual scene as an automobile face-on and decreasing in distance. The object-mean-
ing part of reasoning requires recognition, despite differing perspectives and distances of the perception. Finally, the mind must process the relevance of the visual
scene to the self. If the face-on auto perspective with decreasing distance means that
the auto might hit you, further prior experience then tells you to leap quickly out of
the way or risk being injured or killed. Attention of the mind to the external world
results in interpreting the meaning of the world to the self and determining action
by the self to be taken in the external world. This is the reasoning side of the mental activity.

REASONING

There are three general steps for reason to convert sensory information into a mental output—a representation of an object:

1. A necessary step of *object processing,* which transforms sensory information into a concept of an object
2. A further necessary step of determining *object meaning* by interpreting the relevance of the mental object to the purposes of the mind
3. A necessary step of *decision/action,* in which a decision to act initiates physical action in response to the meaning of the perceived object to the mind.

Historically, all modern theories of reason can be traced back to the German philosopher Immanuel Kant. In 1800, Kant proposed that to create mental conceptions of objects in experience, the mind as pure reason must have two kinds of *prior formatting capability.* By *prior formatting capability,* Kant meant that these capabilities must be built mechanically into the mind for the mind to process any experience; and he used the term *transcendental* for this prior capability. As we saw in the brain, people must be born with mechanically operating eyes, optical nerves, and optical processing portions of the brain for them to experience any sight of objects in the world. Kant was interested in these prior formatting capabilities for the brain to function as a mind. He identified the prior formatting capabilities necessary for formatting data input and data processing, which he called *transcendental aesthetics* and *transcendental logic,* respectively.

The transcendental aesthetic of the mind is the capability of formatting the sense perceptions prior to any sensory experience of the world. The transcendental logic of the mind is the capability of formatting synthesis of perceptions into mental representations of objects as the source of sensory experience. Today we call these prior capabilities of the mind to perceive and represent the world *mechanisms* of the brain. These mechanisms must provide two general capabilities for the brain to function: (1) the capability of formatting sensory perceptions, the transcendental aesthetic, and (2) the capability of formatting mental reconstructions of the sensory perceptions as objects (things outside the mind), the transcendental logic.

Perception begins with sensory input as physical data into the brain, stimulated by an object outside the mind. Then *reason* formulates perceptions into recognizable and relevant objects. The human brain does provide the sensory inputs of vision, hearing, smell, taste, touch, and balance. Next, perceptions must be recognized and formed into objects. A Kantian model of the mind distinguishes perception and reasoning. Kant's model was that perceptions are structured by the transcendental aesthetics of the mind as *space and time.* We recall that space and time do provide the basic reference frame for any scientific description of physical phenomena. For Kant, human senses discriminate (format) sensory experience in the general forms of space and time. The formats of the physiological sensory processing to the mind are the transcendental aesthetics of the mind.

Reasoning determines what the mind does with sensory inputs, or perceptions. Mind assembles sensory data into conceptions, representations of objects—pictures, images, representations, ideas of things existing outside the mind, outside the self—external in the real world. The physical source of these sensory inputs to the mind, Kant called an external *object*. An external world filled with objects is the world about which the mind constructs mental images and concepts. This external world is represented as material objects existing in the transcendental aesthetical forms of space and time. Conversely, the guarantor to the mind that a perceived object is real and external to the mind is that it can be perceived in the forms of space and time. To assemble sensory information into a representation of an object in the world external to the mind, the mind must also have a prior logical processing capability for distinguishing any object as to *quality, quantity, relation,* and *modality.*

The *quality* of an object allows the mind to distinguish one external object from another. For example, there are several differences in the physical quality of electrons and protons. They differ in electrical charge (negative versus positive) and spin ($\frac{1}{2}$ versus 1). They differ in size (the electron has only a fraction of the mass of a proton). They differ in force interactions (the electron interacts with electromagnetic forces, weak nuclear forces, and gravitational forces, whereas the proton interacts with electromagnetic, weak and strong nuclear forces, and gravitational forces). The *quantity* of external objects is the number of objects that are qualitatively similar. What is interesting about this prior category of logic for Kant is that it indicates that the basic element of mathematics, counting, is a prior capability of the mind. Thus, as mathematics is a science, it is a science of the mind rather than one of physical nature (e.g., chemistry or physics). *Relationship* in Kant's transcendental logic is the category of judging representations of objects with respect to reality. Judgments on the existence of objects could be positive (existent) or negative (nonexistent), depending on evidence that the object really did or did not exist. The last prior logic category was *modality,* and by this, Kant meant that the mind also judges how objects exist in time to relation to one another, as, for example, in causality (the existence of an object in time could later cause the existence of another object).

In summary, Kant's transcendental logic asserts that the mind must have a prior capability of assembling processed sensory inputs into a mental concept (that can represent an object external to the mind). Kant argued that the mental concept was a representation or an image of something actually existing outside the mind. This thing-outside-the-mind that supplied the original sensory stimulation to the mind Kant called a "thing-in-itself."

Kant's model of the mind with transcendental aesthetics, logics, and things-in-itself (external objects) does really have a modern basis in the mechanisms of the brain. For example, scientists know that people whose brains are damaged or people born with brains that are mechanistically incomplete do have problems with the ordinary perceptive and cognitive functions performed easily by normal people. Brain areas are localized in different specializations for processing speech, vision, balance, feelings, and so on. Damage to local areas of the brain have specific impacts on the mental ability of the victim. Damage in the speech areas of the brain make speech difficult or impossible. Damage in vision areas of the brain cause blindness. Damage in the connection between the right and left halves of the brain causes the inability of information transmitted through one eye to be received in the opposite acting part of the brain. We know that the right brain lobe controls sensory and motor functions on the left side of the body, while the left brain lobe controls right-side body sensory and motor functions.

The modern computer does follow a scheme similar to Kant's model of mind. The basic idea of a computer is that sense data are fed into the computer's "mind" and that data are processed (reasoned) into an information output from the computer by a stored computer program. To put data into a computer, one must first specify the format in which the data are entered. This data format is a prior condition to the computer receiving input and provides a transcendental aesthetic for the computer's mind. Now once the data are formatted in this transcendental aesthetic, the stored program within the computer computes on those data. This stored program specifying the computing is the transcendental logic of the computer's mind. Finally, the information output of the computer is displayed in a form recognizable by a human as visual displays and/or sound, and this display is the computer's mental representation of an object, a thing-in-itself.

MODERN REASON

Now with the idea of *mind* added into the paradigm of *function,* we can examine the idea of *reason* as a mental operation. In a modern view of reason, we now recognize that there are two general kinds of reason processing in human minds: spatial and linguistic. Sensory information from external objects assembled as mental concepts can be assembled in either spatial or linguistic forms. Spatial forms can be reasoned about in their spatial relationships. Linguistic forms can be reasoned about through language processes.

Spatial Reasoning

We noted earlier that space and time are the representational framework of the paradigm of mechanism. Object processing in the brain allows the mind to create mental descriptions of things existing outside the mind as material objects in space and time. All of the modern physical and life sciences in the disciplines of physics, chemistry, and molecular biology describe existing material objects in the external world of nature of spatial and temporal description. The kinds of material objects so existing are particles, atoms, molecules, and macro structures from molecules, such as rocks and organisms and stars. The guarantor to the mind that a perceived material object is real and external to the mind is that it can be perceived interacting with other material objects in space and time.

Spatial reasoning is conceptual manipulation of geometric and temporal forms. Spatial reasoning is essential to physical design in engineering and architecture. Spatial reasoning is essential to the visual arts, presenting emotional content through the manipulation of geometric forms.

Linguistic Reasoning

In the human mind, there is in addition to spatial reasoning, linguistic reasoning. One unique sensing capability of the human brain has been the evolution of an elaborate language capability for signaling and communicating information between humans. All social animals, those who graze or hunt in herds or packs, have some kind of signaling capability—for signaling danger or coordinating attack or marking territory. Yet the signaling capability of humans through voice and facial expressions and gestures far exceeds in sophistication that of any animals. It has resulted in an internal form of consciousness wherein mental concepts are expressed as words. Humans are the only form of animals where internal dialogues in the form of words are continually being held. The capability of holding internal and external dialogues between self and others in the forms of words is distinctly human.

Language is the basic tool and form of thinking. We think in words as well as in visual images. The thinking in words involves either an internal dialogue with self or an external dialogue with other people. Thinking with other people or computers is usually called *communication* or *information*. What linguistic reasoning does in object processing is (1) to focus sensory perception into a linguistic object, (2) to abstract and generalize experience as linguistic objects from one specific external event and context to general events and contexts, and (3) to formulate the conceptual basis for resultant social cooperation and/or conflict.

Broca's area is a frontal lobe structure in the brain that has been found to be critical in linguistic reasoning. As one of the examples of studies that have found this correlation between brain region and mental function of linguistic processing, consider a study reported in 1997 about the differences in brain region in people who acquired second languages differently. For example, a neuroscientist, Joy Hirsh (of Memorial Sloan–Kettering Cancer Center in New York), and her colleagues there compared linguistic processing in children who had grown up bilingual to adults who had acquired a second language later (Bower, 1997). Using a scientific instrument called functional magnetic resonance imaging (FMRI), they were able to see which part of Broca's area was being used by a subject in processing a foreign language. They found that bilingual children growing up with the two languages process either language in the same brain area. In contrast, adults who had later acquired their second language process the two different languages in different segments of Broca's area.

The fundamental units of language are words, sentences, and message. *Words* are the sounds that correspond to the meaning of unit objects of perception (such as material things, material events, persons, actions, etc.). The set of words in a language constitute the *dictionary* of the language. *Sentences* are the sequences of sounds that correspond to the meaning asserted in combinations of words. *Messages* are sequences of sentences that communicate a unit of thematic meaning between speakers. Using these fundamental units, the meaning conveyed in any language is structured in sentential, grammatical, and logical forms.

Sentences are structures as sets of words grouped into subject and predication. The *subject* of a sentence names the object that is the attention of the sentence.

Subjects can be things, people, events, or actions. *Predication* is the information asserted about the subject. Predication consists of verbs and prepositional objects. Verbs relate the prepositional object to the subject. Verbs can denote existence, properties, belonging, events, or actions about the subject. Prepositional objects complete the description of the verb relationship to the subject.

Grammatical structure orders the sentence in precedence of subject and predication. Grammatical structure also categorizes the function of the sentence as exposition, inquiry, or command. Thus, sentences are grammatically punctuated as a declarative, imperative, or question. Grammar also classifies verbs in temporal phase as present, past, or future. (Grammar may or may not also classify nouns by gender.)

In addition to grammar, a logic of discourse provides sentences with classification in the level of generality, particularity in discourse. The logical structure of language provides for a level of concrete description, a higher level of abstraction of concrete description, to an even higher logical level of general description. It also provides the logical capability of moving up and down these levels by *inference*—deductive inference going from general statements to particular statements or inductive inference going from particular statements to general statements. Logical structure also allows thematic development of the subject of the linguistic discourse through recurrence and development of theme. (We next elaborate on *language* as a paradigm of the *soft side of information technology*.)

> **The paradigm of biological function includes complex ideas of action, behavior, will, instinct, and reason.**

SUMMARY AND PRACTICAL IDEAS

In case studies on

- Reductionist methodology in biology
- First evidence of mechanisms of cancer
- Genome duplications
- Fire ant society
- Mechanisms of the human brain

we have examined the following key ideas:

- Scale and systems of biological mechanisms
- Action
- Behavior
- Evolution theory
- Will
- Mind
- Sensing
- Reasoning

These theoretical ideas can be used practically in understanding the concept of function as a paradigm in biological representation of animate matter. We see that both paradigms of mechanism and function are fundamental and complex in biology. Technology strategy for biotechnology needs to elucidate both the mechanics and functional value of biological behavior.

FOR REFLECTION

Read accounts of the discovery of the DNA molecule and of the later human genome project. How did science in biology evolve into big science, and what does that portend for R&D strategy in the twenty-first century?

23

INFORMATION TECHNOLOGY PARADIGM

INTRODUCTION

We have seen the scientific paradigms underlying physical and biological technologies are *mechanism* and *function*. And we have seen how the paradigm of function leads to the ideas of reasoning and language. Next, we examine the underlying scientific paradigms of information technology: *language* and *logic*.

> ## CASE STUDY:
>
> ### Early Programming Languages
>
> The case of the early history of computer programming provides a clear example of how language is a paradigm idea underlying information technology. After the commercial innovation of the electronic stored-program computer in the 1950s, programming on the machines required writing the programs directly in computer language, in which sentences are expressed in the binary mathematics of one and zeros. For example, Lohr (2001) commented about those early days before the first high-level programming language, Fortran, was invented: "Professional programmers at the time worked in binary, the natural vernacular of the machine then and now—strings of 1's and 0's. Before Fortran, putting a human problem—typically an engineering or scientific calculation—on a computer was an arduous and arcane task. It could take weeks and required special skills. Like high priests in a primitive society, only a small group of people knew how to speak to the machine" (p. 20).
>
> With our human minds, we had to talk to those computers with their machine minds—but in a language that computers understood. The idea of a higher-level language in which to write programs was a good idea, a language closer to the human mind than to the machine mind. The computer could compile the higher-level language program into a binary language program. In the early 1950s, John

Backus, an IBM employee, suggested a research project to improve computer programming capability: "In late 1953, Mr. Backus sent a brief letter to his boss, asking that he be allowed to search for a 'better way' of programming. He got the nod and thus began the research project that would eventually produce Fortran" (Lohr, 2001, p. 20).

Backus put together a team to create a language in which to program calculations on a computer, using a syntax more easily learnable and humanly transparent than was binary programming: "They worked together in one open room, their desks side by side. They often worked at night because it was the only way they could get valuable time on the IBM 704 computer, in another room, to test and debug their code. . . . The group devised a programming language that resembled a combination of pidgin English and algebra. The result was a computing nomenclature that was similar to the algebraic formulas that scientists and engineers used in their work" (Lohr, 2001, p. 20).

This was the essential point of Fortran—that it look to the human mind of the scientist or engineer sort of like a familiar algebraic computation. Learning to program in Fortran could use a familiar linguistic process that scientists or engineers had learned as students, how to formulate and solve algebraic equations. For example, writing a program in Fortran first required a scientific programmer to define the variables that would appear as operands in the program: scalar variable (A), vector variables (A_i), or matrix variables (A_{ij}). Then an algebraic operation could be written that looks like an equation (e.g., $A_i = B_i + C_i$). Any algebraic operation could be repeated automatically by embedding it in repetition loop (called a "DO" loop in Fortran, e.g., For $i = 1$ to 10, Do $A_i = B_i + C_i$). One can see that the successful trick of developing computer programming language lay in making it more familiar to humans.

Fortran was the first high-level language developed specifically to talk to computers. Fortran was released in 1957 as a syntactical computer language to write formulas for performing calculations. Next, Algol was invented in 1958, as was Flow-matic. In 1959, a syntactic language for linguistic inferences, LISP, was invented to assist writing artificial intelligence programs. Pascal was invented in 1972 to improve the structuring of programs. The first object-oriented program language (to further improve structuring of programs) was Smalltalk in 1980. In 1985, C++ was a modification to C to provide a widely used object-oriented programming language. Visual Basic was invented to add ease of graphics in basic programming. Java was invented in 1995 for performing operations independent of computer platform (and transmittable across the Internet).

All programming languages are specialized in linguistic syntax to facilitate communication between the human mind and the computer mind.

COMPUTER LANGUAGES

We can look again at the model of science and engineering interacting, but now we add computers into the communities of scientists and engineers and businesses, as

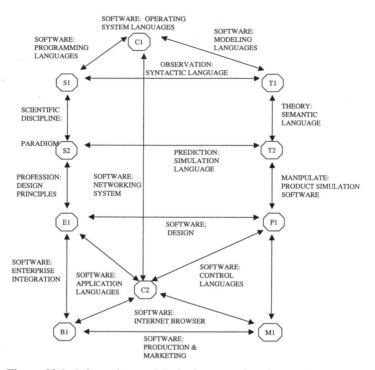

Figure 23.1. Information model of science, engineering, and computers.

shown in Figure 23.1. First look at the lines of the scientists to computers (the connection of S_1 to C_1), and there one can indicate the importance of software for programming, *programming languages,* for humans (such as a scientist) to communicate with a computer. In the connection within a computer (C_1), an operating system software, operating system language, is necessary for the internal working of the computer's mind. In the connection of one computer to another $(C_1$ to $C_2)$, a networking system software, networking system language, is necessary. Both professions (E_1) and businesses (B_1) use software or application languages. The connection of business-to-market networks $(B_1$ to $M_1)$, requires another kind of software, a search capability, as for example a browser language. Thus, several kinds of software languages are necessary for computer communications: programming, operating, networking, and browser languages in specialized software.

TYPES OF SPECIALIZED LANGUAGES

Since language is a scientific paradigm underlying information technology, it will be useful to classify the different types of possible languages. This can be used to identify the linguistic focus of any software. Traditionally, logicians have distinguished between studies of language focused upon the syntactical, semantical, or pragmatical aspects of discourse: "In syntax one is interested exclusively in the signs or expression of the language. . . . In semantics . . . one is concerned . . . with the objects which the signs denote. . . . Finally, in pragmatics, there is reference . . . also to the speakers or users of the language." (Martin, 1959, p. XI) Some logicians further suggested that the scientific community has developed formalized languages around these different aspects: (1) mathematics as a *syntactically specialized language* for quantitative expression, (2) physical theory as a *semantically specialized language* for expression of physical nature, and (3) *pragmatically specialized language* for description of social nature (e.g. economic theory). After the invention of the electronic computer, we can now add a fourth as languages specialized for process control, a *control specialized language.*

A syntactic specialized language is a formal language focused on aiding the form of expression, such as mathematics provides scientists a syntactic language to express the quantitative form of observations. For example, we recall that quantitative expressions in physics (such as $F = ma$) were expressed by different mathematical forms, syntax—force F and acceleration a are in the syntactical form of a mathematical vector and mass m is in the syntactical form of a mathematical scalar.

A semantic specialized language is a formal language focused on aiding the elucidation of a field of experience. For example, we recall that physics theory carefully defines a set of terms and relations to express the commonality of physical experience, such as force, mass, motion, momentum, and energy.

A pragmatically specialized language is a formal language focused on aiding an area of purposeful activity. For example, a word-processing software program, is constructed as a special application language for the application of writing. A control specialized language is a formal language that focuses on aiding the control of a process or system. For example, a computer operating software program is constructed as a special control language for controlling computational operations in a computer.

These four categories can be viewed as a taxonomy defined by two sets of distinctions: *content* and *form,* and *experience* and *value.*

Content and Experience: Semantic Specialized Language

Semantic specialized languages enable sophistication of communication about things in an experiential field of existence, such as things of physical, chemical, or biological phenomena. In Figure 23.1, look vertically at the connection between things of nature (T_1 and T_2) as expressed in scientific theories of nature, and such theory is expressed as a semantically specialized language. The manipulative connection between things of nature and products (T_2 and P_1) can be described in semantically specialized languages of technology. The value connection between market and products (M_1 to P_1) can be described in semantically specialized languages of utility.

Form and Experience: Syntactic Specialized Languages

Syntactic specialized languages enable a sophistication of communication about the forms of communication, as mathematics provides forms for expressing the quantitative aspects of phenomenon. Look again at the information model; the observation connection between scientist and natural things (S_1 to T_1) is facilitated by syntactic languages such as mathematics, which enables expression of the quantitative aspects of things. The programming connection between scientist and computer (S_1 to C_1) is facilitated by syntactic programming languages, such as Fortran, Basic, and C++. The industrial standards connection between engineering and computer (E_1 to C_2) is facilitated by syntactic standards in computerized communications. The search connection between business and computer (B_1 to C_2) is facilitated by syntactic browser languages.

Content and Value: Pragmatic Specialized Language

Pragmatic specialized languages enable sophistication of communication about valuable and purposeful tasks, such as the task of writing as aided by an application software word-processing program. The disciplinary connection between scientist and scientist (S_1 to S_2) or professional connection between scientist and engineer (S_2 to E_1) can be facilitated in communications that build disciplinary and professional knowledge, such as professional conferences, archival scholarly journals, and textbooks. These activities are now being facilitated by application-specialized software that facilitates conferences or online journals and distance education. The enterprise connection between engineering and business (E_1 and B_1) is being facilitated by enterprise integration software and/or intranet software that utilizes application specialized languages.

Form and Value: Control Specialized Languages

Control specialized languages enable sophistication of communication about the control of valuable and purposeful systems of activities, such as modeling and simulation programs. The prediction connections between science and nature (S_2 to T_2) and models that enable a quantitative prediction function are constructed in control specialized languages. So is the design connection between engineering and products (E_1 to P_1) aided by design software that is constructed as a kind of control specialized language. The business-to-market (B_1 to M_1) sales connection is aided by production process control software to produce products and by customer relations software to assist in the sales of product, both of which are constructed as control specialized languages. In the vertical connections between computers (C_1 to C_2), the operating system and network system connections are facilitated by software written as control languages.

CASE STUDY:

XML and Ontology

As the Internet evolved, a computer programming language for Internet Web pages was developed, *hypertext markup language* (html), which became the standard language for communicating text on the Internet. However, the communication of data as quantitative variables required extending markup language to be able to handle quantitative variables in formatted forms. This was called an *extended markup language* (XML). For example, in XML, variable data would be written in the form ($\langle$variable$\rangle$47$\langle$variable$\rangle$), where the term within the angle brackets is the name of the variable and the number between the bracketed terms is the quantitative value of the variable. The form ($\langle$variable$\rangle \cdots \langle$variable$\rangle$) provides a format for inserting quantitative data as different values of the variable.

To communicate the form and meaning of variables between two computers hooked to the Internet, both computers must have access to the form and meaning of the variable. This requires a shared dictionary of the variables in XML, or a common registry. Since there cannot be a completely universal dictionary for all possible variables in the world, registries must be constructed and maintained by a community of computer users who wish share information using a particular variable. Such a registry becomes a kind of shared *ontology* for that community of XML users. The term *ontology* is an old philosophical term that used to mean a science of being or existence. But for computer scientists it began to be used to mean the shared content, or semantics, of a sharing community. For example, as Kim et al. (1995) commented: A shared understanding about a community—information that its members possess—is always applied in solving problems in that community. The terminology used by a community's members can be codified as a community's [ontology]. Ontologies, as explicit representations of shared understanding, can be used to codify [the community's] . . . semantics" (1995, p. 52).

Kim goes on to explain that a large community's use of XML will require that shared registries of an ontology of variables must be developed: "For example, it must be assumed in using XML that the author and reader of $\langle$foo$\rangle$7$\langle$foo$\rangle$ have the same understanding of what $\langle$foo$\rangle$ means. . . . Ontologies can be adopted in situations where the capability to represent semantics is important . . ." (Kim et al., 1995, p. 52).

What is interesting about this case of the emergence of XML as a standard for communicating formatted data on the Internet is the recognition by computer scientists that a community's semantics (registries of ontologies) would be required for XML to be widely used by a large community.

LANGUAGE AS THE TECHNOLOGY OF CULTURE

As we see that language plays a major role in software, we should next review the general idea of language. A convenient and useful modern point of departure in reviewing the topic of language is again that of a philosopher, John Dewey. Dewey thought about the philosophy of language from the scientific perspective of biology, emphasizing that language was a kind of biological technology—a human tool for thinking and communication. In the 1930s, Dewey wrote that language and logic should be seen as a fundamental aspect of the social nature of the human species: "Man is naturally a being that lives in association with others in communities possessing language and transmitting culture. Language in its widest sense . . . is the medium in which culture exists and through which it is transmitted. . . . Language is the record that perpetuates occurrences and rends them amenable to public consideration" (Dewey, 1938, p. 20).

By the term *occurrences,* Dewey meant human experience. Experience is the stuff of life, or *reality* as *existence.* Language records experience and perpetuates the meaning of experience in public forums of humanity. In Dewey's perspective, human culture is as much the "nature" of humanity as are human biological mechanisms (such as DNA). We recall that earlier we saw evolution as being of two kinds: biological and cultural. Human language is the necessary tool for cultural evolution, and cultural evolution has been the big change in the human species for the last 100,000 years.

Language is a biological technology for human culture.

What we can learn from the pragmatic philosophical school is that the concept of *language* provides the basic technology for the cultural *level of human biology.* For example, Dewey argued one should view the evolution of speech in humanity as the evolution of a "natural" kind of communal technology: "The importance of language as the necessary . . . condition of [human] existence and transmission of non-purely organic activities [culture] . . . is that language compels one individual to take the standpoint of other individuals and to see and inquire from a common standpoint . . ." (Dewey, 1938, p. 46).

This is the essential feature of human culture—being able to view the world in a shared perspective of a communicating group—*a common standpoint.* Language enables individuals participating in a linguistic group to share perception, understanding, inquiry, decision, and cooperative action. We could see the importance of this idea of a common standpoint in the important role that computer standards (such for as XML) play in computer technology. Thus, as computer minds were included within modern human culture, languages specialized for computation became a part of the linguistic culture of modern civilization. A technology such as the Internet alters the global capability of creating *common standpoints* around the entire world.

CASE STUDY:

Von Neumann Architecture as Inference

Having looked at language as a paradigm idea underlying information technology, we next look at how logic provides a second paradigm idea underling information technology. Let us briefly recall the computer architecture of the first stored-program computer, which ran computation as cycles of logical steps:

1. Initiate the program.
2. Fetch the first instruction from main memory to the program register.
3. Read the instruction and set the appropriate control signals for the various internal units of the computer to execute the instruction.
4. Fetch the data to be operated upon by the instruction from main memory to the data register.

5. Execute the first instruction upon the data and store the results in a storage register.

6. Fetch the second instruction from the main memory the program register.

7. Read the instruction and set the appropriate control signals for the various internal units of the computer to execute the instruction.

8. Execute the second instruction upon the recently processed data whose result is in the storage register and store the new result in the storage register.

9. Proceed to fetch, read, set, and execute the sequence of program instructions, storing the most recent result in the storage register until the complete program has been executed.

10. Transfer the final calculated result from the storage register to the main memory and/or to the output of the computer.

The von Neumann computer architecture is a *logic for conducting calculations,* the logic of a Turing machine. Any mathematical calculation can be expressed as a sequence of ordered algorithmic steps that transformed initial data into calculated results. In this case we see that logic is a structuring idea, a paradigmatic idea for computers.

LOGIC

In this case of von Neumann architecture, we can see that writing an operating system program (such as MSdos or Windows), the architecture provides a logic for structuring the computational operations of the program. The operating program is written as a language (with terms such as *file* and *run* and relations such as *save* and *recall*). But the architectural logic provides the form of calculation—operation. In general, any kind of language has both a grammar and a form of calculation or inference. The relationship of logic to language is to provide grammar and inferential forms for expression and reasoning (calculation) in the language.

We recall that the *grammar* of a language consists of terms and relations and sentence order. Terms are classified as nouns or adjectives. Relations are classified as verbs, adverbs, propositions, and connectives. Sentence order specifies the order of terms and relations into a meaningful predication of a noun. Sentences also have different modalities: declarative, inquisitive, and imperative.

As the *grammar* of a language, a logic provides a language its form of sentence structure.

We recall that reasoning is an operation of the mind that both constructs and relates mental objects, concepts, and that linguistic reasoning operates with language. Language not only provides modes of expressing experience in sentences but also connections between experiences as *inference*. Inference is the form of reasoning about things as linguistic objects. Linguistically, inference is defined as deriving a conclusion from prior premises.

Inference is a proper form for making valid arguments from premises.

Computer programming language uses the logical process of calculational inference to conduct calculations in the computer.

As a *grammar,* logic provides both proper *forms of sentences* and *inference* in a language.

In providing both grammatical and inferential forms, one can see that logic can be viewed as a kind of language about language, a metalanguage.

As a metalanguage, logic provides a language its proper forms of grammar and inference.

CASE STUDY:

Teaching Logic in the Modern University

Let us look next at the kinds of inferences that modern logic provides to modern languages, both common and specialized languages. We can see the range of inferential logics by looking at the case of teaching logic in the modern university. Logic is taught in several places in a university and in several forms—in the academic departments of languages, philosophy, mathematics, computer science, physical sciences, management, and engineering.

Logic in Ordinary Languages

In departments of languages (such as English, French, Chinese, German, Spanish, etc.), logic is taught as the grammatical structure of the language. We recall that in the grammar of many modern languages, terms are classified as nouns, verbs, adjectives, adverbs, and connectives. Verbs indicate the condition about a noun, and adjectives modify nouns and adverbs modify verbs. Nouns may be classified by gender, and verbs are classified by tense. Sentences are structured as subjects and predication about subjects. Sentences may also include propositional clauses and be connected into complex sentences. Sentences might also be classified as declarative, inquisitive, or imperative. In written languages, further organization of long expositions may occur in paragraphs, subheadings, or chapters. Also in written languages, expositions may have informal structures and plots, such as introduction, episodes, climax, and denouement.

In ordinary languages, a linguistic logic provides the formats for grammar and argument.

Logic in Philosophy

In philosophy, logic is taught not as a grammar but as a form of inference or argument. Inferences may proceed either from particular statements to general statements, or vice versa. For example, in the classical dialogues of Plato about his teacher, Socrates, Plato emphasized logic's importance in probing the assumptions of an intellectual position. This is called the *Socratic method* of questioning a person about assumptions. The direction of the inference lies in going from conclusions to the assumptions of the argument—inductive inference.

Conversely, logic is also taught as a deductive direction of inference from the general to the specific. An example is Aristotle's deductive form of inference, the famous *syllogism,* such as: All humans are mortal, and Socrates is a human, and therefore Socrates is mortal.

In modern terminology, the logic of this argument is that of *set theory,* which states that if a particular object (Socrates) can be identified as belonging to a set of things (human), the deductive inference can be made that all members of the set (humans) share all properties (mortality) of the set. In modern philosophy departments, this set theory is called *propositional calculus.* For example, a compound statement $C = A$ and B (C is composed as the conjunction of statements A and B) is true if and only if statements A and B are both true. As another example, the nonexclusive conjunctive "or" can form the complex statement $C = A$ or B, so that C is true if either A or B is true or both are true. This propositional calculus is a expansion of Aristotle's syllogism, as that form of syllogism can be written in propositional calculus as the complex sentence: If a belongs to set $A,$ all the properties of set A accrue to any member a of the set.

In philosophy, logic is focused on the correct forms of propositional inference.

Logic in Mathematics

Set theory as taught in modern mathematics departments is similar in ideas to the propositional calculus of philosophy departments, but more complete. In mathematics, set theory provides a modern foundation (or metalanguage) for modern mathematical topics. Early in the twentieth century, the mathematicians Alfred North Whitehead and Bertrand Russell provided a modern foundation of mathematics using set theory as a metalanguage for structuring mathematical languages. Afterward, all modern mathematical topics are taught with the basic terms of each topic so defined in the mathematical metatopic of set theory. For example, in the mathematical topic of probability, the basic idea of probability is defined as the ratio of the cardinality (number) of a subset to the cardinality (number) of its inclusive set. This is, if a set S can be partitioned into subsets, the probability of any subset S_i is the ratio S_i/S.

The practical result of this is that one can use any mathematical topic together with any other mathematical topics and be assured that no logical contradictions will occur in their use together.

> **A formal metalanguage assures that all languages structured grammatically by the same metalanguage will be logically consistent from language to language in their grammatical structure.**

Thus, in mathematics, logic is taught as a metalanguage providing the grammar and inference for the quantitative languages of mathematics (e.g., math topics). The mathematical topics of algebra, vectors and matrices, analytical geometry, calculus, and differential equations are all mathematical languages useful to the scientific disciplines, such as physics, as quantitative deductive languages. The mathematical topic of probability and statistics is useful to the scientific disciplines, such as physics, as a quantitative inductive language.

> **In mathematics, logic is a metalanguage for mathematical topics providing logical self-consistency in defining terms among all math topics.**

Logic in Computer Science

In the computer science departments of the late twentieth century, logic was taught in the form of Boolean algebra, which provides a true–false look-up table for conjoined sentences by the set conjunctions "and," "or," "exclusive or," and "negation." However, other kinds of logic play implicit and very important roles in computer science, such as (1) standards and (2) the grammatical structures of computational languages (e.g., Fortran, Basic, Pascal, C, C++). There structures differ according to features such as graphics and object orientation.

In computer science, standards operate as types of metalanguages (structuring logics) for compatibility of a computer science language across applications and vendors of applications.

> **For any language, logic provides the proper forms of the language's grammar, inferences, and standards of communication.**

Logic in the Physical Sciences

In departments of the physical sciences (e.g., physics, chemistry, biology), logic is taught a methodology for observing nature. Nature should be observed with natural observations, measurements, and experiments. Abstractions and generalizations from these observations are used in the construction of theory about the natural phenomena observed. The constructed theory uses the intellectual paradigm of mechanism to describe and explain the physical phenomena.

The logic of a scientific discipline provides the forms for proper methods of observation and paradigms of explanation. These methods guide perception and analysis in the scientific languages of the discipline.

> **In addition to proper forms of grammar, inferences, and standard, logic also provides the proper forms of perception and analysis in the language.**

Logic in Management

In management, logic is taught as a form of reasoning about action, planning, and decisions. Planning is thinking out future action, making concrete what actions one needs to perform in the present to bring about a future state that one desires. Action is the basis of all productive organizations: commercial, governmental, military, or other. We recall that the logic of thinking about action is in the conceptual dichotomy of means and ends. The *means* of an action is the way the action is carried out by an actor, and the *end* of an action is the purpose of the action to the actor. Humans reason and communicate about possible futures as means and ends in order to satisfy both instinctive and learned needs as ends, and to plan action as a means to bring about such futures. The reasoning and communication about future action is called *planning,* and the means–ends description for future action from such reasoning is called a *plan.*

  Action as a *logic* involves the important idea that all action, although planned, will occur in a *present time* of the action. This implies that all preparation for any current action must have preceded it in time. Accordingly, *planning* is *present knowledge* applied to *future action.* An important implication of this is that present knowledge must always be incomplete and not perfectly accurate about the future conditions and environments, in which action will occur. From this arises the basic idea of *risk* in planning. In any present time, complete knowledge of any future time is never possible. Consequently, planning will always provide incomplete knowledge for future action and risk is inherent in any plan. Risk is inherent in any action because planning for that action must have be reasoned about with incomplete knowledge of the future. The quality of the logic of planning is measured by its effectiveness in facilitating action to bring about a planned future and reduce the risk in action toward that future. Action implemented according to a plan is the operations of the plan.

> **In addition to proper forms of grammar, inferences, standards, and perception, logic also provides proper forms of planning and operations.**

Logic in Engineering

In schools of engineering, logic is taught as design principles in the different engineering disciplines. All engineered systems can be designed as unit processes and system integration. For example, the logic of any production consists of a production system and generic unit production processes. A production system is an ordered set of activities that transforms material, energy, and information resources into products. In logic in engineering, generic types of technology systems (such as production systems) provide the logical forms for engineering design.

  We recall that in a technology system, there is a physical morphology (mechanisms) and a schematic logic such that the mechanistic states are ordered in a logical sequence to transform states of nature in a functional pattern. We called this a *schematic logic,* the functional order of transformed states of nature. All logic in engineering is in the form of schematic logic, transforming states of nature in a functional pattern.

> **In addition to proper forms of grammar, inferences, standards, perception and analysis, planning and operations, logic also provides design principles and transformation schemes.**

SOFTWARE STRATEGY

From this case we can see that logic is a complex topic and is taught in different ways in different disciplines in academia. Logic plays the important formal role of structuring reasoning in a language. Reasoning in a language requires proper logical forms for structuring:

- Grammar, inference, and standards
- Perception and analysis
- Planning and operations
- Design and transformations

The logic used in a software development depends on how the language in the software is intended to process these different kinds of activities: inference, perception, planning, design, and so on.

Software development can be facilitated by generic strategies that think first about the language and logic basis of the proposed software. We have seen that language for software can be categorized as principally syntactic, semantic, control, or application:

- Software that is linguistically *syntactic* focuses on *forms*.
- Software that is linguistically *semantic* focuses on *content*.
- Software that is linguistically *control* focuses on *processes*.
- Software that is linguistically *application* focuses on *tasks*.

The first strategic decision in developing a new category of software is to decide the focus of the software, as principally on forms, content, processes, or tasks. This focus will set the linguistic domain of the software:

- Syntactic software must be complete in all the forms of communication and internally consistent among the forms.
- Semantic software must be nominally and relationally complete in all the general things and their relationships in the field of existence.
- Control software must be complete in decision criteria for control of a process and temporally coherent in the sensing of the process.
- Application software must be complete in the tasks of an application and properly coordinated in task sequencing.

The second strategic decision in new software design is to identify the kinds of logical structuring necessary for the software; and we have seen that there are several choices about the roles of logic in structuring software language: grammar, inference, and standards; perception and analysis; planning and operations; and design and transformations.

> **All software design requires strategies about the use of appropriate specialized languages and logics.**

In conclusion, in our review of the paradigms underlying technology strategies, we have seen that of the four kinds of technologies: (1) in material technologies, the mechanism paradigm is primary; (2) in energy technologies, mechanism and logic are primary; (3) in biological technologies, mechanism and function are primary; and (4) in information technologies, mechanism, logic, function, and language paradigms are all primary.

SUMMARY AND PRACTICAL IDEAS

In case studies on

- Early programming languages
- XML and ontology
- Von Neumann architecture as inference
- Teaching logic in the modern university

we have examined the following key ideas:

- Computer languages
- Specialized languages
- Logic as meta-language
- Kinds of logic
- Software strategy

These theoretical ideas can be used practically for understanding the nature of software development as an exercise in language and logic construction.

FOR REFLECTION

Examine an instruction manual for a software program, and identify the dictionary of terms and relations in the grammar of the program. What kinds of inferences are assisted by the program?

24

HIGH-TECH STRATEGY

INTRODUCTION

Not all businesses are high tech. As we have emphasized, businesses are only high-tech when technological innovation improves a business's economic value-adding (EVA) capabilities. For a competitive edge, new EVA capability must be seen by the customer in terms of *higher-quality* and/or *lower-priced* products (services). High-tech businesses remain high tech only as long as they can innovate products or production better than a competitor's. Product planning is the core business strategy for maintaining a high-tech business—product planning in terms of generations of superior products: superior in performance, superior in quality, and lower in cost. We will examine effective high-tech product strategy in the key ideas of: *product-line lifetimes, product-model lifetimes, product planning,* and *product families* and *platforms.*

Electronic Products and Tires

Products can have finite commercial lifetimes. New technologies make older products obsolete. We can see this in two cases of finite commercial life of two types of products, tire chords and consumer electronics products.

Product Lifetimes in Tire Cords

The historical setting is the first half of the twentieth century, and Figure 24.1 shows the finite commercial lifetimes for tires made of various cord materials. At first, tires were produced using cotton fibers in the sidewalls of rubber tires to reinforce the tires. Next, the reinforcing cotton fiber was replaced by rayon fiber, making a stronger tire. Then rayon was replaced by nylon fibers to increase tire strength further. Then nylon fibers were replaced by polyester fibers for another

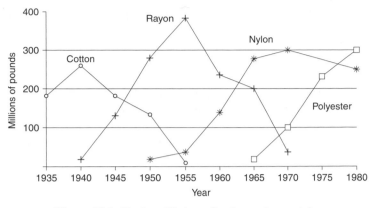

Figure 24.1. Product lifetimes for tire cord materials.

advance in strength. Since each tire cord reinforcement material was stronger than the prior cord material, each generation of tires were stronger.

Consumer Electronic Product Lifetimes

The historical setting was the second half of the twentieth century, and Figure 24.2 shows market volumes of the various types of consumer electronic products as a percent of total value of the consumer electronics industrial sales. One can see that radio, the first consumer electronic product, declined from 100% of the industry's products prior to 1950 toward a small percent by 1985. Of course, some radios are still made and sold, but as a percentage of industrial value, they had become very small. Moreover, as a technology in consumer electronics, the radio function was

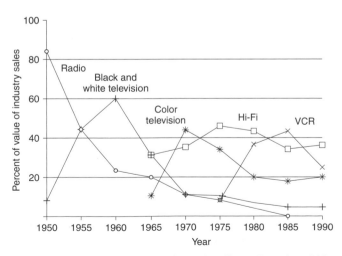

Figure 24.2. Product lifetimes of electronic product lines. (Data from Valery, 1991.)

incorporated within high-fidelity (hi-fi) technology products (e.g., home hi-fi, home entertainment products, car hi-fi, Walkman-like cassette players).

Also, one can see that the technology of black-and-white television was replaced by color TV. Monochrome TV sets rose to a peak at 60% of industrial value in 1960 and declined toward 0% after 1990. Color TV as an industrial technology rose rapidly after 1964 to a peak of industrial value of about 45% in 1971. Thereafter, color TV continued as a mature technology in the industry at about 20% of industrial value by 1990. Color sets were sold to consumers as replacement units or as second or third sets. Hi-fi sales peaked as an industrial technology at about 55% of industrial sales in 1975, steadying to about 40% by 1990. Sales of TV videocassette recorders (VCRs) began in 1975, peaking at just over 40% in 1985.

What we see in this case is that an industry such as consumer electronics may have several generic technology products (radio, monochrome TV, color TV, hi-fi, VCR). Moreover, some of these technologies become obsolete and are replaced by a newer technology (e.g., monochrome TV replaced by color, radio replaced by hi-fi). For technologies that are not replaced, there will be a growth in sales of the products of that technology until their market saturates. Thereafter, sales will continue at a replacement rate and any growth of demographics in customers.

PRODUCT LIFETIMES

Products can be grouped into product models, product models into product families, and product families into product lines. Product lines provide the highest grouping of products. For example, in the U.S. automobile industry (which we studied earlier) different product lines emerged early in its history according to different applications of land transportation: passenger cars, trucks, motorcycles, and tractors. We recall that by the year 2000 in the United States, there were only two domestic auto manufacturers (General Motors and Ford), both of which had passenger car and light truck divisions. The tractor industry in the United States evolved into firms producing only tractors and construction equipment, Caterpillar and International Harvester. Also in the United States in 1990, the large truck firms that had survived were Navistar (formerly a division of International Harvester) and Mack Trucks. In contrast, the Japanese auto industry, which began in the 1930s with Nissan and Toyota, producing trucks for military applications, after World War II began producing passenger cars and light trucks, first for the domestic market and then for the global market. Toyota grew to be a globally dominant firm. Historically no automakers had entered the motorcycle business, but in Japan some motorcycle makers, including Honda and Suzuki, became major producers of passenger cars.

> **A product line is a class of products embodying similar functionality, similar technology, produced by similar production processes for a major market and application segment.**

Product lines can evolve within an industry to serve different broad market niches and different broad applications, which was the case for product lines in the automobile industry. Product lines can also evolve from advancing technology. An example was the computer industry, which organized around mainframes, minicomputers, workstations, personal computers, routers, and servers.

When technical obsolescence occurs in a product line, the product line has a finite lifetime. In the cases of tire chord materials and consumer electronics, we saw that product lines became obsolete by technical performance (technology substitution) or technical integration (product integration). The commercial lifetime of a product line is the time between the introduction of a new product into the marketplace and its withdrawal from the market. Figure 24.3 depicts the general form of a product line lifetime expressed as the market volume of the product line over time. After the new product line is innovated, there is a rapid growth in market volume to a saturation point in the market. This saturation point occurs, as in the concept of the industrial technology life cycle, after product and process innovations in the product line have stopped and the core technologies of the product line are mature. Then the market volume of the product line remains at a level determined by replacement rates and demographics. If a new-technology-based new product line begins substituting for the existing product line, market volume for the older product line begins to decline as it has become technically obsolescent in the marketplace.

> **Product line lifetimes are terminated by technical obsolescence or by technical integration.**

We recall that technological change can be continuous or discontinuous; and we recall that continuous change reinforces industrial structures and discontinuous change creates and alters industrial structures. One form of a discontinuous technological change is the next-generation technology system. When a next generation of a product technology system is innovated, the product line changes, making obsolete the earlier product line and introducing a next-generation product line.

Substituting generations of next-generation product lines will have overlapping curves (as in the tire chord example). The reason for this is that initially, a next-generation product line will be priced higher than the existing generation of product line. The higher price will be required due to the R&D costs of developing the

Figure 24.3. General form of product line lifetime.

product line and the initial capital cost of constructing production facilities and initially higher costs of materials and unit production.

Substitution will begin for applications that demand the higher performance of the new-generation product line that justifies the price premium. After volume production and sales grow and competitive pressures reduce the price premium, the new-generation product line begins more completely to substitute for the existing product line. The existing product line may lower prices to lengthen its product line lifetime or to form a permanent low-priced market niche for low-performance applications.

The preparation of product line lifetime forecasts requires judgments on the extrapolation of the price–performance charts and judgments on the minimum performance required for a class of applications. If for a new generation product line, not all the performance parameters are superior to the older generation product line, one might identify a unique-performance market for continuing the old-generation product line as a market niche.

As we noted, product lines divide into families and into models: product families and product models within a family. Product models are variant designs of a product family in a product line. Product models can be differentiated within a product line according to different kinds of criteria:

- Technical performance
- Technical features
- Dependability
- Price
- Safety
- Fashion

Product lines have finite lifetimes due solely to technical obsolescence, but product models have even shorter lifetimes. Product models can have finite lifetimes due to:

- Technical performance obsolescence
- Technical feature obsolescence
- Maintenance obsolescence
- Cost obsolescence
- Safety
- Fashion changes

Performance or feature obsolescence in a product occurs when its performance and/or features are markedly less at the same price as a competing product. Cost obsolescence occurs in a product when the same performance can be obtained in a competing product at a lower price and the product cannot be produced at a cost to meet that lower price. Safety obsolescence occurs in a product when a competing product offers similar performance and price with improved safety of operation or when government regulations require safer features or operation in a product line. Finally, for products in which technology and costs and safety features are relatively stable, products can still become obsolete due to fashion changes (such as in clothing). Fashion obsolescence occurs in a product in which product competition is not differentiable in performance and price but is differentiable in lifestyle.

CASE STUDY:

First Commercial Computers

It helps in the commercial success of new high-tech products if they are designed for a specific customer and application. Going back to the invention of the computer, we can see this in the case of the first commercial computers products. Historically, there were two electronic computer projects that directly fostered the innovation of commercial computers: (1) a Mauchly and Eckert project, EDVAC, at the University of Pennsylvania and sponsored by the U.S. Army, and (2) a Forrester project, Whirlwind, at the Massachusetts Institute of Technology (MIT) and sponsored by the U.S. Air Force. These both occurred in the middle of the twentieth century when the process of knowledge creation and implementation had *evolved institutionally* into an elaborate societal structure, including (1) research at universities such as at Penn and MIT, (2) research-intensive industrial firms such as IBM, and (3) large amounts of research funds spent by government agencies such as the U.S. Department of Defense.

The UNIVAC Computer

We recall that central to the invention of the first electronic computers were von Neumann, Goedel, Turing, Mauchly, and Eckert (Heppenheimer, 1990). John von Neumann had suggested the idea of the world's first stored-program electronic computer but was influenced by the earlier ideas of Kurt Goedel and Allen Turing. Mauchly and Eckert had invented one of the world's first all-electron-vacuum-tube special-purpose computers. Mauchly had borrowed earlier ideas from John Atanasoff. Mauchly and Eckert were influenced by von Neumann's ideas. Mauchly and Eckert did create the world's first *commercial* stored-program electronic computer product, the UNIVAC.

When Mauchly and Eckert built their first vacuum-tube computer, the ENIAC, a major problem was the need to use of vacuum tubes for active memory in the computer. Today in personal computers, we use integrated circuit chips called DRAMs (dynamic random access memory), with a storage capacity of millions of bits (64M, for example). But for the original inventors, for each memory bit they would have needed two electronic vacuum tubes to store just one memory bit. For example, an electronic sentence in today's personal computers which requires 32 bits would then have required 64 tubes. Electron vacuum tubes were the size of a half a banana, gave off the heat of a light bulb, and burned out frequently. Thus, electron tubes were used for the logic circuits of the first general-purpose computers, but they could be used for the memory circuits.

Eckert and Mauchly invented a memory unit consisting of an electroacoustic delay line in a longitudinal tube filled with mercury. Their idea was to store a string of bits of 1's and 0's morphologically as successive acoustic pulses in a long mercury tube. An electrical transducer at the start end of the tube would pulse acoustic waves into the tube corresponding to a 1 bit, or no pulse corresponding to a 0 bit. As the 1-bit pulses and 0-bit spaces between them traveled the length of the tube, the word they represented would be stored temporarily. When the string of pulses began to reach the end of the mercury tube, the acoustic pulse would be retransduced back into electrical pulses for reading by the computer system, or if it was not to be read would be reinserted electrically back into the front end of the tube as a renewed string of pulses and spaces morphologically expressing the stored word. This storage function could recycle until the word was "read" and a new "word" stored temporarily in the mercury tube.

Then Eckert and Mauchly had a conflict with the University of Pennsylvania over rights to the invention. When the university would not assign them the rights to the invention, they left to build computers on their own. They obtained a $75,000 contract from the Census Bureau to develop a computer and used this contract to start their new company, the Eckert–Mauchly Computer Company. Subsequently, they received a $300,000 contract from the bureau to build the first commercial mainframe computer, which they called the universal automatic computer (UNIVAC). However, even this amount of money was not sufficient for the development costs, and Eckert and Mauchly were forced to solve their financial problems by selling to Remington Rand (a typewriter manufacturer) on February 1, 1950. The mercury storage tube was used in that first UNIVAC built for the Census Bureau, delivered by Remington Rand on March 31, 1951. For permanent storage, the UNIVAC read data and programs from and to magnetic tape and/or punch cards.

Eckert and Mauchly were brilliant engineers but not good managers, for they never really prospered in their business venture. We also recall that because a later dispute over patent rights (in which von Neumann testified against them), they did not gain even valuable intellectual property from any computer patent. Later, Remington Rand sold the computer company to Sperry, which even later was put out of the computer business by the great success of IBM in building a next generation of computers.

The SAGE Computer

The company that really made it big in early computer products was IBM. The story of how IBM got into the computer business goes back to another university-based research project sponsored by a U.S. military agency. While Eckart and Mauchly were building their first commercial computer product, there was another very early computer project at MIT, the Whirlwind project directed by Jay Forrester. Historically, it was as important to the history of commercial computers as was the Eckart–Mauchly UNIVAC project. Forrester's computer made the next breakthrough invention required for real progress in the active memory part of the computer, the ferrite-core main memory. The Whirlwind project enabled IBM's entry into computers and positioned it to become the first mover and dominant player in the new mainframe computer industry.

Jay Forrester graduated in engineering from the University of Nebraska in 1939. He went to MIT as a graduate student in electrical engineering, obtaining a doctorate during World War II and worked in military research at the Servomechanisms Laboratory at MIT. In 1944, Forrester participated in studies for an aircraft analyzer and led the aircraft stability and control analyzer project: "Jay Forrester was described by people in the project as brilliant as well as cool and distant and personally remote in a way that kept him in control without ever diminishing our loyalty and pride in the project. He insisted on finding and hiring the best people according to his own definition, people with originality and genius who were not bound by the traditional approach" (Pugh, 1984, p. 63).

In August 1945, the Servomechanisms Laboratory received a feasibility study contract for the aircraft analyzer from the Naval Office of Research and Invention. Two months later, in October, Forrester attended a conference on advanced computation techniques. He wanted to learn about the ENIAC, which Mauchly and Eckert had built. He was interested in using digital electronic computation in his aircraft stability and control analyzer project. When he saw the digital circuit technologies developed for ENIAC, he thought he could use digital techniques.

With this in mind, Forrester decided to redirect the analyzer project. In January 1946 he went back to the Navy with a new project proposal to design a digital computer and adapt the computer to the aircraft analyzer. The crux of the problem was the main memory subsystem for the computer. The need for response in real time for control in the aircraft simulator made use of both the mercury delay line and the rotating magnetic drum technology, too slow for this application.

Then international events resulted in a reorientation of the project. After the USSR exploded an atomic bomb, the U.S. government decided to build an early warning air defense system. An air defense system engineering committee was created in January 1950 under the chairmanship of George E. Vally of MIT to make technical recommendations for such a system. Vally suggested using the Whirlwind computer for air defense. Forrester's project was redirected to the new objective of air defense.

Still, Forrester had not solved the technical bottleneck in the computer system, active memory, but he did have an idea for a kind of magnetic memory. In April 1949 he saw an advertisement for a new material called Deltamax and thought of using it for a novel three-dimensional magnetic memory array. Deltamax, developed during the World War II by the Germans, was made of 50% nickel and 50% iron rolled very thin. After the war, U.S. naval scientists brought samples of the material, and one of the special machines required to make it, back to the United States. They encouraged a U.S. firm, Arnold Engineering (a subsidiary of Allegheny-Ludlum), to make it. Its important property was its sharp threshold for magnetization reversal when an external magnetic field was applied to it.

Forrester's inventive idea was that he could use the magnetic direction of the metal to store information in binary mathematical form either as a 1 (in one direction of the magnetization in the material) or as a 0 (in the reverse direction of the magnetization in the material). He constructed a rectangular set of magnetic loops as a two-dimensional array for storing data. These loops of small magnetic toroids were constructed of Deltamax tape wrapped around each loop to which were connected two sets of electric wires, to carry signals for writing and reading. One wire could be charged electrically with a "write" signal to magnetize the tape in one direction; the other wire was used to sense a "read" signal, represented by a second magnetization direction. If Forrester's computer wanted to store a 1 data bit in a Deltamax tape loop, the computer would send a write signal to the loop to magnetize it in the proper "1" direction. This magnetic direction would stay stored in the loop until the computer wished to read the data bit, which it could do at any time by sensing the direction of magnetization in the tape through the read wire. If the computer wished to change the data bit stored on that particular loop, all it had to do at any time was to send a new write signal to reverse the direction of magnetization in the loop (which the computer would interpret as a 0 rather than a 1 data bit). Each of Forrester's tiny little ferromagnetic tape loops could store a 1 or 0 as a date bit. Taken together, a lot of these storage loops could provide an active memory for an electronic computer. Assembled in three-dimensional arrays, these tiny toroids would function as a main memory of the first successful mainframe commercial computers. These ferrite-cores memory arrays really made the first computers practical.

When Forrester completed the project, he had built the first computer to use ferrite-core arrays for main memory, with a 16×16 array. We no longer use ferrite-core memories in computers but transistors in integrated-circuit memory chips, storing millions of data bits in a single chip. One can see the vast progress in computer memory knowledge that occurred in the 50 years from 1950 to 2000.

The next stage of the computer project was then at hand. Although the university research project had designed the computers for the U.S. Department of Defence, an industrial firm would be required to manufacture it in volume. This is how and when IBM got into the computer business.

In June 1952, John McPherson of IBM participated in a committee meeting of a professional society and there talked to Norm Taylor of MIT. There Norm Taylor advised that the MIT Digital Computer Laboratory was looking for a commercial concern to manufacture the proposed air defense system. Taylor asked McPherson if IBM was interested, and McPherson responded that IBM would indeed be interested. McPherson returned to IBM headquarters and discussed the project with executives. It was the kind of opportunity that Tom Watson had been looking for in order to rebuild IBM's military products division and to improve electronic technology capabilities. IBM told Forrester of their interest. Forrester and his group were reviewing several companies as potential manufacturers of the Air Force computer. They visited Remington Rand, Raytheon, and IBM, and Forrester chose IBM to build the computers.

With the deal between MIT and IBM concluded, IBM rented office space at a necktie factory in Poughkeepsie, New York and got to work. The Whirlwind II project was renamed SAGE. By the summer of 1953, 203 technical and 26 administrative people were working on the IBM part of the project. The system was to have many digital computers at different sites around the country in continual communication with each other. They were to share data and calculate the paths of all aircraft over the country in order to identify any hostile aircraft. These first computers were to use electronic vacuum-tube logic circuits and ferrite-core memory. IBM would use the design principles from the SAGE project to design and produce their first commercial computer mainframe product line.

At the time of the innovation of the new computer industry, all the pioneering research had been performed at universities: Aiken at Harvard, Mauchly and Eckert at the University of Pennsylvania, and Forrester at MIT. Moreover, their research was funded primarily by the U.S. government: Aiken by the Navy (with IBM assistance), Mauchly and Eckert by the Army, and Forrester by the Air Force.

Next, the transfer of technology into commercial applications occurred through the formation of new firms and through existing firms entering the new industry. Mauchly and Eckert formed a new company, financed on a government contract from the Bureau of Census, and IBM produced the SAGE computer financed by the Air Force.

When IBM built the SAGE computers for the Air Force, it provided IBM with a commercially important technological advantage. IBM innovated production capabilities to build ferrite-core memories, which were the strategic competitive key to the successful early computers.

PRODUCT REALIZATION PROCESS

We recall that in going from research to utility, there are logical steps in the innovation process that connects science to the economic marketplace, beginning with successive steps from science as *fundamental research* to *applied research* to *technology development* and finally, to *commercialization* of a product (Figure 24.4). The cost of product creation increases over time as these various stages are pursued. The stage of technology development is much, much more expensive than prior stages.

In the stage of *fundamental research,* scientific activities develop new knowledge as the discovery, understanding, and manipulation of nature. This is the scientific technology focus of fundamental research. In the case of the first computers, Mauchly and Eckert's invention of the electronic computer can be seen as fundamental research in how to accomplish computation using the technology of electronic vacuum tubes. Mauchly, Eckert, and von Neumann together were constructing the first basic principles of what would become *computer science* as a basis for the new *technology of electronic computers.*

In the next stage of *applied research,* further research activities (both scientific and engineering research) focus on generating additional knowledge about how the functional manipulation of a technology can be improved for a specific need. In the case of the first computers, Forrester's SAGE project used Mauchley and Eckert's earlier invention to develop a military application of computation to track incoming

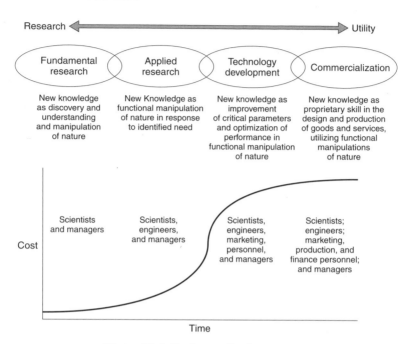

Figure 24.4. Product realization process.

enemy airplanes toward the United States. New knowledge was developed to improve critical computation performance. Forrester invented a new active memory component, the ferrite-core memory, for the computer, significantly improving memory capacity and response time.

In the next stage of *technology development,* more new knowledge is created, focused on optimization of the new technology system. For example, to manufacture the SAGE computers for the U.S. Air Force, IBM had to develop new manufacturing techniques to produce the complex matrices of ferrite-core memories in large numbers. It was IBM's leading large-scale production of computer memory that enabled it to leap out ahead of all competitors.

> **The transition from the applied research stage of product development to the technology development stage is** *a functional prototype* **of the new product that establishes** *industrial standards* **for the product.**

The final stage of the product realization process is the *commercialization* of a new product. The new product is designed and manufactured and marketed. In the case of IBM, IBM needed to supply not only the new electronic computer but also software to run on the computer. Most of IBM's new business was built on being both a hardware and a software producer.

IBM's Next-Generation System/360 Computers

However, even an early start in a new product line does not ensure continuing competitive success. In new technologies there will be new generations of products as the technology continues to progress. In the 1950s, the new transistor would replace the electronic tube in computers. In a strategy meeting in January 1955, IBM executives noted that their customers all wanted machines as fast as possible with as much memory as possible (Strohman, 1990). (Even in early computer applications, the demand for speed and memory was apparent and would remain a dominant performance requirement for new computers.) The issue was how to finance further research for a next generation of high-performance computers. In that meeting the decision was made for IBM to seek government funds to assist in developing a next-generation computer. Next-generation technologies (next-generation knowledges) are expensive and risky. Government funding of the SAGE computers had helped IBM in the design of its first generation of its computers, and the government contract had helped IBM build its first production capabilities for computers. IBM executives saw direct benefit to them in government-sponsored research and would seek to use it for its next generation of computers.

IBM learned that Edward Teller (then the associate director of the University of California Lawrence Radiation Laboratory in Livermore, California) was ready to sponsor a high-performance computer. The Livermore Laboratory was a university-administered, government-sponsored research laboratory for hydrogen bomb research. In 1955, both IBM and Sperry Rand submitted proposals to Teller for the project, but Sperry Rand won the contract.

IBM continued to refine its planned next-generation computer, which they called Stretch, whose logic circuits were to be entirely transistorized. Undaunted, IBM next approached the U.S. National Security Agency (NSA) with its proposal. In January 1956, IBM received a contract for memory development and computer design totaling $1,350,000. Then IBM won a second contract from the Atomic Energy Commission's Los Alamos Laboratory to deliver the Stretch computer to the AEC within 42 months for $4.3 million. As the Stretch project continued, IBM received another contract from the Air Force to prove a transistorized version of its 709 computer.

However, with its radically new transistor technology, the next-generation Stretch project ran into technical difficulties. The project turned out to be costly and to take longer than expected. However, IBM under Tom Watson, Jr. valued its reputation and delivered the computer at the previously agreed upon price, taking a loss. Moreover, the project had achieved only half its speed objective. Still there were commercial benefits from the Stretch technology, being the first IBM computer to be fully transistorized.

Government-sponsored research was not always profitable for IBM, but it kept IBM at the cutting edge of knowledge in the early days of the industry. IBM's ability to translate the research into commercial technology provided IBM with an economic and competitive advantage. But this was not easy, for IBM had already become large enough that it was beginning to have trouble coordinating innovation within its divisions and research laboratories. In 1959, IBM had introduced its advanced 7090 computer and was selling several different computer models, from entry-level to advanced models. Yet these models were not compatible; when a customer wished to move up from entry-level performance, they had to rewrite software applications.

So it happened that IBM's incompatible product models became a severe problem. IBM was then organized as two divisions, the General Products Division (GPD) for the lower-priced machines and the Data Systems Division (DSD) for the higher-priced machines. They designed their own computer lines independently. In addition, two different research laboratories were also designing incompatible computers: the IBM Poughkeepsie Laboratory and the Endicott laboratory. In June 1959, IBM was reorganized into three product divisions: the Data Systems Division for high-end computers, the General Products Division for low-end computers, and the Data Processing Division for sales and service.

The development of the next large computer, the 7070, was thus given to DSD. Responding to this reorganization, the Poughkeepsie Laboratory, now in GPD, began a frantic effort to regain control of the design of the high-end computers. Steve Dunwell of the Poughkeepsie Lab was certain the GPD could design a computer they would call 70AB, which would be twice as fast and cheaper than the 7070 computer being designed by DSD.

We can see that within a large company, strategic rivalry over product strategies can exist between its divisions.

The two research laboratories, Endicott and Poughkeepsie, were internal rivals, with different designers and different judgments on best design. At the time of the reorganization, T. Vincent Learson had been made group executive responsible for the three divisions. When Learson learned of Dunwell's decision, Learson ordered that DSD 7070 be continued, with maximum effort. This meant that the Poughkeepsie Laboratory of GPD was not allowed to try to exceed the 7070's performance, and Dunwell retargeted the 70AB project toward a lower-performance region just above IBM's current 1401 computer and below the projected 7070. GPD renamed the 70AB project the 8000 computer series and formally presented their plan to top management.

But in that presentation, there was one IBM executive who did not like what he heard in the GPD product plan. He was T. Vincent Learson, the new group executive responsible for all three IBM divisions. He was disappointed at the planned performance of the 8000 series. Although some improvements had been planned in circuit technology, to get the machine out fast, older technology was still planned for use in other circuits. Learson was also disturbed at the lack of planed mutual compatibility of computers across IBM. Learson saw that IBM was spending most of its development resources propagating a wide variety of central processors and little development effort was devoted to other areas, such as programming or peripheral devices. After the presentation, Learson decided to force IBM to formulate a coherent product strategy. Learson asked Robert Evans (a PhD in electrical engineering) to review the 8000 series plans, telling him: "If it's right, build it; and if it's not right, do what's right." (Strohman, 1990, p. 35)

Evans decided that it was not right. He wanted a product plan for the entire business, not just the upper line of the GPD for which the 8000 series had been planned. This brought Evans and Brooks (who was responsible for the 8000 series) into direct conflict. As Brooks (who lost the battle) recalled the conflict: " 'Bob [Evans] and I [Brooks] fought bitterly with two separate armies against and for the 8000 series. He was arguing that we ought not to do a new product plan for the upper half of the business, but for the total business. I was arguing that that was a put-off, and it would mean delaying at least two years. The battle . . . went to the Corporate Management Committee twice. We won the first time, and they won the second time—and Bob was right' " (Strohman, 1990, p. 35).

Evans then proceeded to get IBM to plan for a wholly new and completely compatible product line from the low end to the top end. Evans proceeded toward healing the wounds and getting the teams together. He assembled everyone at the Gideon Putnam Hotel in Saratoga Springs, New York, to plan out the new strategy for mutually compatible IBM computers. He assigned Brooks the job of guiding the project, which would be called System/360.

The disagreement had been an honest one about how important to the company a better technology was versus the time to do it. Next-generation product strategy often involves technical disagreements and disagreements about the relative commercial benefits.

Strategy on next-generation products affects the careers of the various individuals involved, and therefore different product strategies can have different impassioned champions in a large firm.

Having bitterly fought a strategy battle, Evans and Brooks still respected each other and continued to work together. Evans's next job was to get the GPD aboard for his total compatibility strategy. For this, Evans needed the help of Donald T. Spaulding, who led Learson's staff group. Spaulding helped Evans by proposing an international top-secret task force to plan every detail of the new product line (System/360) and to make necessary compromises along the way. Spaulding brought GPD into the new strategy by making John Hanstra of GPD chair of the group, with Evans as vice-chair.

This task force of top technical experts represented all the company's manufacturing and marketing divisions. In November and December 1961, they met daily at the Sheraton New Englander Motel in Greenwich, Connecticut. By the end of December they had a technical strategy specifying the requirements of a next generation of IBM computers, the System/360 product line. It took a top-level executive to lead a strategic vision on computer knowledge as compatibility across all product lines. It also required a total management commitment (worked out through internal political battles and eventual consensus) to translate that knowledge vision into a concrete technology plan. That knowledge strategy specified seven basic technical points:

- The central processing units were to handle both scientific and business applications with equal ease.
- All products were to accept the same peripheral devices (such as disks and printers).
- To reduce technical risk to get the products speedily to market, the processors were to be constructed in microminiaturization using transistors (although integrated circuits had been considered).
- Uniform programming would be developed for the entire line.
- A single high-level language would be designed for both business and scientific applications.
- Processors would be able to address up to 2 billion characters.
- The basic unit of information would be the 8-bit byte.

There were problems in the product development, but they were all solved and the first 360 computer was shipped in April 1965: "The financial records speak for the smashing ultimate success of IBM's gamble. In the six years from 1966 through 1971, IBM's gross income more than doubled, from $3.6 billion to $8.3 billion" (Strohman, 1990, p. 40).

PRODUCT PLANNING

High-tech strategy requires long-term product planning for a next-generation product line. Product design is not a "one-shot" activity in a business, but a recurring engineering activity. Businesses sell more than one product, or more than one product model, and need to update product models periodically. A business will vary a product design into a family of products to cover the niches in a business's market. Variation of a product into different models of a product family is a redesign problem, altering the needs and specs of the product model to improve focus on a niche of the market. One product model may be replaced by an improved or redesigned product model. The time from introduction into the market of a product model and its replacement by a newer product model is called a *product-model lifetime*. Since all the product models are made obsolete by a newer generation of a technology, the newer product models are called a *new generation of the product*. Product generations are designed to provide substantial improvement in performance, features, or safety and/or to reduce product cost substantially.

For example, Wheelwright and Sasser (1989) emphasized the importance of long-term product model planning—mapping out the evolution of product models of a line of products. They pointed out that product models evolve from a core product, as one can enhance models for a higher-priced line or strip a model of some features for a price-reduced line. They also argued that from the core product, one should plan its evolution in generations of products. The core product system will express the generic technology system, and higher- or lower-priced versions will differ in the subsidiary technologies of features. Next-generation products differ in dramatically improved performance or new features, and improved technologies differ in features or dramatically reduced cost. One can diagram the planning of a line of product models as branched product models and as generations of products. Wheelwright and Sasser called this a *new product development map*. Such a map is particularly useful for the long-term product development process, when the anticipated technical goals for product performance and features are listed in the product boxes.

Product family planning is especially important to deal with competitive conditions of shortened product life cycles, which can decrease profits. For example, von Braun (1990, 1991) charted shortening product lifetimes at Siemens from 1975 to 1985 at an average of 25% shortening. With shorter lifetimes usually come smaller cumulative product volume sold and hence less return on investment from the R&D needed to introduce the product. This makes the idea of the product family important by extending product lifetimes over a variety of product models and product model improvements. The key to efficient design and production of a product family is to develop common technology platforms for the core technologies of the product. Common product platforms that can be modularized can be varied efficiently to adapt the product line to different market niches.

For example, Sanderson and Uzumeri (1995b) studied the successful innovation and subsequent market dominance of the SONY Walkman product line, concluding that "success in fast cycle industries (e.g., consumer electronics) can depend both on rapid model replacement and on model longevity. Sony was as fast as any of its chief competitors in getting new models to market; an important explanation for the wide variety of models . . . is the greater longevity of its key models."

The key models provided common technology platforms from which to vary models. In general for any product family, Sanderson and Uzumeri (1995b) stressed the importance of developing several generations of product platforms for successive generations of product families, used to (1) vary models for covering market niches with a full product line and to (2) advance the performance of product lines to stay ahead of competitors. Meyer and Utterback (1993) have also emphasized the concept of product platforms to provide the basis for individual product design.

The strategy for keeping product families high-tech and competitive are next generations of product platforms for product families.

SUMMARY AND PRACTICAL IDEAS

In case studies on

- Electronic products and tires
- First commercial computers
- IBM's next-generation system/360 computers

we have examined the following key ideas:

* Product line lifetimes
* Product model lifetime
* Product planning
* Product families and platforms

These theoretical ideas can be used practically for understanding how to plan product strategy for next-generation products.

FOR REFLECTION

Identify a product line with a history of product models. Describe the product, market, and product family evolution in the history. Which companies succeeded in which market niches of the product line? What was the critical performance/feature/price balance that contributed to market success?

25
INTEGRATING TECHNOLOGY AND BUSINESS STRATEGY

INTRODUCTION

We have just looked at the formulation of scientific technology. We have seen this formulation in two parts, scientific and technological. The scientific part of the scientific technology strategy is to perform research *to better understand paradigmatically the phenomena* underlying the technology system. The technological part of scientific technology strategy is to aim research toward invention of alternative configurations, morphologies, or logics of technology systems. We conclude by examining how to integrate technology strategy into business planning.

When a new business begins on a radically innovative technology (new basic or next generation), at first, technology strategy drives business strategy. But as business grows, business strategy needs to drive technology strategy. The business and technology strategy provides the foundation for long-term survival of a business. For example, in the case of Cisco in Chapter 18 we saw how the router technology first drove business strategy for the Bosacks. Later, under Chambers's management, Cisco business strategy drove technology strategy through strategic acquisitions of new technologies.

For any business, there are several goals for its technology strategy in its business planning:

1. *Maintenance of technical competence* in its existing business through:
   a. Incremental improvement of existing products
   b. Next-generation improvements of products and processes
2. *Expansion of markets* in the existing business or launching of new businesses through:
   a. Innovation of new products
   b. Innovation of new processes

3. *Distinctive competitive advantages* from outside a business's industrial sector through:

   a. Vertical integration from core products and skills

   b. Horizontal transfer of core products and skills

To accomplish these ends, corporate research and engineering and information activities need to:

- Deepen the understanding of the science beneath existing business technologies
- Invent or acquire improvements in existing technologies
- Maintain a window upon science for new or substituting technologies
- Invent or acquire new technologies for new products and businesses.

Technology strategy in a business is formulated within a planning process of the annual business planning cycle of the firm. For technology strategy, the setting up such a planning process requires special attention to the contexts of both research and business. The technology planning process must integrate the two different contexts, matching business goals to technology strategy, using the techniques of *technology planning, technology roadmaps,* and *technology core competencies.*

CASE STUDY:

Technology Planning at Eastman Chemicals

As an illustration of a technology planning process, we review the case of planning at the Eastman Chemicals Division of Eastman Kodak. This process was initiated by Harry W. Coover in 1965 after he had been promoted to director of research. Then the Eastman Chemicals Division was successful, with 1965 sales of $368 million and after-tax earnings at 16% of sales. Coover thought that a more systematic approach to research planning could improve operations: "Our approach to research at that time was much like everybody else's. Research was somewhat of an island unto itself with very little interaction with other company functions. And the director of research was not considered an integral part of the top level management team. . . . When I looked forward into the late 1960s and early 1970s I began to wonder: *How can research be made central to determining the company's future?*" (Coover, 1986, p. 12).

This question expressed Coover's determination to make research central to the company's future. His problem was how to create a management process for planning innovation systematically and routinely. The technology planning process that Coover began in the research laboratory focused on three goals: (1) to generate new products, (2) to ensure that all products were technological advanced, and (3) to maintain a knowledge base in the sciences underlying company technologies. We note that these goals cover several of the general goals for technology planning listed above. We also note that the business emphasis of the goals also satisfies the two goals to provide economic value to customers and obtain a competitive rate of return.

The expression of the goals above as the *mission* of the R&D laboratory was important to changing the culture of the laboratory: "We defined our responsibility as both applied and basic [technology and science]. But we maintained that all research must be consistent with current and future corporate directions. . . . We were coming out of our ivory tower and getting involved in the 'business of our business.' That meant a greater reliance on *planning*" (Coover, 1986, p. 13).

This is a fundamental point. A formal set of planning procedures for technology can improve the integration of technology with business strategy.

> **The purpose of the technology planning process is to integrate technical change with business development.**

However, this does not mean that creativity or serendipity in research efforts need be stifled by a ponderous, heavy-handed bureaucratic style of centralized planning: "In research, we usually aren't presented with well-charted territory. No one has clearly defined the new products that need to be invented, or the problems that need to be solved. Research management has the responsibility of developing ways of perceiving and conveying to the research scientists those things that the corporation needs from research" (Coover, 1986, p. 13).

Coover was saying that the real point of a technology planning process is to inculcate a *strategic view throughout the research organization*. Although technology invention cannot be planned in the way that one plans a battle, creative attention of the researcher can be focused. How did Coover focus the attention of researchers on corporate business goals? First he organized projects into a focus on businesses. Coover divided research planning for different strategic business units, both for relevance to current product lines and to potential new businesses for the units. The business context of research projects need not follow exactly the organization of the businesses in a corporation. For technology planning, the conception of business groupings should be to challenge and stimulate research imagination as to the potentially cross-cutting impact of technology change in the corporation as a whole.

The next step in Coover's technology planning process was to have his staff make technology and market projections for each business group—forecasting. The kinds of factors they considered in their projections were (1) technical changes, (2) economic changes, (3) market changes, (4) political and social trends. They considered all factors that might have an impact on the value of a product. Next, they identified what would be required technically to maintain the competitiveness of a product and what new products might be required to maintain or enlarge a market share. They also did a critical analysis of current and potential competitors in the business segment, estimating their strengths and weaknesses. To make these forecasts, they formed teams of five or six people with a member from R&D, marketing, manufacturing, finance, and a staff person skilled in forecasting techniques.

Management responsibility for the studies resided in R&D. Also, in potentially new business areas with which they were unfamiliar, they prepared a business area analysis, attempting to learn and understand a possible target business area. The next step was to organize a panel to consider the studies, called an *opportunities panel,* made up of people from the functions of marketing, production, finance, engineering, and research. Such a panel would then recommend which market segments for new potential businesses of the new technology would offer the greatest potential for the corporation.

After creating forecasts, they organized a special panel to specifically review and identify opportunities. That panel needed to be cross-functional and forward looking. The market segments that were identified for greatest potential then provided market-oriented opportunities on which research could concentrate. The advantage of this procedure was that research areas were focused systematically by adding information from other functional areas of the firm: marketing, manufacturing, and finance. Collectively, their identification of market potential was likely to be much better than if research personnel did it on their own.

The usefulness of these projections was to establish areas of markets in which research needs could be identified. A research need was a description of a valid new product or process. Coover's planning process was clearly intended to facilitate the synthesis of market need and technical opportunity. Coover understood that a planning process should foster focused creativity on the identification of needs: "People throughout the company . . . play a role in describing these needs. . . . we are seeking ideas for *value-added products and next-generation products* that represent the major opportunities of the future" (Coover, 1986, p. 15).

In addition to the ability of this planning approach to make researchers sensitive to and focused on future market needs, Coover had instituted a major programmatic effort to get them out into the marketplace. R&D personnel were assigned to each market segment of major interest, requiring them to be expert not only in their disciplines but also in the markets assigned. Coover then intended to improve the researchers' ability to communicate with senior management.

TECHNOLOGY PLANNING PROCESS IN A STRATEGIC BUSINESS UNIT

As we can see in this case of technology planning, the challenge of a planning process to integrate technology and business strategy is to *match technology push to market pull.* Since the research staff is technically responsible for technology push, being informed by this kind of planning process (which included teams of people from marketing and manufacturing and finance) about market needs provided a way for the research people to integrate into their creativity a focus on market pull. For a new product, the *market need* would describe the kinds of features and performance required for the product to find utility in the marketplace. For a new process, the market need would describe the desirable process characteristics and/or changing economic conditions that necessitate new process techniques. These market need descriptions provided market-oriented and targeted technical opportunities toward which R&D personnel could direct their research and inventive creativity.

> **The technology planning process in a corporate research organization does not plan research projects so much as it *stimulates a business-focused technology strategy in research personnel.***

A business-focused technology strategy provides a research culture within which the individual researchers are motivated to think of and to plan their own R&D projects that are both technologically creative and business relevant. In dealing with the uncertainty, creativity, and inventiveness that characterize technological invention, formal planning processes should facilitate the conditions for creativity, not constrain them. Creativity and invention fundamentally involve surprise and serendipity—the bright idea never before thought!

Although research creativity and invention cannot be planned, the *areas of research focus and the conditions for fostering creativity can be planned.*

CASE STUDY:

Motorola's Technology Roadmaps

When technology-push opportunities envisioned by researchers can provide a *technical pathway* for future progress, these technology-push visions can be used for future product planning as *technology roadmaps*. A historical case of a firm using technology roadmaps in their product planning was the U.S. electronics firm Motorola, in the 1980s: "Motorola is a technology-based company and has, for each of its many businesses, a strategic plan which is often based on the anticipated advancements that will be made in certain technologies. Our products involve the application of scientific principles, directed by our people, to solve the problems of our customers, to improve their productivity, or to let them do things they could not do otherwise" (Willyard and McClees, 1987, p. 13).

Motorola's systematic approach to technology and product planning arose from the growing technological complexity of their products and processes: "Because our products and processes were becoming much more complex over the years, we realized there was the danger that we could neglect some important element of technology. This potential danger gave rise to corporate-wide processes we call the 'Technology Roadmaps'" (Willyard and McClees, 1987, p. 13).

The formal outputs of the process were documents that recorded the information generated in the process: "The purpose of these documents is to encourage our business managers to give proper attention to their technological future, as well as to provide them a vehicle with which to organize their forecasting process" (Willyard and McClees, 1987, p. 13).

Their technology roadmaps facilitated orientation and internal communication among managers in engineering, manufacturing, and marketing. Together they could track the evolution and anticipation in their technology/product/market planning. Their purposes in the roadmap process was to (1) forecast the progress of technology, (2) obtain an objective evaluation of Motorola's technology capabilities, and (3) compare Motorola's capabilities to those of competitors. Each technology roadmap document was composed of several sections (Willyard and McClees, 1987): (1) description of business; (2) technology forecast; (3) technology roadmap matrix; (4) quality; (5) allocation of resources; (6) patent portfolio; (7) product descriptions, status reports, and summary charts; and (8) minority report.

The "description of business" section covered the following areas: business mission, strategies, market share, sales history and forecast, product-life-cycle curves, product plan, production-experience curve, and competition.

The *business mission* statement provided an answer to the question: What is expected for the business to achieve greater success? The *strategies* part summarized the timing and application of resources to a product line to generate a competence distinctive from competition. The *market share* part summarized the present shares of worldwide markets by Motorola and major competitors and trends of change. The *sales history and forecast* part displayed forecasts based on strategic assumptions. The *product-life-cycle curves* expressed the anticipated length of time from conception of a product through development to market introduction, growth, decline, and withdrawal from the market. (They used the product-life-cycle histories from past products on which to base educated guesses about current or planned products.) The *product plan* part summarized the milestones for product development and introduction and redesign. The *production-experience curve* projected the planned rate of decline of production costs over the volumes and lifetime of the product. The *competition* part emphasized understanding competitors' current strengths and weaknesses and directions of their technological development. (Motorola summarized that information in a competitor/technology matrix.) For technology forecasts, Motorola used the technology S-curve format and summarized their product plan as sets of milestones and product characteristics.

In a 1989 interview, George Fisher, then CEO of Motorola, discussed Motorola's technology planning process: " 'We are such a major force in the radio communications business that we virtually define the standard. The basics of the business are not all that complex, although the technical details are quite demanding. In portable equipment . . . there are four critical forces: size, weight, current drain [driving function of the size and weight of the battery] and cost. So in our formal technology planning, for particular products, we develop 5-year and 10-year technology roadmaps of those forces' " (Avishai and Taylor, 1989, p. 110).

What Motorola was projecting were the key technical performance parameters underlying their product line of portable radio communications. Because the technologies were being planned, the opportunities for new products to embody the technical changes could also be planned. At that time they led in smaller-size wireless phones, then at 11 ounces. The point of their formal product development process was to ensure that technology planning and product planning were closely coupled: " 'Most of us who have been brought up in the portable equipment business—serving the needs of people and machines on the move—have developed pretty refined instincts as to what the next steps have to be technologically. When we go to customers with a new product, we don't expect many surprises. . . . This is a business we know very, very well" (Avishai and Taylor, 1989, p. 110).

This last comment underlies the central point of innovating technology in new products—one does not want to be badly surprised when the new product is introduced to a customer. Bad surprises are of two kinds: (1) a product that does not focus on a customer need, or (2) a product that focuses on a customer need too early. For the first problem, Fisher emphasized the importance of having Motorola's technologists visit customers frequently. Motorola personnel were encouraged to visit their customers. But even talking to customers still required an interpretation of what the customer really would want if new technology made it possible: " 'Customers will always pull you in directions of interest to them. As a technology leader, sometimes we have to show people what we can do. The portable telephone is a good example. Motorola jumped way out in front on the car phone. . . . We didn't have thousands of people saying, 'We want portable telephones that look a certain way and have particular kinds of features.' We presented it to the world' " (Avishai and Taylor, 1989, p. 109).

At the beginning of the 1990s, Motorola was one of the U.S. firms that was performing well in global competition. But later a Finnish firm, Nokia, competed well against Motorola in cellular telephones. Also we recall that in 2000 Motorola failed at a new business venture, Iridium.

TECHNOLOGY ROADMAP

A technology roadmap for a business (or a division of a business) is a set of documents summarizing the product development and manufacturing development plans of the division. In those documents are included:

1. Technology and market forecasts
2. Product and competitor benchmarks and manufacturing competitive benchmarks
3. Product-life-cycle estimates
4. Divisional profitability gap analysis
5. Product development maps
6. New product project plans
7. Manufacturing improvement project plans
8. Divisional marketing plans
9. Divisional business plans

The technology roadmap of a business expresses the integration of technology planning with business strategy at the business level in a current business and product–line.

CASE STUDY:

Technology and Diversification at Ethyl Corporation

A complication in technology strategy occurs when firms diversify into different businesses. Then several different kinds of businesses need to be integrated with different technology strategies. This is easier to do when business diversification is driven from technological progress. We look at the case of a firm whose diversification was guided by technology strategy, the Ethyl Corporation. Ethyl began as a joint venture between General Motors and Exxon to produce a leaded antiknock gasoline additive from a new technology that had been invented at General Motors for high-compression gasoline engines but which General Motors itself did not wish to produce. In 1923, Ethyl first sold antiknock additive for gasoline, and for nearly 40 years it was a one-product company.

In a strategy to diversify, Albemarle Paper Manufacturing Co acquired Ethyl in 1962. Floyd D. Gottwald, Jr. was executive vice-president of the newly merged firm, then named Ethyl. Gottwald had joined Albemarle in 1943 as a chemist and advanced to be president of Albemarle: "When Albemarle purchased Ethyl in 1962, Ethyl was clearly a one-product company with a wealth of pent-up talent restless to exert itself. Under the previous joint owners, GM and Exxon, there had been virtually no opportunity for commercialization of the many possibilities that had emerged from 40 years of research on improving or finding a better antiknock. For good reasons of their own, the previous owners had preferred to keep Ethyl a one-product company. Our change of perspective in 1962, as we sought to diversify, could not have been more dramatic" (Gottwald, 1987, p. 27).

The continuing diversification challenge related to what further businesses to acquire. Ethyl chose to diversify toward areas in which they had a strong underlying research base. The Ethyl research program had grown out of their original focus on lead antiknock compounds, whose original rights they had acquired from General Motors (Moser, 1987).

Ethyl developed two families of technologies, one branching from lead antiknock chemistry and a second branching from aluminum alkyl chemicals. Technology strategy as research branching allowed Ethyl to innovate on technical strength and focused their original business acquisitions, which further stimulated research branching.

As Ethyl began acquiring businesses, they were originally focused by their research base in chemicals and paper. In 1963 they acquired Visqueen film for its plastics products; in 1967 they acquired Oxford Paper; and in 1968, IMCO Container. Ethyl had also an interesting research program in aluminum, which eventually didn't work out but in the meantime motivated them to acquire the William L. Bonnel Company in 1966 and Capital Product Corporation in 1970. Ethyl continued acquiring businesses in the 1970s: "As we reached the mid-1970s and Ethyl's success seemed assured, our acquisition program entered phase two—broadening the base. In 1975, we acquired the Edwin Cooper Division of Burma Oil to give added strength to our existing lube additive lines. Harwicke Chemical, purchased in 1978, expanded our insecticide business to include synthetic pyrethroids. In 1980, we acquired Saytech, Inc. to extend Ethyl's basic bromine position into flame retardants" (Gottwald, 1987, p. 27).

The diversification program of the 1960s dramatically altered Ethyl's businesses, and timing of the diversification was fortunate. Later in the 1970s, the U.S. government legislated lead additives out of gasoline as being a health hazard. The diversification program at Ethyl had looked at many other businesses which it chose not to acquire. Those had no relation to Ethyl's research strengths. Ethyl was founded on a core technology product (antiknock lead compounds) and later built its diversified businesses by technology strategy branching out from core technical competencies.

BUSINESS DIVERSIFICATION

Over the long term, all firms must change in order to survive—since neither competition, markets, nor technology are ever completely unchanging. As in the case of Ethyl, when business diversification is based on technological and research capability, the likelihood of commercial success is high. A proven strategy for diversification is to build businesses branching from the research strengths of the corporation. But this is not always possible, and certainly never quick. The quick way for a firm to diversify is through buying other businesses and leads to many kinds of diversified firms.

A *diversified firm* is a set of businesses, and the businesses owned by a firm can be said to be the business portfolio of the firm (Steele, 1989). Each business in the firm's portfolio will be organized as a division of the firm or as a corporation in its own right; and for planning purposes in the firm, each business is often called a *strategic business unit* (SBU). Using technological innovation in the business strategy of a firm is complicated by business diversification in the firm.

Some firms are single businesses with several product lines (such as Cisco Systems or Intel), and these are *single-business firms*. But many firms contain a set of diversified businesses. Some diversified firms diversify on a core technological competency (such as Ethyl), and this is a *technology-branching diversification*. Other businesses may diversify about the industrial value chain of the core business (such as General Motors), and this is *vertical-integration diversification*. Most businesses diversify without any integrating strategy (such as General Electric), and this is *conglomerate diversification*.

Since each business (SBU) uses certain core technologies in its products and production, the kinds of technologies used by a diversified firm can be numerous. Therefore, in diversified firms, technology strategy is complicated by the numbers and ranges of relevant technologies to the business portfolio of the firm.

CASE STUDY:

Technology Planning at NEC

Technology strategy for a single business differs from technology strategy for an entire diversified corporation. Business or divisional-level technology planning then needs to fit into a larger technology planning process for the entire corporation. As an example we will review technology planning in the NEC Corporation. Michiyuki Uenohara described the technology planning process in place at NEC in 1991. Uenohara was an executive advisor to NEC Corporation in Tokyo and chairman of the board of trustees of the NEC Research Institute for Advanced Management Systems and of the NEC Research Institute in Princeton, New Jersey. He had received a B.E degree from Nihon University and an M.S. and Ph.D. from Ohio State University. At first he went to work for Bell Laboratories, and then joined NEC in 1967, subsequently managing NEC's Central Research Laboratories. He was elected to NEC's board, with responsibility for corporate research and development. NEC focused on technology for its core businesses, attending to the links from research to product and market: "The key to competing successfully in industry is overall strength—each link in the chain from basic research to development to production to marketing must be as strong as the next. I believe that overall strength is the secret of the success of Japanese industry in general and of the NEC Corporation in particular. . . . Our core business areas are communications equipment and systems, computers and industrial systems, electron devices and home electronics. These businesses contributed respectively 25, 44, 18, and 13 percent of our sales in 1989" (Uenohara, 1991, p. 18).

This selection of businesses was integrated by a strategic vision of functionality in an information-oriented society: "NEC is contributing to the betterment of the highly information-oriented society that is to come. We strongly believe that information, including software, will play an important role in this future society, but hardware innovation is basic to the improvement of information productivity" (Uenohara, 1991, p. 18).

Then NEC had over 190 companies which formed the NEC group. Total employment was around 160,000, with 35,000 engineers. NEC had distributed the research and development (R&D) activities throughout the company and affiliated companies. The corporate R&D laboratory was comprised of the Research and Development Group, the Computer and Communications Software Development Group, and the Production Engineering Group. The corporate laboratory was engaged primarily in long-term R&D for future products—the products after the next products. The divisional laboratories focused on the next products. Overall, NEC spent 10% of sales on R&D. The corporate laboratories received 65% of funds from corporate headquarters, 33% from manufacturing groups, and 2% from government. It is also interesting to note that software and production were singled out as both corporate research and divisional responsibilities.

NEC's businesses were clustered into 10 manufacturing groups, with four groups in communications and two groups in electronic devices (one in home electronics and one for special projects). Together these constituted 60 product divisions and 25 divisional laboratories. Thus, the management of technology strategy in this large, diversified firm required sophisticated planning and implementation procedures. Cooperation between the corporate research laboratories and the divisional laboratories of the strategic business units occurred in joint projects: "Generic technologies are developed far ahead of manufacturing division's needs in each professional laboratory. As soon as basic technologies are reasonably well developed, new core product models that effectively utilize such basic technologies are proposed to various manufacturing divisions. After hard negotiation and evaluation, joint development projects are initiated and development resources concentrated from both laboratories and manufacturing divisions. Laboratories provide key technologies and brains, and manufacturing divisions provide product development skills and established peripheral technologies" (Uenohara, 1991, p. 20).

But to make such joint projects possible, long-range planning was required so that the generic technologies could be focused on next-generation products: "Educating R&D managers to the point where they are able to identify the correlations between the market and the generic technologies and then getting them to develop these technologies requires a time-consuming effort and long-range planning. . . . Considerable information must be gathered about future market trends, potential product ideas for business growth, and science and technology trends" (Uenohara, 1991, p. 21).

The key concept in this long-range planning is the notion of core technologies to the corporation as a whole: "The heart of the strategy is the core technology program initiated by me [Uenohara] 17 years ago. The core technology program is developed from extensive analyses of both market and the technology. A major analysis is done every 10 years for a two-year period, and minor modifications are made every year if necessary" (Uenohara, 1991, p. 21).

The formulation of the core technologies strategy is the responsibility of corporate research management, assisted by a small corporate technology planning office: "The corporate laboratory top management is responsible for establishing the core technology program. . . . However, since it requires extensive analysis and evaluation, the R&D planning office handles most of the process. The office members consist of a few managers, who are laboratory general manager candidates, a small number of analytical experts, and most senior researchers from every laboratory who hold two jobs concurrently—research leader and planner" (Uenohara, 1991, p. 20).

This case again emphasizes that technology strategy is a management responsibility, housed in corporate research laboratory management, assisted by an analytical staff, but composed primarily of *research leaders from every divisional laboratory*. The planning of core technology competencies requires involving both corporate research staff and SBU product development staff in the formulation of technology strategy. The is an example of the general principle about planning that planning works well only when it is both a top-down and a bottom-up process.

In that planning office, the research leaders from the divisions worked together with senior researchers from corporate research to forecast both technology and market opportunities. One-half of the researchers/planners were laboratory manager candidates and the remainder were specialist candidates. They analyzed the strength of all the strategic business units. They then analyzed science and technology trends and listed the most important technology areas. Finally, they correlated the technologies to the business units. These correlations were used to identity NEC's key strategic technologies.

One sees in this case a focus on forecasting both market and technology trends and identifying matrices (correlation tables) between key strategic technologies and market/product needs. In a large, diversified firm such as NEC, one may end with a large number of key strategic technologies from such a planning exercise; and these must be rationalized: "Strategic key technologies that should be pursued are grouped into a limited number of core technologies, each of which can be created and nurtured by a professional group. The strategic importance, minimum number of committed key basic technologies, important core products, and interdisciplinary relationships among other core technologies are defined for each core technology. The core technology program is presented to laboratory managers and divisional managers before the five-year plan is developed at each division" (Uenohara, 1991, p. 21).

The core technology program at NEC was important for providing the strategic vision of NEC as an information company. The number of core technologies for a diversified firm will probably be large, but it is finite in number. In 1975, NEC defined 27 core technologies. In 1983, they included software technologies and increased the number to 30 core technologies. In 1990, they were again increased to 34 core technologies. Thus because of the numbers of core technologies and the numbers of businesses in a diversified firm, the activity of managing strategic technologies is a complex activity requiring formal attention and procedures by management. A way to deal with complexity is to create conceptual hierarchies. Just as key strategic technologies were grouped into sets of core technologies, core technologies can be further grouped into strategic technology domains: "A group of core technologies demands a common strategy for better communication and collaboration. Hence, we have defined six strategic technology domains (STDs)" (Uenohara, 1991, p. 22).

> **A strategic technology domain is an area of rapidly changing research that underlies a group of strategic technologies.**

CORE TECHNOLOGY COMPETENCY IN A DIVERSIFIED FIRM

The idea of core technologies provides central competencies to businesses and to groups of businesses in a diversified corporation. Corporate commitments to a strong corporate technology capability is necessary to fund the long-term research and engineering necessary for such a capability. Corporations that invest in and manage innovation successfully over the long term have been said to have a core competence in technology.

Didrichsen (1972) pointed out that historically some corporations have shown broad technological competence in a scientific area (such as DuPont in chemistry), while others have had a kind of branching technological competence (such as 3M in adhesives and surface coatings). The case of Ethyl was a branching technology competency. Later, Prahalad and Hamel (1990) popularized this notion that strategic competency was important for diversified corporations: "During the 1980s, the top executives were judged on their ability to restructure, declutter, and delayer their corporations. In the 1990s, they'll be judged on their ability to identify, cultivate, and exploit the core competencies that make growth possible—indeed, they'll have to rethink the concept of the corporation itself" (p. 79).

As we saw in the example about NEC, core technology competency planning within a diversified corporation is complicated. There core technologies are organized as strategy technology domains, which provide the basis both for (1) the long-range planning of the divisional laboratories in the SBUs, and (2) the long-range planning of corporate research.

The problem in corporate technology strategy requires both the identification of corporate core technology competencies and planning of long-range research programs for technical progress in the competencies. These projects result in core products (or processes or services) for the strategic business units to use competitively in their products and services to their customers.

> **Because of large numbers of different businesses and relevant technologies, integrating technology strategy with business strategy in a diversified firm is complicated.**

Figure 25.1 shows that this complexity can be handled systematically in a diversified firm's strategic planning—for core technology competencies. Corporate-level business plans must be both the beginning and ending points in a cyclic planning process, as indicated in the top right of the scheme, the corporate business plan.

On the left side of the scheme lies the economic context of the businesses of the firm—in the form of industrial value chains in which the businesses of the corporation participate. Also on the left side are various techniques that can assist the technology planning required to strategically identify core competencies for the

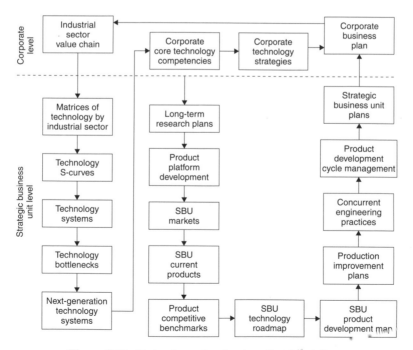

Figure 25.1. Technology planning in a diversified firm.

corporation: (1) technology by industrial-sector matrix, (2) technology S-curves, (3) technology systems, (4) technology bottlenecks, and (5) visions of next-generations of technology systems.

On the right side are indicated the steps in the product development cycles of the SBUs which utilize technology and core products, processes, or services of the corporation in each SBU business plan: (1) product development maps, (2) production improvement plans, (3) concurrent engineering practices, (4) product development cycle management, and (5) strategic business units plans.

The center of a strategic technology process for a diversified corporation turns on (1) core technology competencies of the entire firm and (2) corporate core products (or processes or services). These give SBUs competitive advantages by (1) market focus of strategic business units, and (2) competitive benchmarking for both the corporation and its business units. In the middle are the procedures that connect technology forecasting to technology implementation:

- Formulation of corporate core technology competencies
- Long-range research plans
- Corporate core products development
- Review of markets of strategic business units
- Review of current products of strategic business units
- Competitive benchmarking
- Formulation of corporate technology strategy
- Formulation of technology roadmaps for each strategic business unit

We can walk through the complicated scheme of Figure 25.1 to see the many kinds of issues that need to be addressed in the various aspects of formulating technology and business strategy:

1. *Corporate business plans*
   - The businesses of a firm should be planned within groupings of coherent industrial sectors, indicating the generic product lines of the sector and connections between customers and suppliers.
   - The dominant theme of corporate-level business strategy should be reconciling long-term survival with short-term profitability.
   - Survival requires both corporate prosperity and corporate change; short-term profitability requires competitiveness and productivity.
   - Technology strategic at the corporate-level should be focused on the necessity and opportunity for technical change and the contributions of such change to long-term corporate prosperity.
2. *Industrial value chains*
   - The concept of industrial system value chains facilitates a diversified firm to understand how to include technology within its corporate-level strategies through a regrouping of its businesses into relations within industrial value chains.

• For the current businesses of the firm, each industrial sector relation (from resources to final customer) should be described in detail.

3. *Technology/industrial sector matrices*
   • Generic-level technological change (either as core technologies or as substituting technologies) that affects any of the industrial sectors should be described.
   • Lists should be made of the generic core and substituting technologies against the industrial sectors to identify all the potential industrial relevancies of technical changes.

4. *Technology S-curves*
   • For the relevant technologies identified in the technology/sector matrix, it is important to anticipate rates of change in core and substituting technologies.
   • In particular, one should identify new and/or rapidly changing technologies, as distinct from slowly changing and mature technologies.
   • Appropriate and generic technology performance measures should be constructed to express the potential impact of technological progress on functional applications.
   • These should be projected against time, identifying natural limits and the reasons for such limits.

5. *Technology systems*
   • Although the rates of change of technologies are summarized in generic and key performance parameters, the actual changes in technologies occur in system aspects of the technologies.
   • It is necessary to describe the relevant technologies as systems in order to connect technology forecasting with technology planning.

6. *Technology bottlenecks*
   • The current technical bottlenecks to progress can be identified, wherein if overcome, the overall performance of the technology system would be improved.
   • Such bottlenecks are the points for planning research to improve technologies.

7. *Next-generation technology systems*
   • Most progress will be incremental within current technology systems, but it is also important to look further into the future to envision a next generation of a technology system.
   • Plans for incremental improvement toward a next generation of technology that will not result in an eventual technical dead-end (and an ultimately failed business).

8. *Corporate core technology competencies*
   • Core technology competencies express the corporation's ability to integrate technologies basic to the critical features of products and production generic to several businesses of the corporation.

- Core competencies identify the technological distinctiveness of the corporation (with respect to competitors) in the diverse businesses and markets of the firm's strategic business units.
- The attitude of senior management about technology is expressed in the core competencies of the firm and realized in the culture for technological change created and reinforced by the CEO team.
- If simply ignored or treated superficially by the CEO team, the diversified firm may miss opportunities for technological synergism between its businesses.

9. *Corporate technology strategies*
   - Corporate technology strategies express the directions of technological change forecasted in the strategic technologies of the corporation.
   - Such strategies identify the impact of strategic technologies changes on the businesses and the competitive position of the firm.
   - Such strategies also identify the long-term capital investments required to exploit technological changes as a corporate first-mover.

10. *Long-term research plans*
    - Within the commitment to corporate core technology competencies, a firm should plan its long-term research to acquire, maintain, and advance core competencies.
    - A corporate research laboratory is required to organize research for core competencies that cut across business units.
    - The long-term research program plans should include:
      —Areas of research thrusts
      —Methodological approaches in these thrusts
      —Technical goals and targets
      —Technical risks and tests of technical feasibility
      —Plans for incorporating successful results into core products, processes, or services
      —Profit gap analyses of the impact of research on products
      —Resources and schedules required for progress
      —Organization and management responsibility for implementation of progress

11. *Core corporate products/processes/services*
    - Core products, processes, and services are the tangible embodiment of core competencies as components, materials, production processes, software systems, and design capabilities used by strategic business units in their products, production, or services.
    - Core products, processes, and services enable business units to compete with technological factors beyond their particular industrial sector.

12. *Strategic business unit markets*
    - SBU markets implement the direct competitiveness aspect of corporate strategy.
    - SBU competitive strategy attends to the options chosen to offer attractive value to customers of the SBU market focus.
    - Factors such as product attributes, cost, service, and segmented customer focus target SBU markets.

13. *Strategic business unit current products*
    - The current product lines of strategic business units are reviewed to the extent that they may exploit the corporate core products, processes, or services.

14. *Competitive benchmarks*
    - In addition to forecasting technology change, a firm also needs to forecast:
      —The commercial impact of such change on the competitive environment of the firm
      —Anticipation of changes in current or new markets due to technical progress (in core and substituting technologies by industrial sector)
      —Strengths and weaknesses of competitors and any changes in competitive factors of industrial organization that technical progress may facilitate
      —Scenarios for possibilities of restructuring competition (through market niching or new business ventures or product substitution or vertical integration within and between industrial value chains)

15. *Strategic business unit technology roadmap*
    - Each SBU creates a technology roadmap for the unit, indicating the long-range product or process development of the unit.

16. *SBU product development map*
    - The implementation of research results will be in products, which together with customer needs, provide the basis for the development and evolution of product lines.
    - Each strategic business unit should plan their long-term product development as a map, depicting core products, product branching, next-generation models, and new product lines.
    - The product development map facilitates research cooperation with the corporate research lab and the use of their core product, process, and services developments.

17. *Production improvement plans*
    - The implementation of research results will also be in production processes and systems, which provide the capabilities of mass producing high quality products at low cost, with quick responsiveness.

18. *Concurrent engineering practices*
    - The computerization of engineering design and manufacturing is increasingly providing enabling technologies for coordinating product design and

manufacturing—a necessity for companies that wish to establish rapid market response companies

19. *Product development cycle management*
   * All the tools and procedures for anticipating, planning, and implementing technological innovation will be of no use unless the firm's product development cycles are well managed.

20. *Strategic business unit plans*
   * Business plans of SBUs should integrate business planning with technology planning, as technology roadmaps.
   * Technology roadmaps should include:
   —Technology and market forecasts
   —Product and competitor benchmarks and manufacturing competitive benchmarks
   —Product-life-cycle estimates
   —Divisional profitability gap analysis
   —Product development maps
   —New product project plans
   —Manufacturing improvement project plans
   —Divisional marketing plans
   —Divisional business plans

In summary, a formal planning procedure for formulating technology strategy and integrating it into product and business planning should provide a systematic method of utilizing technology as a competitive edge. Managing strategic technologies can develop core technical competencies that provide competitive advantages to the businesses of the firm—beyond the ordinary boundaries of their respective industries. Commercially successful innovation is both complex and clever. Successful technology strategy is complex, but successful business strategy is clever. Technology strategy is complex because of the complexity of relating utility to nature. Business strategy is clever by focusing the right product for the right customer at the right price. This is the point of integrating technology strategy with business strategy:

Successful technological innovation requires both complexity and cleverness— complexity in technology and cleverness of business.

SUMMARY AND PRACTICAL IDEAS

In case studies on

* Technology planning at Eastman Chemicals
* Motorola's technology roadmaps

- Technology and diversification at Ethyl Corporation
- Technology planning at NEC

we have examined the following key ideas:

- Business goals for technology strategy
- Planning to match technology push to market pull
- Technology roadmap
- Technology core competency

These theoretical ideas can be used practically for understanding how technology strategy can be integrated into business planning through a technology planning process.

FOR REFLECTION

Identify a high-tech corporation that has a corporate laboratory. List the corporate businesses and identify the kinds of technologies the corporate laboratory is likely to address. Are these all physical or biological or information technologies, or a combination?

BIBLIOGRAPHY

Abelson, Philip H. 1996. "Pharmaceuticals Based on Biotechnology," *Science,* Vol. 273, August 9, p. 719.

Abernathy, William J. 1978. *The Productivity Dilemma,* Johns Hopkins University Press, Baltimore.

Abernathy, William J., and K. B. Clark. 1985. "Mapping the Winds of Creative Destruction," *Research Policy,* Vol. 14, No. 1, pp. 2–22.

Acton, Jan Paul, and Lloyd S. Dixon. 1992. "Superfund and Transaction Costs: The Experiences of Insurers and Very Large Industrial Firms," working paper, Rand Institute for Civil Justice, Santa Monica, CA.

Afuah, Allan N., and Nik Bahram. 1995. "The Hypercube of Innovation," *Research Policy,* Vol. 24, pp. 51–76.

Ahl, David H. 1984. "The First Decade of Personal Computing," *Creative Computing,* Vol. 10, No. 11, pp. 30–45.

Allen, Thomas J., and Ralph Katz. 1992. "Age, Education, and the Technical Ladder," *IEEE Transactions on Engineering Management,* Vol. 39, No. 3, August, pp. 237–245.

Amatucci, Francis, and John H. Grant. 1991. "Eight Strategic Decisions That Weakened Gulf Oil," *Long Range Planning,* Vol. 26, No. 1, pp. 98–110.

Angier, Natalie. 1991. "Molecular 'Hot Spot' Hints at a Cause of Liver Cancer," *New York Times,* April 4, pp. A1, A21.

Angwin, Julia, 2000. "Business-Method Patents, Key to Priceline, Draw Growing Protest," *Wall Street Journal,* October 3, p. B4.

Armst, Catherine, Judith H. Dorbrzynski, and Bart Ziegler. 1993. "Faith in a Stranger," *Business Week,* April 5, pp. 18–21.

Armstrong, Larry, and Peter Elstrom. 1996. "Inside Apple's Boardroom Coup," *Business Week,* February 19, pp. 28–30.

Avishai, Bernard, and William Taylor. 1989. "Customers Drive a Technology-Driven Company: An Interview with George Fisher," *Harvard Business Review,* November–December, pp. 107–116.

Ayres, Robert U. 1989. "The Future of Technological Forecasting," *Technology Forecasting and Social Change,* Vol. 36, No. 1–2, August, pp. 49–60.

Ayres, Robert U. 1990. "Technological Transformations and Long Waves: Parts I and II," *Technology Forecasting and Social Change*, Vol. 37, pp. 1–37, 111–137.

Badiru, Adedeji. 1992, "Computational Survey of Univariate and Multivariate Learning Curve Models," *IEEE Transactions on Engineering Management*, Vol. 39, No. 2, May, pp. 176–188.

Bailetti, Antonio, and John R. Callahan. 1995. "Managing Consistency between Product Development and Public Standards Evolution," *Research Policy*, Vol. 24, pp. 913–931.

Bailetti, Antonio J., John R. Callahan, and Pat Dipietro. 1994. "A Coordination Structure Approach to the Management of Projects," *IEEE Transactions on Engineering Management*, Vol. 41, No. 4, November, pp. 394–402.

Barboza, David. 2000. "Iridium, Bankrupt, Is Planning a Fiery Ending for Its 88 Satellites," *New York Times*, April 11, pp. C1–C2.

Bard, Jonathan F., Annura deSilva, and Andre Bergevin. 1997. "Evaluation Simulation Software for Postal Service Use: Technique versus Perception," *IEEE Transactions on Engineering Management*, Vol. 44, No. 1, February, pp. 31–42.

Bardeen, John. 1984. "To a Solid State," *Science 84*, November, pp. 143–145.

Bartimo, Jim. 1984. " 'Smalltalk' with Alan Kay," *InfoWorld*, June 11, pp. 58–61.

Bastable, Marshall J. 1992. "From Breechloaders to Monster Guns: Sir William Armstrong and the Invention of Modern Artillery, 1954–1880," *Technology and Culture*, Vol. 33, pp. 213–247.

Baugh, S. Gayle, and Ralph M. Roberts. 1994. "Professional and Organizational Commitment among Engineers: Conflicting or Complementing?" *IEEE Transactions on Engineering Management*, Vol. 41, No. 2, May, pp. 108–114.

Becker, Robert H., and Laurine M. Speltz. 1983. "Putting the S-Curve to Work," *Research Management*, September–October, pp. 31–33.

Bell, James R. 1984. "Patent Guidelines for Research Managers," *IEEE Transactions in Engineering Management*, Vol. 31, pp. 102–104.

Beltramini, Richard. 1996. "Concurrent Engineering: Information Acquisition between High Technology Marketeers and R&D Engineers in New Product Development," *International Journal of Technology Management*, Vol. 11, No. 1–2, pp. 58–66.

Berenson, Alex. 2002. "Tweaking Numbers to Meet Goals Comes Back to Haunt Executives," *New York Times*, June 29, pp. A1–B3.

Berger, A. 1975. "Factors Influencing the Locus of Innovation Activity Leading to Scientific Instrument and Plastics Innovation," unpublished S.M. thesis, MIT Sloan School of Management, Cambridge, MA.

Berkowitz, Leonard. 1993. "Getting the Most from Your Patents," *Research-Technology Management*, March–April, pp. 26–30.

Betz, Frederick. 1991. "Next-Generation Technology and Research Consortia," *International Journal of Technology Management*, special issue, pp. 298–310.

Betz, Frederick. 1997. "Industry/University Centres in the USA: Connecting Industry to Science," *Industry and Higher Education*, Fall.

Betz, Frederick, and Ian Mitroff. 1972. "Representational Systems Theory," *Management Science*, May, pp. 1242–1252.

Bosomworth, Charles E., and Burton H. Sage, Jr. 1995. "How 26 Companies Manage Their Central Research," *Research-Technology Management*, May–June, pp. 32–40.

Bowen, H. Kent, Kim B. Clark, Charles A. Holloway, and Steven C. Wheelwright. 1994. "Development Projects: The Engine of Renewal," *Harvard Business Review,* September–October, pp. 110–120.

Bower, Joseph L., and Thomas M. Hout. 1988. "Fast-Cycle Capability for Competitive Power," *Harvard Business Review,* November–December, pp. 110–118.

Boyden, J. 1976. "A Study of the Innovation Process in the Plastics Additives Industry," unpublished S.M. thesis, MIT Sloan School of Management, Cambridge, MA.

Brainerd, J. G., and T. K. Sharpless. 1984. "The ENIAC," *Proceedings of the IEEE,* Vol. 72, No. 9, September, pp. 1202–1205.

Braudel, Fernand. 1979. *The Wheels of Commerce,* Harper & Row, New York.

Brick, Michael. 2002. "What Was the Heart of Enron Keeps Shrinking," *New York Times,* April 6, pp. B1–B2.

Bricklin, Dan, and Bob Frankston. 1984. "Viscalc '79," *Creative Computing,* Vol. 10, No. 11, pp. 122–123.

Bridenbaugh, Peter R. 1992. "Credibility between CEO and CTO—A CTO's Perspective," *Research-Technology Management,* November–December, pp. 27–33.

Brittain, James E. 1984. "Introduction to 'Hopper and Mauchly on Computer Programming,' " *Proceedings of the IEEE,* Vol. 72, No. 9, September, pp. 1213–1214.

Broad, William J. 1997. "Incredible Shrinking Transistor Nears Its Ultimate Limit," *New York Times,* February 4, pp. C1–C5.

Brown, Ken, and Henny Sender. 2002. "Did Andersen Act Properly? Firm Is Fired," *Wall Street Journal,* January 18, pp. C1–C15.

Bunnell, David. 2000. *Making the Cisco Connection,* Wiley, New York.

Burrows, Peter. 1996. "The Man in the Disk Driver's Seat," *Business Week,* March 18, pp. 71–73.

Business Week. 1984a. "Biotech Comes of Age," January 23, pp. 84–94.

Business Week. 1984b. "Apple Computer's Counterattack against IBM," January 16, pp. 78–82.

Byrne, John A. 2002. "Why Chainsaw Al Opened His Wallet," *Business Week,* January 28, p. 8.

Cabral-Cardoso, Carlos, and Roy L. Payne. 1996. "Instrumental and Supportive Use of Formal Selection Methods in R&D Project Selection," *IEEE Transactions on Engineering Management,* Vol. 43, No. 4, November, pp. 402–410.

Chandler, Alfred. 1990. "The Enduring Logic of Industrial Success," *Harvard Business Review,* March–April, pp. 130–140.

Chase, Richard B., and Robert H. Hayes. 1991. "Beefing Up Operations in Service Firms," *Sloan Management Review,* Fall, pp. 15–16.

Chemical & Engineering News. 1983. "Monsanto's Richard Mahone: Ready to Take on the 1980s," September 26, pp. 10–13.

Chester, Arthur N. 1994. "Aligning Technology with Business Strategy," *Research-Technology Management,* January–February, pp. 25–32.

Chiesa, Vittorio. 1996. "Managing the Internationalization of R&D Activities," *IEEE Transactions on Engineering Management,* Vol. 43, No. 1, February, pp. 7–23.

Christensen, Clayton. 2000. *The Innovator's Dilemma,* HarperBusiness, New York.

Christensen, Clayton M., and Michael Overdorf. 2000. "Meeting the Challenge of Disruptive Change," *Harvard Business Review,* March–April, pp. 67–76.

Clark, Douglas W. 1990. "Bugs Are Good: A Problem-Oriented Approach to the Management of Design Engineering," *Research-Technology Management,* May–June, pp. 23–27.

Cole, Tim. 1988. "Science at Sea," *Popular Mechanics,* September, p. 28.

Coover, Harry W. 1986. "Programmed Innovation: Strategy for Success," *Research-Technology Management,* November–December, pp. 12–17.

Copper, Robert G. 1990. "New Products: What Distinguishes the Winners?" *Research-Technology Management,* November–December, pp. 27–31.

Cortese, Amy. 2000. "Venture Capital, Withering and Dying," *New York Times,* October 21, Section 3, pp. 1–11.

Costello, D. 1983. "A Practical Approach to R&D Project Selection," *Technology Forecasting & Social Change,* Vol. 23, 1983.

Cringley, Robert. 1997. "High Tech Wealth," *Forbes,* July, pp. 296–304.

Crow, Kenneth A. 1989. "Ten Steps to Competitive Design for Manufacturability," presented at the 2nd International Conference on Design for Manufacturability, sponsored by CAD/CIM Alert, Management Roundtable, Washington, DC.

Cusumano, Michael A. 1985. *The Japanese Automobile Industry: Technology and Management at Nissan and Toyota,* Harvard University Press, Cambridge, MA.

Cusumano, Michael A. 1998. "Manufacturing Innovation: Lessons from the Japanese Auto Industry," *Sloan Management Review,* Fall, pp. 29–40.

Cyret, Richard M., and Praveen Kumar. 1994. "Technology Management and the Future," *IEEE Transactions on Engineering Management,* Vol. 41, No. 4, November, pp. 333–334.

Debresson, Chris. 1995. "Predicting the Most Likely Diffusion Sequence of a New Technology through the Economy: The Case of Superconductivity," *Research Policy,* Vol. 24, pp. 685–705.

Dewey, John. 1938. *Logic: The Theory of Inquiry,* Henry Holt and Company, New York.

Didrichsen, Jon. 1972. "The Development of Diversified and Conglomerated Firms in the United States, 1920–1970," *Business History Review,* Vol. 46, Summer, p. 210.

Drucker, Peter E. 1990. "The Emerging Theory of Manufacturing," *Harvard Business Review,* May–June, pp. 94–102.

Duggan, Paul, and Lois Romano, 2000. "Enron Official Shaken in Days before Suicide," *Washington Post,* February 9, pp. A1–A8.

Dulkey, N. C., and Q. Helmer. 1963. "An Experimental Application of the Delphi Method in the Use of Experts," *Management Science,* Vol. 9, pp. 458–467.

Dunning, John H. 1994. "Multinational Enterprises and the Globalization of Innovative Capacity," *Research Policy,* Vol. 23, pp. 67–88.

Economist. 1997. "Venture Capitalists," January 25, pp. 20–22.

Edleheit, Lewis S. 1995. "Renewing the Corporate R&D Laboratory," *Research-Technology Management,* November–December, pp. 14–18.

Eichenwald, Kurt. 2002a. "Guilty Plea Seen in the Shredding of Enron Records," *New York Times,* April 9, pp. A1–C5.

Eichenwald, Kurt. 2002b. "Andersen Guilty in Effort to Block Inquiry on Enron," *New York Times,* June 16, pp. A1–A18.

Eichenwald, Kurt, and Diana B. Henriques. 2002. "Entron Buffed Image to a Shine Even As It Rotted from Within," *New York Times,* February 10, pp. A1–A26.

Estrom, Peter. 1996. "PDA May Always Mean 'Pretty Darn Average,'" *Business Week,* June 24, p. 110.

Ettlie, John E., and Henry W. Stoll. 1990. *Managing the Design-Manufacturing Process,* McGraw-Hill, New York.

Evans, Michael W., Alex M. Abela, and Thomas Beltz. 2002. "Seven Characteristics of Dysfunctional Software Projects," *CrossTalk,* April, pp. 16–19.

Foray, Dominique, and Arnulf Grubler. 1990. "Morphological Analysis, Diffusion and Lock-Out of Technologies: Ferrous Casting in France and the FRG," *Research Policy,* Vol. 19, pp. 535–550.

Ford, David, and Chris Ryan. 1981. "Taking Technology to Market," *Harvard Business Review,* March–April, pp. 117–126.

Forrester, Jay W. 1961. *Industrial Dynamics,* MIT Press, Cambridge, MA.

Fortune. 1993. "Information Technology Special Report: How to Bolster the Bottom Line," Autumn, pp. 15–28.

Foster, Richard N. 1982. "A Call for Vision in Managing Technology." *Business Week,* May 24, pp. 24–33.

Foster, Richard, Lawrence Linden, Roger Whiteley, and Alan Kantrow. 1985. "Improving the Return of R&D II," *Research Management,* Vol. XXVII, No. @, pp. 13–22.

Freeman, Christopher. 1974. *The Economics of Industrial Innovation,* Penguin Books, New York.

Frosch, Robert A. 1996. "The Customer for R&D Is Always Wrong!" *Research-Technology Management,* November–December, pp. 22–27.

Funk, Jeffrey L. 1993. "Japanese Product-Development Strategies: A Summary and Propositions about Their Implementation," *IEEE Transactions on Engineering Management,* Vol. 40, No. 3, August, pp. 224–236.

Fusfeld, Hebert I. 1995. "Industrial Research—Where It's Been, Where It's Going," *Research-Technology Management,* July–August, pp. 52–56.

Gates, William. 1984. "A Trend toward Softness," *Creative Computing,* Vol. 10, No. 11, pp. 121–122.

Gemmill, Gary, and David Wilemon. 1994. "The Hidden Side of Leadership in Technical Team Management," *Research-Technology Management,* November–December, pp. 25–32.

Geppert, Linda. 1994. "Industrial R&D: The New Priorities," *IEEE Spectrum,* September, pp. 30–41.

Godfrey, A. Blanton, and Peter J. Kolesar. 1988. "Role of Quality in Achieving World Class Competitiveness," in Martin K. Starr (ed.), *Global Competitiveness: Getting the U.S. Back on Track,* W.W. Norton, New York.

Gold, Bela, William S. Pierce, Gerhard Rosegger, and Mark Perlman. 1984. *Technological Progress and Industrial Leadership,* Lexington Books, Lexington, MA.

Gottwald, Floyd D. 1987. "Diversifying at Ethyl," *Research Management,* May–June, pp. 27–29.

Graham, Alan, and Peter Senge. 1980. "A Long-Wave Hypothesis of Innovation," *Technological Forecasting and Social Change,* Vol. 17, August, pp. 283–312.

Grajek, Tim. 2002. "Prosecutors Move Up the Enron Food Chain," *Washington Post,* August 25, p. H2.

Gray, Steven, B. 1984. "The Early Days of Personal Computers," *Creative Computing,* Vol. 10, No. 11, pp. 6–14.

Gunther, Marc. 1998. "The Internet Is Mr. Case's Neighborhood," *Fortune,* March 30, pp. 69–80.

Gunther, Marc. 2000. "These Guys Want It All," *Fortune,* February 7, pp. 71–78.

Gwynne, Peter. 1993. "Directing Technology in Asia's 'Dragons,'" *Research-Technology Management,* March–April, pp. 12–15.

Gwynne, Peter. 1996. "The CTO as Line Manager," *Research-Technology Management,* March–April, pp. 14–18.

Haeffner, Erik. 1980. "Critical Activities of the Innovation Process," pp. 129–144 in B. A. Vedin (ed.) *Current Innovation,* Almqvist & Wiksell, Stockholm.

Hall, Dennis G., and Rita M. Hall. 2000. "Chester F. Carlson: A Man to Remember," *Optics & Photonics News,* September, pp. 14–18.

Hamilton, David P. 1991. "Can Electronic Property Be Protected?" *Science,* Vol. 253, July 5, p. 23.

Hart, Stuart L. 1997. "Strategies for a Sustainable World," *Harvard Business Review,* January–February, pp. 67–76.

Hauben, Michael. 1993. "History of ARPANET," *http://www.dei.ise.ipp.pt/docs/arap.html* (Haruden@columbia.edu), December 2.

Hayes, Robert, and David Gavin. 1982. "Managing As If Innovation Mattered," *Harvard Business Review,* May–June, pp. 70–79.

Hazelrigg, George A. 1982. "Windows for Innovation: A Story of Two Large-Scale Technologies," report 82-180-1 submitted to the National Science Foundation, Washington, DC, December 1.

Heppenheimer, T. A. 1990. "How Von Neumann Showed the Way," *Invention and Technology,* Vol. 6, No. 2, Fall, pp. 8–17.

Hewlett, Richard G., and Francis Duncan. 1974. *Nuclear Navy, 1946–1962,* University of Chicago Press, Chicago.

Holdern, Constance. 2002. "Single Gene Dictates Ant Society," *Science,* Vol. 294, November 16, p. 1434.

Holstein, William J. 2002. "Samsung's Golden Touch," *Fortune,* April 1, pp. 89–94.

Jantsch, Erich. 1967. *Technological Forecasting in Perspective,* Organization for Economic Cooperation and Development, Paris.

IRI. 1994. "First Annual Industrial Research Institute R&D Survey," *Research-Technology Management,* January–February, pp. 18–24.

Jaffe, Greg. 2000. "Pentagon to Pay Iridium for Use of Its Network," *Wall Street Journal,* December 7, p. A18.

Joint Committee on Atomic Energy (JCAE). 1974. *Congress of the United States Hearings of October 30–31 and November 13, 1963, on Nuclear Propulsion for Naval Surface Vessels,* U.S. Government Printing Office, Washington, DC.

Jones, Casper. 2002. "Software Cost Estimation in 2002," *Crosstalk,* June, pp. 4–8.

Judson, Horace Freeland. 1979. *The Eighth Day of Creation,* Simon & Schuster, New York.

Kahn, Brian. 1990. "The Software Patent Crisis," *Technology Review,* April, pp. 53–58.

Kell, Douglas, and G. Rickey Welch. 1991. "Perspective: No Turning Back," *Times Higher Education Supplement,* September 13, p. 15.

Kerwin, Kathleen, James Muller, and David Welch. 2002. "Bill Ford's Long, Hard Road," *Business Week,* October 7, pp. 89–92.

Kerzner, Harold. 1984. *Project Management: A Systems Approach,* Van Nostrand Reinhold, New York.

Kim, Henry M., Mark S. Fox, and Michael Gruniger. 1995. "An Ontology of Quality for Enterprise Modelling," *Proceedings of the 4th IEEE Workshop on Enabling Technologies.*

Kimura, Tatsuya, and Makto Tezuka. 1992. "Managing R&D at Nippon Steel," *Research-Technology Management,* March–April, pp. 21–25.

King, Nelson, and Ann Majchrzak. 1996. "Concurrent Engineering Tools: Are the Human Issues Being Ignored?" *IEEE Transactions on Engineering Management,* Vol. 42, No. 2, May, pp. 189–201.

Kirkpatrick, David. 1997. "Intel's Amazing Profit Machine," *Fortune,* February 17, pp. 60–72.

Klevorick, Alvin K., Richard C. Levin, Richard R. Nelson, and Sidney G. Winter. 1995. "On the Sources and Significance of Interindustry Differences in Technological Opportunities," *Research Policy,* Vol. 24, pp. 185–205.

Klimstra, Paul D., and Ann T. Raphael. 1992. "Integrating R&D and Business Strategy," *Research-Technology Management,* January–February, pp. 22–28.

Knight, K. E. 1963. "A Study of Technological Innovation: The Evolution of Digital Computers," unpublished Ph.D. dissertation, Carnegie Institute of Technology, Pittsburgh, PA.

Kocaoglu, D. F. 1994. "Special Issue on 40 Years of Technology Management," *IEEE Transactions on Engineering Management,* Vol. 41, No. 4, November, pp. 329–330.

Kokubo, Atsuro. 1992. "Japanese Competitive Intelligence for R&D," *Research-Technology Management,* January–February, pp. 33–34.

Kuhn, Arthur J. 1986. *GM Passes Ford, 1918–1938,* Pennsylvania State University Press, University Park, PA.

Kumar, Uma, and Vinod Kumar. 1992. "Technological Innovation Diffusion: The Proliferation of Substitution Models and Easing the User's Dilemma," *IEEE Transactions on Engineering Management,* Vol. 39, No. 2, May, pp. 158–168.

Kunnathur, Anand S., and P. S. Sundararaghavan. 1992. "Issues in FMS Installation: A Field Study and Analysis," *IEEE Transactions on Engineering Management,* Vol. 39, No. 4, November, pp. 370–377.

Kuwahara, Yutaka, and Yasutsugu Takeda. 1990. "A Managerial Approach to Research and Development Cost-Effectiveness Evaluation," *IEEE Transactions on Engineering Management,* Vol. 37, No. 2, May, pp. 134–138.

Labaton, Stephen, and Richard A. Oppel. 2002. "Testimony of Enron Executives Is Contradictory," *New York Times,* February 8, pp. A1–C4.

Labich, Kenneth. 1984. "Monsanto's Brave New World," *Fortune*, April 30, pp. 57–98.

Larson, Charles F. 1996. "Critical Success Factors for R&D Leaders," *Research-Technology Management,* November–December, pp. 19–21.

Lawton, Thomas C. 1999. "The Limits of Price Leadership: Needs-Base Positioning Strategy and the Long-Term Competitiveness of Europe's Low Fare Airlines," *Long Range Planning,* Vol. 32, No. 6, December, pp. 573–586.

Layton, Christopher. 1972. *Ten Innovations,* Crane, Russak, New York.

Lee, Denis M. S. 1992. "Job Challenge, Work Effort, and Job Performance of Young Engineers: A Causal Analysis," *IEEE Transactions on Engineering Management,* Vol. 39, No. 3, August, pp. 214–226.

Leishman, Theron R., and David A. Cook. 2002. "Requirements Risks Can Drown Software Projects," *CrossTalk,* April, pp. 4–7.

Leonard-Barton, Dorothy. 1992. "The Factory as a Learning Laboratory," *Sloan Management Review,* Fall, pp. 23–38.

Leonhardt, David. 2002. "Tell the Good News. Then Cash In," *The New York Times,* April 7, Section 3, p. 1.

Liesman, Steve. 2002. "SEC Accounting Cop's Warning: Playing by Rules May Not Ward Off Fraud Issues," *Wall Street Journal,* February 12, pp. C1–C8.

Lohr, Steve. 1997. "Creating Jobs," *New York Times Magazine,* January 12, pp. 15–19.

Lohr, Steve. 2001. "The PC? That Old Thing?" *New York Times,* August 19, Section 3, pp. 1–12.

Lubar, Steven. 1990. "New, Useful and Nonobvious," *Invention and Technology,* Spring–Summer, pp. 8–16.

MacCormack, Alan David, Lawrence James Newman III, and Donald B. Rosenfield. 1994. "The New Dynamics of Global Manufacturing Site Location," *Sloan Management Review,* Summer, pp. 69–80.

MacLachlan, Alexander. 1995. "Trusting Outsiders to Do Your Research: How Does Industry Learn to Do It?" *Research-Technology Management,* November–December, pp. 48–53.

Magaziner, Ira C., and Mark Patinkin. 1989. "Cold Competition: GE Wages the Refrigerator War," *Harvard Business Review,* March–April, pp. 114–124.

Marcua, Regina Fazio. 1994. "The Right Way to Go Global: An Interview with Whirlpool CEO David Whitwam," *Harvard Business Review,* March–April, pp. 135–145.

Margretta, Joan. 1997. "Growth through Global Sustainability: An Interview with Monsanto's CEO, Robert B. Shapiro," *Harvard Business Review,* January–February, pp. 78–90.

Marquis, Donald G. 1960. "The Anatomy of Successful Innovations," *Innovation,* November. Reprinted in *Readings in the Management of Innovation,* M. L. Tushman and W. L. Morre, eds., Pitman, Marshfield, MA, 1982.

Marshall, Eliot. 1991. "The Patent Game: Raising the Ante," *Science,* Vol. 253, July 5, pp. 20–24.

Martin, R. M. 1959. *Toward A Systematic Pragmatics,* North-Holland Publishing Co., Amsterdam.

Martino, Joseph P. 1983. *Technological Forecasting for Decision Making,* 2nd ed., North-Holland, New York.

Mathis, James. 1992. "Turning R&D Managers into Technology Managers," *Research-Technology Management,* January–February, pp. 35–38.

McHugh, Josh. 1997. "Laser Dudes," *Forbes,* February 24, pp. 154–155.

McNamee, Mike, Amy Borrus, and Christopher Palmeri. 2002. "Out of Control at Andersen," *Business Week,* April 8, pp. 32–33.

Meadows, Donnella. 1996. "Our 'Footprints' Are Treading Too Much Earth," *Charleston (S.C.) Gazette,* April 1.

Mensch, Gerhard. 1975. *Stalemate in Technology,* Ballinger, Cambridge, MA.

Meyer, Christopher, and Ronald E. Purser. 1993. "Six Steps to Becoming a Fast-Cycle-Time Competitor," *Research Technology Management,* September–October, pp. 41–48.

Meyer, Marc H., and James M. Utterback. 1993. "The Product Family and the Dynamics of Core Capability," *Sloan Management Review,* Spring, pp. 29–48.

Meyer, M. H., and J. M. Utterback. 1995. "Product Development Cycle Time and Commercial Success," *IEEE Transactions on Engineering Management,* Vol. 42 No. 4, November, pp. 297–304.

Miller, William. 1999. *Fourth Generation R&D,* Wiley, New York.

Mims, Forrest M. 1984. "The Altair Story," *Creative Computing,* Vol. 10, No. 11, pp. 17–27.

MIT 50K, 1996. *1996–1997 MIT 50K Entrepreneurship Competition,* Massachusetts Institute of Technology, Cambridge, MA.

Morgenson, Gretchen. 2000a. "First Jolts of the Internet Shakeout," *New York Times,* January 16, Section 3, p. 1.

Morgenson, Gretchen. 2000b. "Analysts Talk and Amazon.com Shares Reel," *New York Times,* June 24, pp. B1–B3.

Morgenson, Gretchen. 2002. "As Pressure Grows, Option Costs Come Out of Hiding," *New York Times,* May 19, Section 3, p. 1.

Morone, Joseph. 1993. "Technology and Competitive Advantage—The Role of General Management," *Management,* Vol. 36, March–April, pp. 16–25.

Morrison, Ann M. 1982. "Apple Bites Back," *Fortune,* February 20, pp. 86–100.

Moser, Roger A. 1987. "New Program Development at Ethyl," *Research Management,* May–June, pp. 30–32.

National Academy of Engineering and National Research Council. 1983. *The Competitive Status of the U.S. Phamaceutical Industry, 1983,* National Academy Press, Washington, DC.

National Research Council. 1987. *Management of Technology: The Hidden Competitive Advantage,* National Academy Press, Washington, DC.

National Research Council. 1991. *Manufacturing Studies Board: Improving Engineering Design,* National Academy Press, Washington, DC.

National Science Board. 2002. *Science & Engineering Indicators,* Superintendent of Documents, U.S. Government Printing Office, Washington, DC.

Nevins, Allen, and Frank E. Hill. 1954. *Ford: The Times, the Man, the Company,* Charles Scribner's Sons, New York.

New York Times. 2002. "Enron's Collapse: The Congressional Hearings," January 25, pp. C8–C9.

Nobeoka, Kentaro, and Michael A. Cusumano. 1995. "Multiproject Strategy, Design Transfer, and Project Performance," *IEEE Transactions on Engineering Management,* Vol. 42, No. 4, November, pp. 397–409.

Nocera, Joseph. 1995. "Cooking with Cisco: What Does It Take to Keep a Hot Stock Sizzling?" *Fortune,* December 25, pp. 114–122.

Nocera, Joseph, and Tim Carvell. 2000. "50 Lessons," *Fortune,* October 30, pp. 136–137.

Norris, Floyd. 2002a. "For Chief, $200 Million Wasn't Quite Enough Cash," *New York Times,* January 22, p. C1.

Norris, Floyd. 2002b. "Capital Scorn: Communists to Accountants," *New York Times,* January 25, p. C1.

Norris, Floyd. 2002c. "Can Investors Believe Cash Flow Numbers?" *New York Times,* February 15, p. C1.

Nussbaum, Bruce. 2002. "Can You Trust Anybody Anymore?" *Business Week,* January 28, p. 31.

Olby, Robert, 1974. *The Path to the Double Helix,* Macmillan, London; University of Washington Press, Seattle, WA.

Oppel, Richard A., Jr. 2001. "Former Head of Enron Denies Wrongdoing," *New York Times,* December 22, pp. C1–C3.

Oppel, Richard A., Jr. 2002a. "Enron Official Says Many Knew about Shaky Company Finances," *New York Times,* February 15, pp. A1–C6.

Oppel, Richard A., Jr. 2002b. "Options Foe Is Not So Lonely Now," *New York Times,* April 7, Section 3, p. 2.

Oppel, Richard A., Jr., and Riva D. Atlas. 2001. "Hobbled Enron Tries to Stay on Its Feet," *New York Times,* December 4, pp. C1–C8.

Oppel, Richard A., Jr., and Joseph Kahn. 2002. "Enron's Ex-chief Harshly Criticized by Senate Panel," *New York Times,* February 13, pp. A1–C8.

Osborne, Adam, and John Dvorak. 1984. "Hypergrowth: Adam Osborne's Upcoming Book Tells His Side of the Story," *InfoWorld,* July 9, July 16, and July 23.

Osburn, William F. 1937. *Technological Trends and National Policy,* Research Report, National Research Council, Washington, DC.

Patel, Pari. 1996. "Are Large Firms Internationalizing the Generation of Technology?" *IEEE Transactions on Engineering Management,* Vol. 43, No. 1, February, pp. 41–47.

Pennisi, Elizabeth. 2001. "Genome Duplications: The Stuff of Evolution?" *Science,* Vol. 294, December 21, pp. 2458–2460.

Perry, Tekla S., and Paul Wallich. 1985. "Design Case History: The Commodore 64," *IEEE Spectrum,* March, pp. 48–58.

Polmar, Norman, and Thomas B. Allen. 1982. *Rickover,* Simon & Schuster, New York.

Porter, Michael. 1985. *Competitive Advantage,* Free Press, New York.

Porter, Michael. 1995. *Competitive Advantage of Nations,* McGraw-Hill, New York.

Porter, Michael E., and Class van der Linde. 1995. "Green and Competitive: Ending the Stalemate," *Harvard Business Review,* September–October, pp. 120–134.

Portugal, Franklin H., and Jack S. Cohen. 1977. *A Century of DNA,* MIT Press, Cambridge, MA.

Prahalad, C. K., and Gary Hamel. 1990. "The Core Competence of the Corporation," *Harvard Business Review,* May–June, pp. 79–91.

Pratt, Stanley E., and Jane K. Morris. 1984. *Pratt's Guide to Venture Capital Sources,* Venture Economics Inc., Wellesley, MA.

Pugh, Emerson. 1984. *Memories That Shaped an Industry: Decisions Leading to IBM System/360,* MIT Press, Cambridge, MA.

Purdon, William B. 1996. "Increasing R&D Effectiveness: Researchers as Business People," *Research-Technology Management,* July–August, pp. 48–56.

Quinn, James Brian. 1985. "Managing Innovation: Controlled Chaos," *Harvard Business Review,* May–June, pp. 73–84.

Quinn, James Brian, and Penny C. Paquette. 1988. "Ford: Team Tarus," Case Study, Amos Tuck School, Dartmouth College, Hanover, NH.

Quinn, James Brian, Jordan J. Baruch, and Karen Anne Zien. 1996. "Software-Based Innovation," *Sloan Management Review,* Summer, pp. 11–24.

Ray, George F. 1980. "Innovation and the Long Cycle," in B. A. Vedin (ed.), *Current Innovation,* Almqvist & Wiksell, Stockholm.

Rebello, Kathy, Peter Burrows, and Ira Sager. 1996. "The Fall of an American Icon," *Business Week,* February 5, pp. 34–42.

Reid, T. T. 1985. "The Chip," *Science 85,* February, pp. 32–41.

Robb, Walter. 1992. "Don't Change the Engineers—Change the Process," *Research-Technology Management,* March–April, pp. 8–9.

Roberts, Edward B., 1991. "High Stakes for High-Tech Entrepreneurs: Understanding Venture Capital Decision Making," *Sloan Management Review,* Winter, pp. 9–20.

Robertson, Arthur L. 1984. "One Billion Transistors on a Chip?" *Science,* Vol. 223, January 20, pp. 267–268.

Robertson, Nat C. 1992. "Technology Acquisition for Corporate Growth," *Research-Technology Management,* March–April, pp. 26–30.

Robinson, A. L. 1984. "One Billion Transistors on a Chip?" *Science,* Vol. 223, January 20, pp. 267–268.

Romero, Simon, and Riva D. Atlas. 2002. "Want a Piece of WorldCom? Contact the Company's Bankers," *New York Times,* June 29, pp. B1–B2.

Rosenberg, Nathan, and Richard R. Nelson. 1994. "American Universities and Technical Advance in Industry," *Research Policy,* Vol. 23, pp. 323–348.

Rosenthal, Stephen R., and Anil Khurana. 1997. "Integrating the Fuzzy Front End of New Product Development," *Sloan Management Review,* Winter, pp. 103–118.

Roth, Daniel. 1999. "Dell's Big New Act," *Fortune,* December 6, pp. 152–156.

Rowe, Gene, George Wright, and Fergus Bolger. 1991. "Delphi: A Reevaluation of Research and Theory," *Technology Forecasting and Social Change,* Vol. 39, pp. 235–251.

Rowen, Robert B. 1990. "Software Project Management under Incomplete and Ambiguous Specifications," *IEEE Transactions on Engineering Management,* Vol. 37, No. 1, February, pp. 10–21.

Rubenstein, Albert H. 1989. *Managing Technology in the Decentralized Firm,* Wiley, New York.

Ryans, John K., Jr., and William L. Shanklin. 1984. "Positioning and Selecting Target Markets," *Research Management,* Vol. XXVII, No. 5, pp. 583–607.

Sanderson, Susan, and Mustafa Uzumeri. 1995a. "Managing Product Families: The Case of the Sony Walkman," *Research Policy,* Vol. 24, pp. 761–782.

Sanderson, Susan, and Mustafa Uzumeri. 1995b. "A Framework for Model and Product Family Competition," *Research Policy,* Vol. 24, pp. 583–607.

Sanderson, Susan, and Mustafa Uzumeri. 1995c. "Managing Product Families: The Case of the Sony Walkman," *Research Policy,* Vol. 24, pp. 761–782.

Scarre, Chris. 1998. *Past Worlds: Atlas of Archaeology,* HarperCollins, London.

Schmidt, Susan. 2002. "Senators Fill Lay's Silence," *Washington Post,* February 13, pp. A1–A8.

Schmidt, Robert L., and James R. Freeland. 1994. "Recent Progress in Modeling R&D Project-Selection Processes," *IEEE Transactions on Engineering Management,* Vol. 39, No. 2, May, pp. 189–201.

Schmitt, Roland W. 1985. "Successful Corporate R&D," *Harvard Business Review,* May–June, pp. 124–129.

Schroeder, Michael. 2002. "As Enron's Derivatives Trading Comes into Focus, Gap in Oversight Is Spotlighted," *Wall Street Journal,* January 28, pp. C1–C15.

Schwartz, John. 2001a. "Business on Internet Time," *New York Times,* March 20, pp. C1–C2.

Schwartz, John. 2001b. "Congress Begins an Investigation, Raising New Questions, and Silence," *New York Times,* February 10, pp. A1–A26.

Segré, Emilio G. 1989. "The Discovery of Nuclear Fission," *Physics Today,* July, pp. 38–43.

Serapio, Manuel G. 1995. "Growth of Japan–U.S. Cross-Border Investments in the Electronics Industry," *Research-Technology Management,* November–December, pp. 42–47.

Service, Robert F. 1997. "Making Single Electrons Compute," *Science,* Vol. 275, January 17, pp. 303–304.

Shackil, Albert F., 1981. "Design Case History: Wang's Word Processor," *IEEE Spectrum,* Vol. 18, No. 1, pp. 29–34.

Shanklin, William L. 1983. "Supply-Side Marketing Can Restore 'Yankee Ingenuity,'" *Research Management,* May–June, pp. 20–25.

Shenhar, Aaron. 1988. "From Low to High Tech Project Management," Tel Aviv University, Tel Aviv, Israel, prepublication copy.

Sherman, Stratford. 1994. "Is He Too Cautious to Save IBM?" *Fortune,* October 3, pp. 78–90.

Shrayer, Michael. 1984. "Confessions of a Naked Programmer," *Creative Computing,* Vol. 10, No. 11, pp. 130–131.

Solberg, James. 1992. "Why Does It Take So Long, Cost So Much?" Seminar, Purdue University, West Lafayette, IN.

Sorenson, Charles E. 1956. *My Forty Years with Ford,* W.W. Norton, New York.

Steele, Lowell W. 1989. *Managing Technology: The Strategic View,* McGraw-Hill, New York.

Stevenson, H.H., and D.E. Gumpert. 1985. "The Heart of Entrepreneurship," *Harvard Business Review,* March–April, pp. 85–94.

Strohman, James E. 1990. "The Ancient History of System/360," *Invention and Technology,* Winter, pp. 34–40.

Stewart, Thomas A. 1993. "The King Is Dead," *Fortune,* January 11, pp. 34–40.

Stuewer, Roger H. 1985. "Bringing the News of Fission to America," *Physics Today,* October, pp. 49–56.

Suarez, Fernando F., Michael A. Cusumano, and Charles H. Fine. 1995. "An Empirical Study of Flexibility in Manufacturing," *Sloan Management Review,* Fall, pp. 25–32.

Suh, Nam. 1990. *Axiomatic Design,* Oxford University Press, Oxford.

Sugiura, Hideo. 1990. "How Honda Localizes Its Global Strategy," *Sloan Management Review*, Fall, pp. 77–82.

Swamidass, Paul M., and M. Dayne Aldridge. 1966. "Ten Rules for Timely Task Completion in Cross-Functional Teams," *Research-Technology Management,* July–August, pp. 12–13.

Taguchi, Genichi, and Don Clausing. 1990. "Robust Quality," *Harvard Business Review,* January–February, pp. 65–75.

Takeschi, Bob. 2002. "E-Commerce Report," *New York Times,* July 22, p. C6.

Tedishi, Bob. 2000. "How to Lure Prestigious Beauty Goods to Cyberspace," *New York Times,* January 17, pp. C1–C8.

Thayer, Ann M. 1996. "Market, Investor Attitudes Challenge Developers of Biopharmaceuticals," *Chemical & Engineering News,* August 12, pp. 13–21.

Thorne, Barbara. 1997. "Overview of Biopharmaceutical Development at Bristol–Meyers Squibb Co.," presentation at Biotechnology Process Engineering Center, MIT, Cambridge, MA, February 25.

Titus, George J. 1994. "Forty-Year Evolution of Engineering Research: A Case Study of DuPont's Engineering Research and Development," *IEEE Transactions on Engineering Management,* Vol. 41, No. 4, November, pp. 350–353.

Tripping, James W., Eugene Zeffren, and Alan R. Fusfeld. 1995. "Assessing the Value of Your Technology," *Research-Technology Management,* September–October, pp. 22–39.

Tully, Shawn. 1993. "The Real Key to Creating Wealth," *Fortune,* September 20, pp. 38–50.

Tushman, Michael, and Philip Anderson. 1986. "Technological Discontinuities and Organizational Environments," *Administrative Science Quarterly,* Vol. 31, pp. 439–465.

Twiss, Brian. 1968. *Managing Technological Innovation,* Longman Group, Harlow, Essex, England.

Twiss, Brian. 1968. *Management of Research and Development,* Longmans, New York.

Tyre, Marcie J., and Wanda J. Orlikowski. 1993. "Exploiting Opportunities of Technological Improvement in Organizations," *Sloan Management Review,* Fall, pp. 13–26.

Tzidony, Dov, and Beno Zaidman. 1996. "Method for Identifying R&D-Based Strategic Opportunities in the Process Industries," *IEEE Transactions on Engineering Management,* Vol. 43, No. 4, November, pp. 351–355.

Uneohara, Michiyuki. 1991. "A Management View of Japanese Corporate R&D," *Research-Technology Management,* November–December, pp. 17–23.

Urban, Glen, and John Hauser. 1993. *Design and Marketing of New Products,* 2nd ed., Prentice Hall, Upper Saddle River, NJ.

Uttal, Bro. 1981. "Xerox Xooms toward the Office of the Future," *Fortune,* May 18, pp. 105–106.

Uttal, Bro. 1983. "The Lab That Ran Away from Xerox," *Fortune,* September 5, pp. 97–102.

Uttal, Bro, Alan Kantrow, Lawrence H. Linden, and B. Susan Stock. 1992. "Building R&D Leadership and Credibility," *Research-Technology Management,* May–June, pp. 15–23.

Utterback, James. 1978. "Management of Technology," pp. 137–160 in Arnoldo C. Hax (ed.), *Studies in Operations Management,* North-Holland, Amsterdam.

Utterback, James M., and Fernando F. Suarez. 1993. "Innovation, Competition, and Industry Structure," *Research Policy,* Vol. 22, pp. 1–21.

Veraldi, Lew. 1988. "Team Taurus: A Simultaneous Approach to Product Development," speech given at the Thayer School of Engineering, Dartmouth College, Hanover, NH.

Verity, John W., Thane Person, Deidre Depke, and Evan I. Schwartz. 1991. "The New IBM," *Business Week,* December 16, pp. 112–118.

Vesper, Karl H. 1980. *New Venture Strategies,* Prentice Hall, Upper Saddle River, NJ.

Von Braun, Christoph-Friedrich, 1990. "The Acceleration Trap," *Sloan Management Review,* Fall, pp. 49–58.

Von Braun, Christoph-Friedrich. 1991. "The Acceleration Trap in the Real World," *Sloan Management Review,* Summer, pp. 43–52.

Von Hippel, Eric. 1976. "The Dominant Role of Users in the Scientific Instrumentation Innovation Process," *Research Policy,* Vol. 5, No. 3, pp. 212–239.

Von Hippel, Eric. 1982. "Appropriability of Innovation Benefit as a Predictor of the Source of Innovation," *Research Policy,* Vol. 11, No. 2, pp. 95–115.

Von Hippel, Eric, and Marcie J. Tyre. 1995. "How Learning by Doing Is Done: Problem Identification in Novel Process Equipment," *Research Policy,* Vol. 24, pp. 1–12.

Walker, Mark. 1990. "Heisenberg, Goudsmit, and the German Atomic Bomb," *Physics Today,* January, pp. 52–60.

Ward, E. Peter. 1981. "Planning for Technological Innovation: Developing the Necessary Nerve," *Long Range Planning,* Vol. 14, April, pp. 59–71.

Wheelwright, Steven C., and W. Earl Sasser, Jr. 1989. "The New Product Development Map," *Harvard Business Review,* May–June, pp. 112–125.

Whiteley, Roger L., Alden S. Bean, and M. Jean Russo. 1994. "Meet Your Competition: Results from the 1994 IRI/CIMS Annual R&D Survey," *Research-Technology Management,* July–August, pp. 51–55.

Whitlock, Craig, and Susan Schmidt. 2002. "Enron Testimony Clashes," *Washington Post,* February 8, pp. A1–A14.

Willard, Charles H., and Cheryl W. McClees. 1987. "Motorola's Technology Roadmap Process," *Research Management,* September–October, pp. 13–19.

Wilson, Donald K., Roland Mueser, and Josepth A. Raelin. 1994. *Research-Technology Management,* July–August, pp. 51–55.

Wilson, Pete. 1988. "The CPU Wars," *Byte,* May, pp. 213–234.

Wise, G. 1980. "A New Role for Professional Scientists in Industry: Industrial Research at General Electric, 1900–1911," *Technology and Culture,* Vol. 21, pp. 408–415.

Wohleber, Curt. 1993. "Straight Up," *Invention & Technology,* Winter, pp. 26–38.

Wolf, Michael F. 1983. "Mervin J. Kelly: Manager and Motivator," *IEEE Spectrum,* December, pp. 71–75.

Wolff, Michael F. 1984. "William D. Coolidge: Shirt-Sleeves Manager," *IEEE Spectrum,* May, pp. 81–85.

Workman, John P. 1995. "Engineering's Interactions with Marketing Groups in an Engineering-Driven Organization," *IEEE Transactions on Engineering Management,* Vol. 42, No. 2, May, pp. 129–139.

Zachary, G. Pascal. 1995. "Vannevar Bush on the Engineer's Role," *IEEE Spectrum,* July, pp. 65–69.

Zirger, B. J., and Janet L. Hartley. 1996. "The Effect of Acceleration Techniques on Product Development Time," *IEEE Transactions on Engineering Management,* Vol. 43, No. 2, May, pp. 143–152.

Zorpette, Glenn. 1987. "Breaking the Enemy's Code," *IEEE Spectrum*, September, pp. 47–51.

Zuckerman, Laurence. 2001. "The B-52's Psychological Punch," *New York Times,* December 8, p. A17.

Zwicky, F. 1948. "The Morphological Analysis and Construction," in *Studies and Essays, Courant Anniversary Volume,* Interscience, New York.

INDEX

in major device fabrication sector, 131
in materials refining sector, 128
in parts sector, 130
product design factor in, 244
and redesign time reduction, 109
in resource extraction sector, 124
in retail service sector, 134
speed for, 279–281
in subsystems sector, 130
Component invention, 73
Components sector, 129–131
Computers. *See also* Information technologies;
    Personal computers
architecture of, 91–93
first commercial, 427–431
IBM System/360 computers case study, 433–436
information model of science, engineering, and, 410
invention of, 85–91
laptop, 129–130
logics in, 96
and model of mind, 404
PDP-8, 93–95
programming languages, 408–410
von Neumann architecture case study, 414–415
Computer Associates, 333
Computer science, 418, 431
Conceptual world, 9
Concurrent engineering, 254–257
Conditions, internal and external, 169
Conference on Biohazards in Cancer Research, 31
Conglomerate diversification, 447
Conner, Finis, 176
Conner Peripherals, 176
Constraints, 106
Construction (as production), 282
Content (in languages), 411, 412
Continuous technological progress, 74. *See also*
    Incremental innovation
Control languages, 412
Control plane, 268, 272–273, 275
Control software, 420
Control systems, 72, 74–76
Coolidge, William, 224–226, 228–229
Coover, Harry W., 440–442
Copernicus, 65
Copyrights, 350
Core business competencies, 50, 449–456
Core technologies, 159–160, 167, 169, 449–456
Corey, Robert, 29
Corporate research. *See also* Industrial research and
    development
evaluation of, 233–235
laboratories for, 231–233, 235
objectives of, 230–231
Corporate shares, 334–335
Correctness (in product development), 110–113
Cosmic-level scale of space, 386
Cost, trade-offs with performance and, 111
Cost advantages, 311
Cost obsolescence, 426
Cost of equity, 267
Country-to-city market, 308
CPU, *see* Central processing unit

Crescendo Communications, 322
Crick, Francis, 29, 30
Cro-Magnons, 7
CSX, 267
Cultural evolution, 4, 6–8, 13–14, 202, 203
Culture:
    of gamesmanship, 333
    language as technology of, 413–414
    as mode of societal association, 17
Cummings, James, 88
Cunningham, Leland, 88
Curie, Irène, 205, 206
Customer(s):
    connection of business to, 102
    and market niche, 105
Customer application systems, 77
Customer needs:
    and commercial success, 103–105
    and delay in product planning, 111
    with next-generation products, 367
    in product development cycle, 108
Cyclic incremental innovation process, 108
Cymer Inc. case study, 171–173, 189

D'Agostino, Oscar, 205
Daiichi Seiyaku, 37
Daimler-Benz, 166
Darwin, Charles, 27, 395
Da Vinci, Leonardo, 63
Davis, Albert, 224
Davis, John, 88
Dayton, Russell, 21
Debt, 325
Debt to equity ratio, 333
Debugging, 249–251
DEC, *see* Digital Equipment Corporation
Decimal number system, 92
Decision/action step (reasoning), 402
Decision-event model, 349
Decision making, 239, 249, 367
Decision-process approach, 349
DeConcini, Dennis, 351
De Forrest, Lee, 378
Delbruck, M., 29
Delegation (by leaders), 175
Dell, Michael, 129, 130, 276, 277
Dell Computer, 129–130, 264, 265, 276–277
Delphi method, 153–154
Dependability, 104
Depreciation, 333
Descriptive MOT approach, xi
Design, 243–244. *See also* Engineering
    best-of-breed, 258–259
    commercial risks in, 105
    commercial success and time for, 112
    competitive benchmarking for, 105–106
    decisions during process of, 109
    definition of, 246
    of early microprocessors, 364–366, 368–369
    of new products, 106, 209–210
    problem solving in, 244–245
    product design standards, 163–164
    technical risks in, 104–105